FAIRNESS 2017: FAIR NExt generation ScientistS

Journal of Physics: Conference Series Volume 1024

Sitges, Spain
28 May – 3 June 2017

ISBN: 978-1-5108-6344-6
ISSN: 1742-6588

Printed from e-media with permission by:

Curran Associates, Inc.
57 Morehouse Lane
Red Hook, NY 12571

CURRAN ASSOCIATES INC. proceedings.com

Some format issues inherent in the e-media version may also appear in this print version.

Printed by Curran Associates, Inc. (2018)

Institute of Physics
Dirac House, Temple Back
Bristol BS1 6BE UK

Phone: 44 1 17 929 7481
Fax: 44 1 17 920 0979

techtracking@iop.org

Additional copies of this publication are available from:

Curran Associates, Inc.
57 Morehouse Lane
Red Hook, NY 12571 USA
Phone: 845-758-0400
Fax: 845-758-2633
Email: curran@proceedings.com
Web: www.proceedings.com

Table of contents

Volume 1024

FAIRNESS 2017: FAIR NExt generation ScientistS 2017

28 May to 3 June 2017, Sitges, Spain

Accepted papers received: 3 May 2018
Published online: 29 May 2018

Preface

FAIRNESS 2017: FAIR NExt generation ScientistS 2017

Peer review statement

Papers

Open-charm physics: from e^+e^- to $\bar{p}p$ machines

Elisabetta Prencipe on behalf of the PANDA collaboration.....1

Excitation of baryon resonances in charge-exchange reactions of heavy nuclei

J L Rodríguez-Sánchez, J-C David, A Boudard, J Cugnon, S Leray and D Mancusi.....6

Holographic vector mesons in a dilaton background

R. Zöllner and B. Kämpfer.....11

Investigation into the charge collection efficiency of prototype microstrip sensors for the CBM Silicon Tracking System

I. Momot, M. Teklishyn, A. Lymanets, O. Bertini and J. Heuser.....16

Recent theoretical advances regarding α-spectroscopy

A Dumitrescu and DS Delion.....21

Measurements and understanding of fundamental properties of hot and dense nuclear matter with HADES

Georgy Kornakov for the HADES Collaboration.....26

AstroBox2E: a detection system for very low energy β-delayed proton decay

I.C. Stefanescu, A. Spiridon, L. Trache, E. Pollacco, A. Saastamoinen and B. Roeder.....31

Higgs $\rightarrow \tau\tau$ Branching Ratio Measurement at CEPC

YU Dan.....36

Calculating hadron properties from dynamical hadronization in the Functional Renormalisation Group

J. París-López, R. Alkofer, A. Maas, W. Mian, M. Mitter, J. Pawlowski and H. Sanchis-Alepuz.....42

S- and *D*-wave vector charmonia

A Krassnigg and T Hilger.....47

Thermal dileptons as QCD matter probes at SIS

F Seck, T Galatyuk, R Rapp and J Stroth.....52

Effects of $\Lambda\Lambda - \Xi N$ mixing in the decay of ${}^{6}_{\Lambda\Lambda}He$

J. Maneu, A. Parreño and A. Ramos.....57

Performance of a prototype Straw Tube Tracker for the $\bar{P}$ANDA experiment

A Apostolou, J Messchendorp, N Kalantar-Nayestanaki, J Ritman and P Wintz.....62

Inclusive Prompt Photons from the Color Glass Condensate at NLO

Sanjin Benić, Kenji Fukushima, Oscar Garcia-Montero and Raju Venugopalan.....67

QCD Critical point from a black hole engineered EoS

Israel Portillo.....72

Determination of electric dipole transitions in heavy quarkonia using potential non-relativistic QCD

Jorge Segovia and Sebastian Steinbeißer.....77

Calculations of $\bar{p}$-nuclear states within the Paris $\bar{N}N$ potential model

J. Hrtánková and J. Mareš.....82

Constraints on the hadronic spectrum from Lattice QCD

P Parotto.....87

Extraction of Polarization Parameters in the $\bar{p}p \to \Omega\Omega$ Reaction

E Perotti.....92

The NLO Chiral Lagrangian from the meson-baryon interaction in the $S = -1$ sector.

A Feijoo, V K Magas and A Ramos.....97

The EXPERT project: part of the Super-FRS Experiment Collaboration

V. Chudoba for the EXPERT project.....102

Heavy quark potentials in effective string theory

Sungmin Hwang.....107

Infrastructures for Distributed Computing: the case of BESIII

J Pellegrino.....112

Higher Order Moments of Proton Number Fluctuations in Au+Au Collisions at 1.23 AGeV studied with HADES

Melanie Szala for the HADES collaboration.....117

Spinodal instability of baryon-rich quark matter

Feng Li and Che Ming Ko.....122

Recent results on small μ_B QCD from the lattice

A. Pásztor.....127

Design studies for the NeuLAND VETO detector

C A Douma, K Boretzky, I Gasparic, N Kalantar-Nayestanaki, D Kresan, J Mayer, C Rigollet for the R^3B collaboration.....132

Systematic errors in transport calculations of shear viscosity using the Green-Kubo formalism

J B Rose, J M Torres-Rincon, D Oliinychenko, A Schäfer and H Petersen.....137

Quark susceptibilities in a generalized quasiparticle model

T. Steinert and W. Cassing.....142

XYZ states at BESIII

Riccardo Farinelli.....147

Precision Laser Spectroscopy Technique for Exotic Radioactive Beams at CERN-ISOLDE

Xiaofei Yang the COLLAPS and CRIS collaboration.....152

Electromagnetic decays of the neutral pion investigated in the Dyson-Schwinger formalism

Esther Weil, Gernot Eichmann, Christian S. Fischer and Richard Williams.....157

$\Lambda N \rightarrow NN$ EFT potentials and hypertriton non-mesonic weak decay

Axel Pérez-Obiol, David R. Entem and Andreas Nogga.....162

Λ_c^+ physics at BESIII

Weiping Wang on behalf of the BESIII collaboration.....167

Light tetraquarks in a Dyson-Schwinger/Bethe-Salpeter approach

Paul C. Wallbott, Gernot Eichmann and Christian S. Fischer.....173

Two-pole structure of the $D_0^*(2400)$

P. Fernandez-Soler, M. Albaladejo, F.-K. Guo and J. Nieves.....178

Charm physics at BESIII

Rupan Guo on behalf of the BESIII Collaboration.....183

EnsarRoot: The framework for simulation and data analysis for ENSAR

Pablo Cabanelas, Héctor Alvarez-Pol, Elisabet Galiana, Yago González-Rozas for the SATNuRSE workpackage of ENSAR2.....188

Properties of open and hidden charm mesons in light quark matter

Martin Cleven, Volodymyr K. Magas and Angels Ramos.....193

Construction of Multi Wire Proportional Chambers for the CBM Transition Radiation Detector

Florian Roether.....198

Polarization reconstruction algorithm for a Compton polarimeter

M Vockert, G Weber, U Spillmann, T Krings and Th Stöhlker.....203

Inverse magnetic catalysis in nonlocal chiral quark models

V P Pagura, D Gomez Dumm, S Noguera and N N Scoccola.....208

Magneto-hydrodynamic simulations of Heavy Ion Collisions with ECHO-QGP

G Inghirami, L Del Zanna, A Beraudo, M Haddadi Moghaddam, F Becattini and M Bleicher.....213

IOP Conf. Series: Journal of Physics: Conf. Series **1024** (2018) 011001 doi:10.1088/1742-6596/1024/1/011001

FAIRNESS 2017 was the fifth edition in a series of workshops designed to bring together excellent international young scientists with research interests focused on physics at FAIR (Facility for Antiproton and Ion Research) and was held on May 28- June 03 2017 in Sitges, Spain. The topics of the workshops cover a wide range of aspects in both theoretical developments and current experimental status, concentrated around the four scientific pillars of FAIR. FAIR is a new accelerator complex with brand new experimental facilities, that is currently being built next to the existing GSI facility close to Darmstadt, Germany.

The spirit of the conference is to bring together young scientists, e.g. young non-tenured scientists, postdocs and advanced PhD students to present their work, to foster active informal discussions and build up of networks. Every participant in the meeting with the exception of the organizers gives an oral presentation, and all sessions are followed by an hour long discussion period. During the talks, questions are anonymously collected in a box to stimulate discussions. The broad physics program at FAIR is reflected in the wide range of topics covered by the workshop:

- Atomic and plasma physics, biophysics, material sciences and applications
- Nuclear structure, astrophysics and reactions
- Physics of hot and dense nuclear matter, QCD phase transitions and critical point
- Hadron Spectroscopy, Hadrons in matter and Hypernuclei
- Experimental programs APPA, CBM, HADES, PANDA, NUSTAR, as well as BES, NICA and the RHIC beam energy scan

For these different areas one invited speaker was selected to give a longer introductory presentation. The write-ups of the talks presented at FAIRNESS 2017 are the content of this issue of Journal of Physics: Conference Series and have been refereed according to the IOP standard for peer review. This issue constitutes therefore a collection of the forefront of research that is dedicated to the physics at FAIR.

June 2017,

Organizers of FAIRNESS 2017: Marco Destefanis, Tetyana Galatyuk, Diana Nicmorus, Claudia Ratti, and Laura Tolos.

Support for holding the conference was provided by:

HELMHOLTZ ASSOCIATION

EMMI

FAIR

IOP Conf. Series: Journal of Physics: Conf. Series **1024** (2018) 011002 doi:10.1088/1742-6596/1024/1/011002

Peer review statement

All papers published in this volume of *Journal of Physics: Conference Series* have been peer reviewed through processes administered by the proceedings Editors. Reviews were conducted by expert referees to the professional and scientific standards expected of a proceedings journal published by IOP Publishing.

IOP Conf. Series: Journal of Physics: Conf. Series **1024** (2018) 012001 doi:10.1088/1742-6596/1024/1/012001

Open-charm physics: from e^+e^- to $\bar{p}p$ machines

Elisabetta Prencipe

Forschungszentrum Jülich, Leo Brandt Strasse, 52428 Jülich, DE
on behalf of the $\overline{\text{P}}$ANDA collaboration

E-mail: e.prencipe@fz-juelich.de

Abstract. This report intends to summarize the most interesting experimental results in open-charm physics achieved by the past e^+e^- experiments, and show some new theoretical developments, in particular for D_s mesons, for the cross section calculations in $\bar{p}p$ annihilation processes. After the observation of the $D^*_{s0}(2317)^+$, more than 10 years ago, still questions remain opened due to the lack of precision measurements. The measurement of the width of narrow states below the $D^{(*)}K$ threshold in this respect is essential for a better understanding: a technique to measure that will be shown, and an overview on new future facilities where this measurement can be achieved will be given. In particular, we will focus on the role of the future $\overline{\text{P}}$ANDA experiment at FAIR, and the unique contribution of this project in the measurement of the $D_s^{(*)}$ very narrow widths.

1. Introduction

Since 2003 unexpected observations and evidence of new resonant states questioned the validity of the Quark Potential Model (QPM), which well explains *e.g.* most of the D_s spectrum, but not the mass shift of the $D^*_{s0}(2317)^+$ and the $D_{s1}(2460)^+$ from the theoretical prediction of Godfrey and Isgur [1]. The peculiar behaviour of these two states is actually not very understandable, because $D_s^{(*)}$ mesons are binding states of a c-quark and a s-quark. The s-quark can be considered light enough to fit the QPM theoretical calculations, and effectively it was experimentally demonstrated so until the discovery of the $D^*_{s0}(2317)^+$ [2]. In the same year of the $D^*_{s0}(2317)^+$ discovery a new state decaying to $J/\psi\pi^+\pi^-$, the so-called X(3872) [3] was announced. This was also a puzzling observation, unexpected and incredibly narrow. After 14 years tens of these new states populate the charm- and charmonium(like)- spectrum. The measurement of their mass, width, and decay products is essential to draw a pattern in order to understand their nature, especially since the charged Z charmonium states have been announced. A great contribution was given by the past experiments BaBar and Belle, and recently from BES III and LHCb. However, still questions remain unanswered due to limitations in mass resolution and photon detection, which nowadays do not allow to discriminate among the theoretical models developed after the discovery of those states.

2. Cross section predictions

The study of charm physics is interesting from both points of view, weak and strong interactions. This short report concentrates on the study and development of the latter.

First Belle [4] then BaBar [5] have studied the cross sections of the processes $e^+e^- \to D^{(*)}\bar{D}^{(*)}$ via ISR, resulting in the same conclusion: for energy values in the center-of-mass region $\sqrt{s}$ >4.2

GeV the cross section is lower than 4 nb, and slightly decreases with increasing $\sqrt{s}$. BaBar also performed the the cross section measurements of the $e^+e^- \to D_s^{(*)} D_s^{(*)}$ processes via ISR [6], ending with measurements lower than 1 nb. No other measurements are published. One could in principle ask the question what the value of the equivalent cross sections in $\bar{p}p$ annihilation is, and if this measurement could add information for better understanding the $D^{(*)}$ and $D_s^{(*)}$ spectum.

Cross section predictions in $\bar{p}p$ annihilations for the cs spectrum are difficult due to the presence of the s-quark, because it is definitively lighter than the c-quark, but not light enough to use the same approximations when developing theoretical calculations for heavy-light systems: divergences, which are difficult to cure, occur. As a general remark, the cross section $\sigma(\bar{p}p) \to \bar{D}D$ is expected to be lower than 100 nb: this conclusion rises from the old studies of $\bar{p}p \to$ charged mesons [7]. However, the most quoted theoretical papers to predict $\bar{p}p$ annihilation into D mesons are not in good agreement with each other: Braaten and Artoisenet [8] make use of NLO and LO approaches, and rate the cross section in the range 0.002-0.09 nb at $\sqrt{s}$ ~4 GeV in inclusive charm production in $\bar{p}p$ collisions, and in the range of 10-100 nb at $\sqrt{s}$ ~6 GeV in the same processes; Khodjamirian, Klein, Mannel and Wang [9] range $\sigma(\bar{p}p \to D^+D^-)$ in the interval 10^{-6}-10^{-2} nb at p_{lab}=6 GeV/c, and it decreases smoothly up to p_{lab}=20 GeV/c. However, they show a huge gap between the prediction for $\bar{p}p \to D^+D^-$ and $\bar{p}p \to D^0\bar{D}^0$ cross sections, in favor of the latter. Recent theoretical developments by Heidenbauer and Krein [10] have come to a different conclusion: using the SU(4) symmetry and then assuming that all coupling constants are fixed, and adding the neutron in the baryon loop exchange, they quote the $\bar{p}p \to D^0\bar{D}^0$ and $\bar{p}p \to D^+D^-$ cross sections in a similar range, *e.g.* 10^{-3} nb at p_{lab}=6.4 GeV/c, and 10^{-1} nb at p_{lab}=7.5 GeV/c. In the same paper for the first time a prediction of the $\bar{p}p \to D_s^+D_s^-$ cross section is given: depending on the model, quark model or baryon exchange approach, a range of 10-100 nb can be extrapolated for the cross section prediction of the $\bar{p}p \to D_s^+D_s^-$ process. Nothing is calculated for $\bar{p}p \to$ excited D_s mesons, for which even with this approach no predictions can be made.

In summary, from a theroretical point of view still no predictions exist for the excited D_s mesons in $\bar{p}p$ annihilation, but with the approach followed in Ref. [10] at least the cross section for the ground D_s states is provided. What is needed then is to perform measurements using a $\bar{p}p$ experiment that can cover the p_{lab} range of 6-10 GeV/c.

3. The $\overline{\text{P}}$ANDA experiment at FAIR

The $\overline{\text{P}}$ANDA (antiProtons ANnihiltion at DArmstadt) [11] experiment at the Facility for Antiprotons and Ion Research (FAIR) is a future project that is going to be realized in different phases, depending on the availability of detectors and infrastructure: *Phase 0* starting in 2018 with the commissioning of detector components; *Phase I*, the start-up phase, with a limited detector setup; and *Phases II-III* utilizing the full detector setup with varying luminosities.

The scheme of the detector with full setup is shown in Fig. 1. The $\overline{\text{P}}$ANDA detector will be a 4π coverage detector, operating with an antiproton beam with momentum of up to p=15 GeV/c, and a fixed target. It will be realized in two main parts: a central spectrometer inside a solenoidal magnetic field (2 T), and a dipole spectrometer (bendig power: 2 T·m). Due to the high boost (β_{cms} ≥0.8) many tracks and photons will be in the forward detector. Background and signal will have the same signature, therefore a complete real-time event reconstruction needs to be performed. The interaction rate will be enormous, up to 20 MHz. We expect a small signal over background (S/B) ratio in our analyses, typically between 10^{-4} and 10^{-7}, so proper techniques to reject the background, online and offline, have to be developed. The $\overline{\text{P}}$ANDA detector will have an excellent photon reconstruction, a good tracking system, and a high PID (particle identification) separation power.

The $\overline{\text{P}}$ANDA physics program is wide and ambitious. In particular, the measurement of the

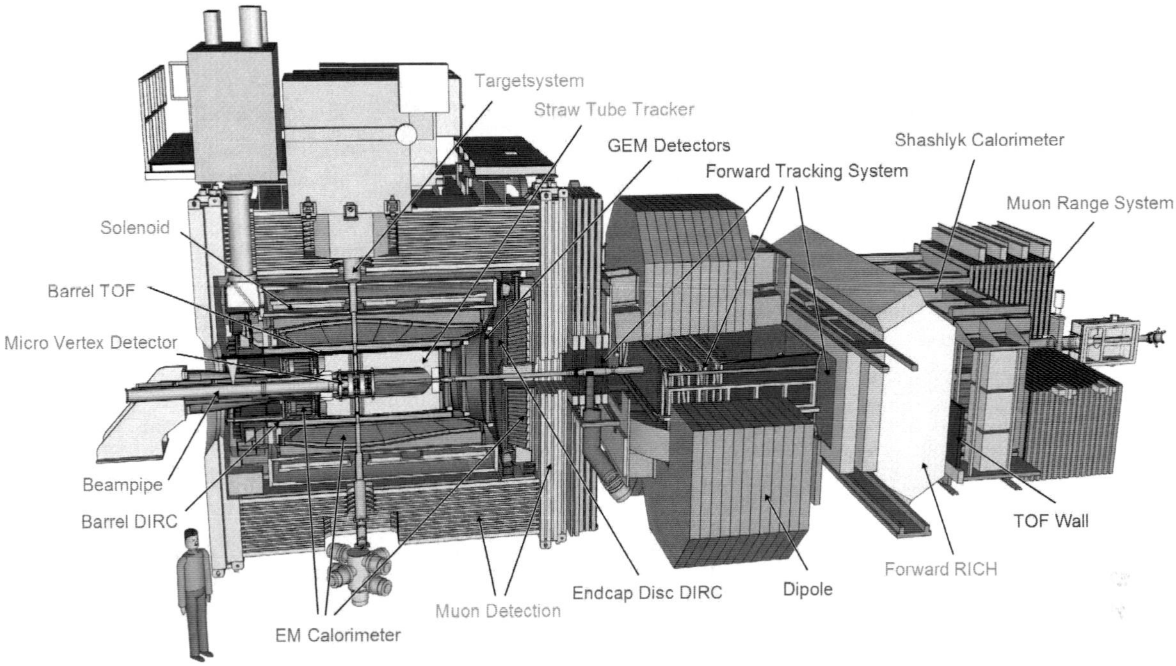

Figure 1. The $\overline{\text{P}}$ANDA detector

open-charm cross sections can start during the first phase of data taking. The measurement of the width of narrow states such as the $D^*_{s0}(2317)^+$, which represents one of the most ambitious challenges of the $\overline{\text{P}}$ANDA physics program, can be performed only with the complete setup of the experiment: due to the expected low values of the $D_s^{(*)} D_s^{(*)}$ cross sections, high statistics is needed.

4. Challenges in D_s meson spectroscopy

The technique to measure the width of narrow states in $\overline{\text{P}}$ANDA is the energy scan. This is a known technique to observe resonant states in $\bar{p}p$ annihilation processes. In any case, before performing the threshold energy scan of the $\bar{p}p \to D_s^- D^*_{s0}(2317)^+$ process to measure the width of the $D^*_{s0}(2317)^+$, the measurement of the cross section is needed and it is prioritized. This would give a hint on how many events per day we can collect for each energy-scan point, and how long this experiment should run.

In order to perform measurements in the open-charm sector, we need a good tracking system, and an excellent photon reconstruction. Our Monte Carlo (MC) simulations suggest that the photon energy resolution is below 2% at $E = 1$ GeV [12], and the photon energy detection threshold is ≤ 30 MeV [13].

Among all channels of interest to which $\overline{\text{P}}$ANDA can make an original contribution, we point here the attention of the reader to the $\bar{p}p \to D_s^- D^*_{s0}(2317)^+$ process, which is a channel of multiple interests:

- cross section measurement;
- width measurement;
- chiral symmetry breaking studies.

We need to run the antiproton beam in $\overline{\text{P}}$ANDA at $p_{lab} \geq 8.802$ GeV/c to study such a process: this is the threshold value to generate the $D_s D^*_{s0}(2317)^+$ pair. We need to produce always a couple of D_s mesons in such a process, because they are charged particles and we want to analyze the process through $\bar{p}p$ annihilation. Then additional values should be collected below and above that value, to reconstruct the excitation function of the cross section of the process under investigation. A proposal to collect 15 energy-scan values is planned.

The measurement of the $D^*_{s0}(2317)^+$ width is crucial to understand the nature of this resonant state. Several theoretical papers have been published, interpreting the $D^*_{s0}(2317)^+$ as a pure cs meson, if its width will be in the order of 10 keV [14]; tetraquark, if the width is in the order of 10-100 keV [15]; molecular state if the width will be ~133 keV [16].

The measurement of the $D^*_{s0}(2317)$ width will be performed by scanning the energy of the $D_s D^*_{s0}(2317)$ system in 100-keV-steps. The mathematical expression of the excitation function of the cross section of such a process is:

$$\sigma(s) = \frac{|\mathcal{M}|^{\in}}{64 \cdot \pi \cdot p_1 \cdot s} \Phi(E),$$

where $\Phi(E)$ is the spectral function, that assumes a very easy expression under the condition that $\sqrt{s}$ - M_{D_s} - $M_{D^*_{s0}(2317)}$ / $\Gamma(D^*_{s0}(2317)) \gg 1$. In this case, the excitation function of the cross section of the mentioned process depends only on the $D^*_{s0}(2317)$ width, the mass of the 2 mesons, and the momentum in the $D^*_{s0}(2317)$ rest frame. By scanning the energy of the $D_s D^*_{s0}(2317)$ system in 15 100-keV-steps, it is possible to draw the plot shown in Fig. 2. Here different width input values deliver different shapes of the excitation function of the cross section. The ability of $\overline{\text{P}}$ANDA to perform this study with precision, and minimizing the error bars inevitably affecting every measurement, point by point, should make it possible to distinguish among different theoretical hypotheses.

Table 1 summarizes the expected number of produced events per day under some cross section input values, for the initial and the final setups of the experiment.

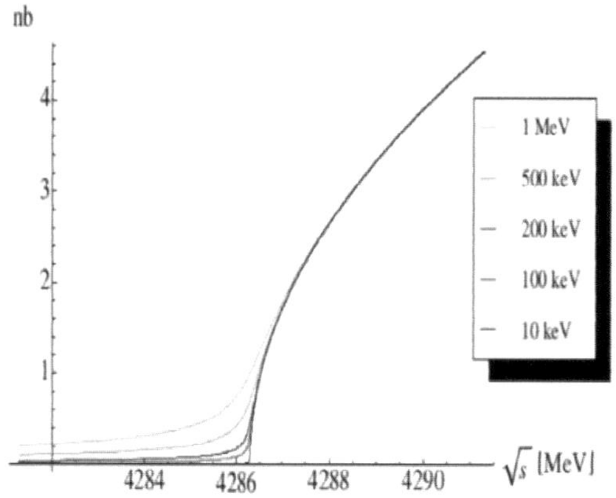

Figure 2. Excitation function of the cross section for the $\bar{p}p \rightarrow D^-_s D^*_{s0}(2317)^+$ process. Different color lines correspond to different input values of the $D^*_{s0}(2317)^+$ width [17].

If the $D^*_{s0}(2317)^+$ width is meaured in the order of 100 keV, then the hypothesis of it being a molecular state will be confirmed, and all other hypotheses will be ruled out. If indeed $\overline{\text{P}}$ANDA will not succeed to measure the $D^*_{s0}(2317)^+$ width, the most precise upper limit to the $D^*_{s0}(2317)^+$ width will be fixed: this would imply that the hypothesis of the $D^*_{s0}(2317)^+$ being

Table 1. Summary of the produced events per day for the $\bar{p}p \to D_s^- D_{s0}^*(2317)^+$ process. The average luminosity for the startup mode corresponds to $10^{31}\text{cm}^{-2}\text{s}^{-1}$; the average luminosity for the final setup in high luminosity mode corresponds to $10^{32}\text{cm}^{-2}\text{s}^{-1}$.

Input σ (nb)	Produced events per day (startup mode)	Produced events per day (full luminosity)
20	17280	172800
10	8640	86400
5	4320	43200
2	1728	17280
1	864	8640

a molecular state can be rejected. The known upper limit from PDG [18] to the $D_{s0}^*(2317)^+$ width is 3.8 MeV.

5. Summary

In summary, the study of charm physics is important because nowadays open-questions remain, which the currently running projects cannot address. Among these, the measurement of the $D_{s0}^*(2317)^+$ width represents a challenge, due to the high background level compared to the signal expectation, and it is one of the highlights of the $\overline{\text{P}}$ANDA physics program. The unknown cross section of the $\bar{p}p \to D_s^- D_{s0}^*(2317)^+$ process does not allow to make clear predictions: data are needed. It will be not easy to perform such a measurement, but the high detector resolution of $\overline{\text{P}}$ANDA allows to perform the energy scan in ~100-keV-steps. Full simulations for the decay process under investigation are ongoing [19], and preliminary results are promising.

References

[1] Godfrey S and Isgur N 1985 *Phys. Rev. D* **32** 189
[2] Aubert B *et al* 2003 *Phys. Rev. Lett.* **90** 242001
[3] Choi S K *et al* 2003 *Phys. Rev. Lett.* **91** 262001
[4] Pakhlova G *et al* 2007 *Phys. Rev. Lett.* **98** 092001
[5] Aubert B *et al* 2009 *Phys. Rev. D* **79** 092001
[6] Lees J P *et al* 2010 *Phys. Rev. D* **82** 052004
[7] Flaminio V *et al* 1984 *CERN-HERA* 84-01
[8] Braaten E and Artoisenet P 2009 *Phys. Rev. D* **79** 114005
[9] Khojdamiriam A *et al* 2012 *Eur. Phys. J* A **48** 31
[10] Haidenbauer J and Krein G 2014 *Phys. Rev. D* **89** 114003
[11] Erni W *et al* 2009 arXiv:0903.3905 [hep-ex]
[12] Singh B *et al* 2015 *Eur. Phys. J* A **51** 107
[13] Singh B *et al* 2016 *Eur. Phys. J* A **52** 325
[14] Godfrey S (2003) *Phys. Lett.* B **570** 180
[15] Cheng H Y *et al* 2006 *Phys. Rev. D* **73** 034004
[16] Clevens M *et al* 2014 *Eur. Phys. J* A **50** 149
[17] Mertens M (2010) PhD thesis: *Der $\overline{\text{P}}$ANDA Micro Vertex Detector: Entwicklung eines Labormessystems, Simulation der MVD-Betriebsparameter sowie Untersuchungen zur Auflösung der Breite des* $D_{s0}^*(2317)$ (RUB)
[18] Patrignani C *et al Particle Data Group* 2017 *Chin. Phys.* C **40** 100001
[19] Prencipe E 2015 *Pos Bormio2015* 044

IOP Conf. Series: Journal of Physics: Conf. Series **1024** (2018) 012002 doi:10.1088/1742-6596/1024/1/012002

Excitation of baryon resonances in charge-exchange reactions of heavy nuclei

J L Rodríguez-Sánchez[1], J-C David[1], A Boudard[1], J Cugnon[2], S Leray[1], D Mancusi[3]

[1] IRFU, CEA, Université Paris-Saclay, F-91191 Gif-sur-Yvette, France
[2] AGO department, University of Liège, B-4000 Liège, Belgium
[3] SERMA, CEA, Université Paris-Saclay, F-91191 Gif-sur-Yvette, France

E-mail: joseluis.rodriguez.sanchez.phys@gmail.com

Abstract. The Liège intranuclear-cascade model (INCL) has been improved using a refined description of the matter and energy densities in the nuclear surface. Hartree-Fock-Bogoliubov calculations with the Skyrme interaction were used to obtain a more realistic description of the proton and neutron density profiles. We find that the new approach, together with a realistic modeling of the de-excitation process of the nuclear pre-fragments, improves the description of the production cross sections of the heaviest nuclear residues produced by charge-exchange processes in spallation reactions, where the excitation of baryonic resonances plays an important role.

1. Introduction

In the last decade, different experiments were performed at GSI using the fragment separator FRS to study the production of charge-exchange residues in spallation reactions [1]. A spallation reaction is a process in which a projectile (proton or neutron) with a kinetic energy from $\sim$100 MeV to several GeVs interacts with a heavy nucleus (e.g. lead) and causes the emission of a large number of hadrons (mostly neutrons) or some light fragments. In these reactions, the charge-exchange residues are populated when the projectile exchanges one proton or one neutron with the target. This charge-exchange process can be produced by quasi-elastic nucleon-nucleon interactions exciting electromagnetic transitions, such as the Gamow-Teller, or by inelastic nucleon-nucleon collisions where baryonic resonances, e.g. the Δ resonance, play an important role [2].

The structure of baryons, such as the Δ and Roper (N*) resonances, and their excitation spectrum is one of the unsolved issues of strong interaction physics. Recently, the early appearance of Δ-isobars in dense nuclear matter has inspired many studies relevant to neutron stars [3] and heavy-ion collisions [4]. In particular, very compact stellar configurations are reached due to the introduction of Δ-isobars [3]. However, the in-medium properties of baryon resonances are not well understood. Up to now, the in-medium effects of Δ-isobars have been studied in heavy-ion collisions at kinetic energies above the production threshold using pion-nucleus and nucleon-nucleus reactions in direct kinematics [5, 6], where there is not a good control over the production of residual fragments and thus over the number of collisions. The production of these resonances is also very important to understand the pion production in heavy-ion collisions [7] because their main decay channel is going into a nucleon and a pion.

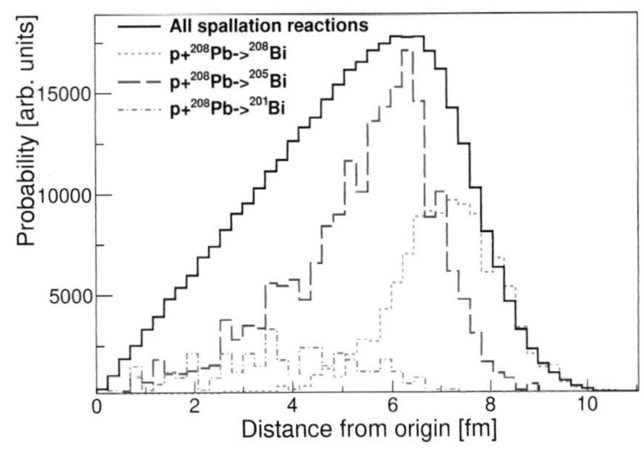

Figure 1. Radial distributions calculated with the INCL model for different nuclear residues produced in proton-induced charge-exchange reactions on ^{208}Pb at 1 GeV. The solid line represents the radial distribution for all possible interactions. The probability for the charge-exchange residues indicated in the figure were multiplied by a factor of 150.

Charge-exchange collisions produced in proton-induced reactions on nuclei are found as an optimum tool to investigate the in-medium properties of baryonic resonances [8]. Recently, it has been demonstrated that the use of the inverse kinematics technique, together with state-of-the-art detection systems, allows a full identification in atomic and mass number of the nuclear residues produced by charge-exchange reactions [1]. In addition, the measurement of the momentum distributions permits the separation of the quasi-elastic and inelastic contributions, as observed in the investigation of isobaric charge-exchange reactions [2, 9], providing a complete identification of the baryonic resonances excited in the nuclear collision. This kind of measurement could be extended to other nuclear residues produced by charge-exchange reactions to investigate the in-medium properties of the baryonic resonances as illustrated in Fig. 1, where the radial distribution calculated with the Liège intranuclear-cascade model (INCL) [10] is displayed for different nuclear residues. As can be seen, the heaviest residues are produced at the nuclear periphery (low nuclear density) while the lighter ones are excited with smaller impact parameters almost in the center of the nucleus (high nuclear density). Therefore, the investigation of the momentum distributions of these nuclear residues could help to understand how the excitation of the baryonic resonances changes with the nuclear density.

In this work, we will present some of the first steps performed in INCL to improve the description of this kind of reaction.

2. Model description

Spallation reactions at high energy ($\geq$ 100A MeV) are commonly described by the coupling of intranuclear cascade (INC) and de-excitation models [11, 12, 13, 14], where the first stage of the reaction can be described as an avalanche of independent binary collisions. The INC scheme can be derived from the usual nuclear transport equations under suitable approximations [15] and its numerical solution can be efficiently tackled on today's computers. The INC model is essentially classical, with the addition of a few suitable ingredients that mimic genuine quantum-mechanical features of the initial conditions and of the dynamics: for instance, target nucleons are endowed with Fermi motion, realistic space densities are used, and elementary nucleon-nucleon collisions are subject to Pauli blocking.

In this work we use the latest C++ version of the Liège intranuclear-cascade code (INCL) [10], which is equivalent to the reference fortran INCL4.6 version [16]. In this model the hadron-nucleus or nucleus-nucleus reactions are modelled as a sequence of binary collisions between the nucleons (hadrons) present in the system. The nucleus is represented by a potential well according to the Woods-Saxon distribution [17]. Nucleons move along straight trajectories until they undergo a collision with another nucleon or until they reach the surface, where they eventually escape. The latest version of the INCL also includes isospin- and energy-dependent nucleon potentials calculated according to optical models [16], isospin-dependent pion potentials [7], and the Pauli principle is considered by means of statistical blocking factors [17]. Cluster emission is also possible via a dynamical phase-space coalescence algorithm [16]. Pions and Δ resonances are supposed to appear and disappear through the $NN \Leftrightarrow N\Delta$ and $\Delta \Leftrightarrow \pi N$ reactions. For πN interactions, we use experimental cross sections including the resonant and nonresonant contributions while the cross sections for $N\Delta$ interactions are taken as for nucleon-nucleon collisions with some correction factors in order to describe pion absorption on nuclei (see Ref. [17] for more details).

In previous studies using INCL, the authors were interested in the isotopic spallation production cross sections [16, 17], where the neutron and proton densities were taken as Woods-Saxon profiles with parameters set to reproduce the charge density measured by electron scattering. However, peripheral reactions, where a few number of nucleons are removed, are expected to be more sensitive to the relative densities of neutrons and protons at the surface of the nucleus. Therefore, in the present manuscript, the neutron and proton densities are taken from microscopic single-particle wave functions from Hartree-Fock-Bogoliubov calculations performed with the HFBRAD code [18], considering the Skyrme Sly5 interaction [19] since it provides a reasonable description of the proton and neutron density radii [20]. The HFBRAD calculations were performed for all nuclei between the proton and neutron drip lines and were included in INCL following our previous work using a basic shell model [21].

At the end of the intranuclear cascade process, an excited remnant is left. This nucleus typically relaxes by emitting low-energy particles or, if possible, by fissioning. The time scale for the second stage is typically much longer than that for the first one, which justifies the fact that de-excitation is not described by INC but by a different class of models which rely on statistical assumptions about the properties of the excited remnant. It is required to couple INC to a de-excitation model if one wishes to describe the production of reaction residues.

For the de-excitation step, we use the ABLA07 model [22], which describes the de-excitation of a nucleus emitting γ-rays, neutrons, light-charged particles, and intermediate-mass fragments (*IMFs*) according to Weisskopf's formalism [23]. For a more realistic description of the de-excitation, the separation energies and the emission barriers for charged particles are considered according to the atomic mass evaluation from 2003 [24] and the Bass potential [25], respectively. In addition, de-excitation by fission is also included according to a dynamical picture described in Ref. [26].

3. Comparison with experimental data

The new ingredients of INCL should only affect the production of nuclear fragments close to the projectile, in particular, those produced by charge-exchange reactions at the surface of the nucleus where the excitation of resonances plays an important role [1, 2, 8]. In Fig. 2 we compare our standard INCL calculations (solid lines) with experimental data (solid circles) of nuclear residues produced by charge-exchange collisions in proton-induced reactions on ^{136}Xe (a) and ^{208}Pb (b) at kinetic energies of 1 GeV per nucleon. Calculations based on the new approach considering HFBRAD density profiles are also shown (dashed lines). In both calculations we assume only the excitation of Δ resonances which is a good approximation at energies of 1 GeV per nucleon because the probability of exciting other resonances is lower. As can be seen in the figures, the new approach improves the description of the production cross sections of intermediate nuclear residues produced through charge-exchange reactions. In addition, we also

IOP Conf. Series: Journal of Physics: Conf. Series **1024** (2018) 012002 doi:10.1088/1742-6596/1024/1/012002

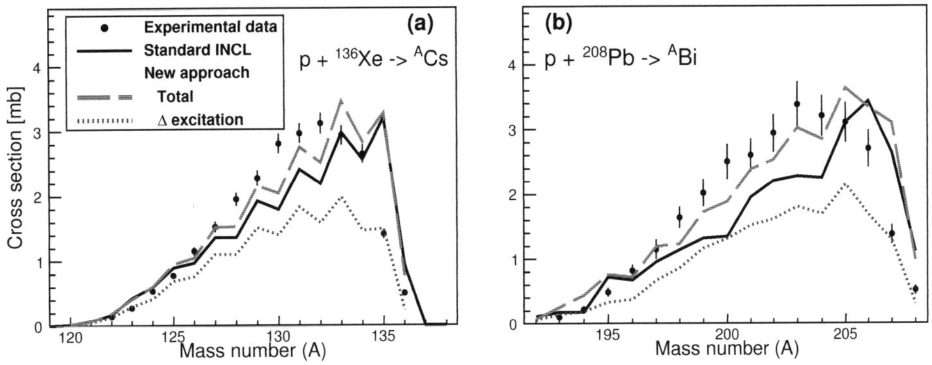

Figure 2. Isotopic distributions of nuclear residues produced in charge-exchange reactions by the excitation of baryonic resonances: p + ^{136}Xe $\rightarrow$ ^{A}Cs [27] (a) and p + ^{208}Pb $\rightarrow$ ^{A}Bi [1] (b) at 1 GeV. The experimental data is compared with the standard INCL (solid lines) and HFBRAD (dashed lines) calculations. The contribution from Δ excitation is also shown (dotted lines).

show the contribution from the excitation of Δ resonances (dotted lines), which corresponds to 50% of the total distribution.

In Fig. 3 we also compare our calculations with experimental data of total charge-exchange cross sections that correspond to the sum over all the nuclear residues produced by charge-exchange reactions that lead to fragments with atomic number $Z = Z_{projectile} + 1$. As can be observed, the new approach (dashed line) increases the total cross sections and brings the prediction in a better agreement with the experimental data. The cross sections from the excitation of Δ resonances are also displayed (dotted line). In this case, one can note that the probability of exciting the Δ resonance increases with the kinetic energy, going from ~15% at 0.5 GeV/A to ~60% at 1.4 GeV/A. This tendency is expected because at low kinetic energies the inelastic collisions disappear [7]. In general, the refined treatment of the surface considerably improves the predictions for these cross sections.

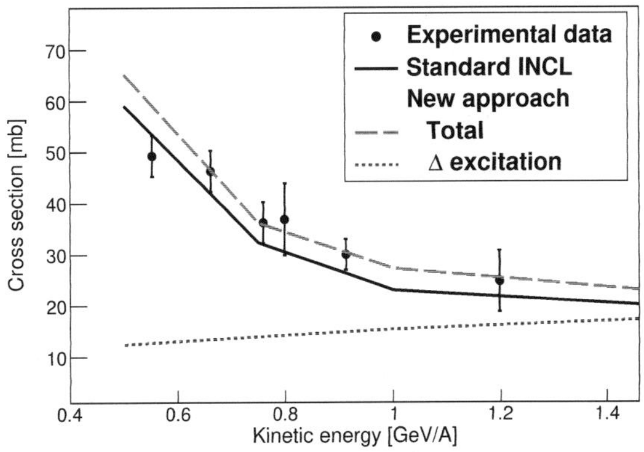

Figure 3. Total production of Hg in the spallation reaction p + ^{197}Au as a function of the projectile kinetic energy per nucleon. Experimental data taken from Ref. [28]. The solid line corresponds to standard INCL calculations while the dashed line represents INCL calculations using the Hartree-Fock-Bogoliubov calculations to define the surface of the nucleus. The contribution from Δ excitation is also shown (dotted line).

4. Conclusions

The initial conditions of the Liège intranuclear-cascade model (INCL) have been improved within Hartree-Fock-Bogoliubov calculations to take into account the presence of proton and neutron skins following the prescriptions given in our previous work [21]. In order to obtain a realistic description of the final nuclear residues produced by charge-exchange reactions, INCL was coupled to the de-excitation model ABLA07. The use of Hartree-Fock-Bogoliubov calculations to shape the surface of the nucleus, together with the capacity of INCL model to describe nucleon-nucleon collisions, allows us to reproduce better the cross sections of the heaviest nuclear residues produced in proton-induced reactions on medium and heavy targets. In particular, we have seen that the new approach improves the description of the cross sections of nuclear residues produced in charge-exchange reactions in which baryonic resonances, e.g. the Δ resonance, are excited with a probability around 50% at projectile energies of 1 GeV per nucleon.

5. Acknowledgments

This work has been partially supported by the project EU-ENSAR2-FP7-654002.

References

[1] Kelić A et al. 2004 *Phys. Rev. C* **70** 064608
[2] Bachelier D et al. 1986 *Phys. Lett. B* **172** 23
[3] Drago A et al. 2014 *Phys. Rev. C* **90** 065809
[4] Feng Z-Q 2016 *Phys. Rev. C* **94** 054617
[5] Kaufman S B et al. 1979 *Phys. Rev. C* **20** 2293
[6] Chiba J et al. 1991 *Phys. Rev. Lett.* **67** 1982
[7] Aoust Th and Cugnon J 2006 *Phys. Rev. C* **74** 064607
[8] Vidaña I et al. 2016 *EPJ Web of Conferences* **107** 10003
[9] Benlliure J et al. 2016 *Il Nuovo Cimento C* **39** 401
[10] Mancusi D, Boudard A, Cugnon J, David J-C,Kaitaniemi P and Leray S 2014 *Phys. Rev. C* **90** 054602
[11] David J-C 2015 *Eur. Phys. J. A* **51** 68
[12] Rodríguez-Sánchez J L et al. 2015 *Phys. Rev. C* **91** 064616
[13] Alcántara-Núñez J et al. 2015 *Phys. Rev. C* **92** 024607
[14] Paradela C et al. 2017 *Phys. Rev. C* **95** 044606
[15] Bunakov V E and Matvejev G V 1985 *Z. Phys. A* **322** 511
[16] Boudard A, Cugnon J, David J-C, Leray S and Mancusi D 2013 *Phys. Rev. C* **87** 014606
[17] Boudard A, Cugnon J, Leray S and Volant C 2002 *Phys. Rev. C* **66** 044615
[18] Bennaceur K and Dobaczewski J 2005 *Comput. Phys. Commun.* **168** 96
[19] Chabanat E et al. 1998 *Nucl. Phys. A* **635** 231
[20] Terashima S et al. 2008 *Phys. Rev. C* **77** 024317
[21] Mancusi D, Boudard A, Carbonell J, Cugnon J, David J-C and Leray S 2015 *Phys. Rev. C* **91** 034602
[22] Kelić A, Ricciardi M V and Schmidt K-H 2008 ABLA07 - *towards a complete description of the decay channels of a nuclear system from spontaneous fission to multifragmentation Joint ICTP-IAEA Advanced Workshop on Model Codes for Spallation Reactions* (Trieste, Italy: IAEA) report INDC(NDC)-0530 p 181
[23] Weisskopf V F and Ewing D H 1940 *Phys. Rev.* **57** 472
[24] Wapstra A H, Audi G and Thibault C 2003 *Nucl. Phys. A* **729** 129
[25] Bass R 1980 *Proceedings of the Symposium on Deep-Inelastic and Fusion Reactions with Heavy Ions*, Springer Verlag, Berlin
[26] Jurado B, Schmitt C, Schmidt K-H, Benlliure J and Junghans A R 2005 *Nucl. Phys. A* **747** 14
[27] Napolitani P et al. 2007 *Phys. Rev. C* **76** 064609
[28] Binns W R et al. 1989 *Phys. Rev. C* **39** 1785

IOP Conf. Series: Journal of Physics: Conf. Series **1024** (2018) 012003 doi:10.1088/1742-6596/1024/1/012003

Holographic vector mesons in a dilaton background

R. Zöllner, B. Kämpfer
Helmholtz-Zentrum Dresden-Rossendorf, PF 510119, D-01314 Dresden, Germany
and
Institut für Theoretische Physik, TU Dresden, D-01062 Dresden, Germany

Abstract. Within a holographic framework, we consider vector mesons riding on a gravity-dilaton background. The latter one is determined directly from a Schrödinger equivalent potential which delivers a proper ρ meson Regge trajectory. The mapping on the dilaton potential yields a thermodynamic phase structure with a first-order transition.

1. Introduction

Vector mesons couple directly to dileptons which are interesting as penetrating probes in relativistic heavy-ion collisions, in particular in a range of beam energies, where transiently large net baryon densities are achieved [1]. Basically, in-medium vector meson properties are accessible via spectral functions in a heat bath from QCD. But the evaluation of QCD as theory of strong interaction in the non-pertubative regime faces some problems, e.g. the sign problem at non-zero chemical potential μ or real-time phenomena hampered by the Euclidean action formulation. One alternative is to employ suitable models which catch some QCD properties. Among such models is the holographic approach which aims at mapping the four-dimensional QCD living in a Minkowski space to a five-dimensional gravitational theory living in an asymptotic anti-de Sitter (AdS) spacetime. Here, we employ the AdS/QCD correspondence [2, 3, 4], conjectured to emerge from its root - the AdS/CFT correspondence -, and explore a set-up, where the Regge trajectory of vector mesons serves as input (section 2). We translate that input into a dilaton potential by exploiting Einstein equations (section 3) to fix the gravity-dilaton background, where the vector meson modes are considered in the probe limit. The gravity-dilaton dynamics adjusted in such a way needs modifications to accommodate the QCD thermodynamics on the same footing. This finding seems to require either the treatment of hadrons beyond the probe limit or/and the inclusion of further fields beyond the one-component dilaton (section 4).

2. Holographic vector mesons

To catch the physics and related scales of vector mesons (V) as a QCD incarnation we use the standard action [5] in Einstein frame

$$S_V \propto \int \mathrm{d}z\, \mathrm{d}^4x\, \sqrt{g} F^2 \tag{1}$$

and the ansatz for the infinitesimal line element squared in five-dimensional Riemann spacetime

$$\mathrm{d}s^2 = e^{A(z)-\frac{2}{3}\Phi(z)} \left(f(z)\mathrm{d}t^2 - \mathrm{d}\vec{x}^2 - \frac{\mathrm{d}z^2}{f(z)} \right), \tag{2}$$

IOP Conf. Series: Journal of Physics: Conf. Series **1024** (2018) 012003 doi:10.1088/1742-6596/1024/1/012003

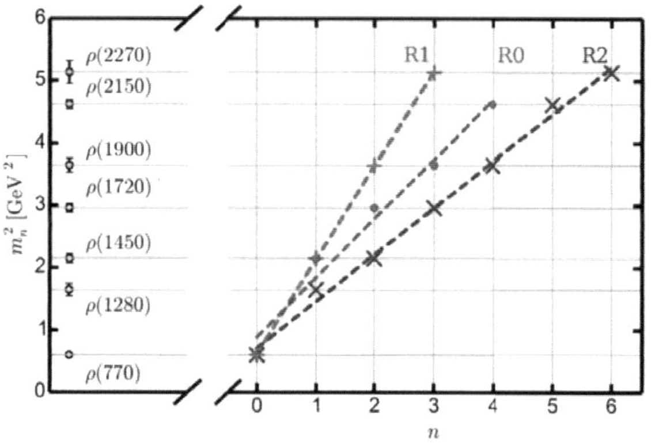

	R0	R1	R2
α [GeV^2]	0.89	0.62	0.71
β [GeV^2]	0.95	1.51	0.75
$-b$	0.253	2.540	0.056
L^{-1} [MeV]	487	614	433

Figure 1. Left: Various assignments of $J^{PC} = 1^{--}$ ρ meson masses squared [7], displayed on the l.h.s.,to radial quantum numbers n (R0: green dots [5], R1: red plusses [8], R2: blue crosses [9]; for other options, cf. [10], and [11] for the iso-scalar vector channel). The tentatively attributed Regge trajectories are depicted by dashed lines in the same colour code as the assignment symbols. Right: Parameters α and β of $m_n^2 = \alpha + \beta n$ for the trajectories depicted in the left, as well as b and L^{-1} of the potential U_0 (5) determining m_n^2 from (3).

where F^2 is the squared field strength tensor of a $U(1)$ vector field, A is a warp factor, f the blackness function, and Φ denotes the dilaton field. The equation of motion follows, after some manipulations (cf. [6]), as one-dimensional Schrödinger type equation

$$\left(\partial_\xi^2 - (U_T - m_n^2)\right)\psi = 0, \tag{3}$$

where $\partial_\xi = (1/f)\partial_z$ and $U_T(z) = U_0 f^2 + \frac{1}{2}Sff'$ with

$$U_0(z) = \frac{1}{2}S' + \frac{1}{4}S^2, \quad S \equiv \frac{1}{2}A' - \frac{2}{3}\Phi', \tag{4}$$

and a prime means derivative w.r.t. z. Considering first the vacuum case, $f = 1$ corresponding to zero temperature [5, 6], the ansatz (cf. [11] for instance)

$$U_0 = \frac{3}{4}\frac{1}{z^2} + 4\frac{b}{L^2} + \left(\frac{z}{L}\right)^2 \frac{1}{L^2} \tag{5}$$

delivers - with L as scale parameter - the spectrum $m_n^2 = 4(n+b+1)/L^2$ of normalisable modes of (3) with $n = 0, 1, 2, \ldots$ to be identified as radial quantum number. The table in Fig. 1 lists optimal parameter choices which catch nicely several options for arranging the first few ρ meson states as Regge trajectories (see left panel and upper part of the table in Fig. 1).
The requirement to have asymptotically an AdS spacetime, $\lim_{z\to 0} A(z) = -2\ln(z/L)$, gives constraints on U_0: The near-boundary expansions

$$S(z) = \frac{1}{z}\sum_{i=0}^{\infty} s_i \frac{z^i}{L^i} \Rightarrow \frac{1}{2}S' + \frac{1}{4}S^2 = \frac{1}{z^2}\sum_{i=0}^{\infty} u_i \frac{z^i}{L^i} \tag{6}$$

yield from (4, 5) $u_2 = 0$ (which implies $b = 0$) and $s_{2l} = 0$ since $u_{2l+1} = 0$ ($l = 0, 1, 2, \ldots$). Inspection of the table in Fig. 1 reveals that $b \neq 0$ could be required for certain Regge trajectories. (The case $b = 0$ recovers the original soft wall model [5], where $\alpha = \beta = 4/L^2$, corresponding to $1/L = 487$ MeV, have been advocated.) A simple remedy is suppressing the impact of b at the boundary by $U_0 \to \hat{U}_0(z) = \frac{3}{4z^2} + \frac{b}{L^2}(1 - e^{-z^4/L^4}) + \frac{z^2}{L^4}$, since the potential $\hat{U}_0$ still generates approximately linear Regge trajectories.

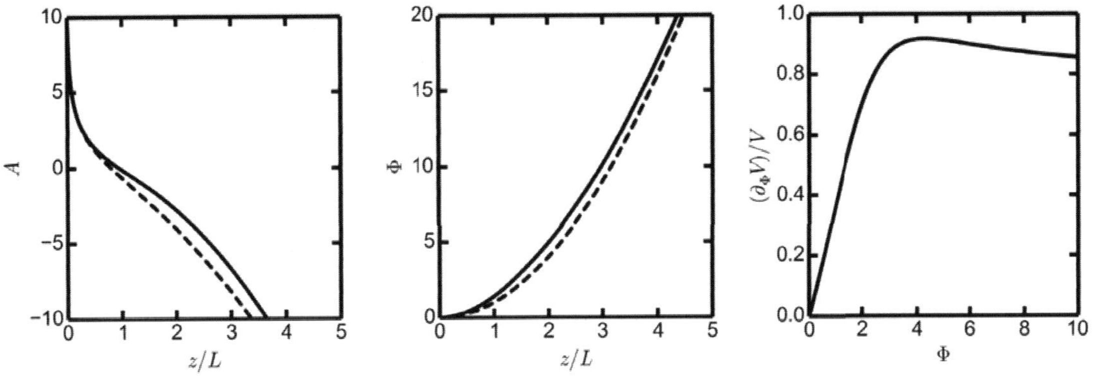

Figure 2. The warp factor A (left) and the dilaton profile Φ (middle) as functions of z/L. The dilaton potential is represented by $(\partial_\Phi V)/V$ as a function of Φ (right). The dashed curves are the profiles of the soft wall model.

3. Dilaton background

Our next aim is to relate that useful U_0 to the dilaton profile $\Phi(z)$ and the warp factor $A(z)$, cf. (4). To do so we consider the vector meson dynamics in the probe limit, i.e. we seek for a suitable approach to A and Φ without back reaction. One option is provided by the Einstein-dilaton model with action [3, 4, 12]

$$S_\Phi = \frac{1}{k}\int \mathrm{d}z\, \mathrm{d}^4x \sqrt{g}\left[R - \frac{1}{2}\partial_M\Phi\partial^M\Phi - V(\Phi)\right], \tag{7}$$

where the key quantity is the dilaton potential $V(\Phi)$, R denotes the Einstein-Hilbert action and $k = 1/(16\pi G_5)$, resulting with $a(z) \equiv \exp\{A(z)\}$ in the field equations

$$f''a + \frac{3}{2}a'f' = 0, \tag{8}$$

$$a''a - \frac{3}{2}a'^2 + \frac{1}{3}\Phi'^2a^2 = 0, \tag{9}$$

$$\left(a'^2 - \frac{1}{6}\Phi'^2a^2\right)f + \frac{1}{2}a'af' + \frac{1}{3}a^3V = 0. \tag{10}$$

Inserting the near-boundary expansions

$$a(z) = \frac{1}{z^2}\sum_{i=0}^{\infty} a_i\frac{z^i}{L^i}, \quad \Phi(z) = \sum_{i=0}^{\infty}\phi_i\frac{z^i}{L^i}, \quad f(z) = \sum_{i=0}^{\infty} f_i\frac{z^i}{L^i} \tag{11}$$

one obtains $a_0 = f_0 = 1$ and $a_{1,2,3} = f_{1,2,3} = 0$ as well as $\phi_{0,1} = 0$. In particular, $\phi_2 \neq 0$, i.e. a quadratic dilaton profile in leading order is permitted iff $m_\Phi^2L^2 = -4$. Such a profile is used in [13] as input to elucidate subsequently $V(\Phi)$ and the Regge trajectory. The case $m_\Phi^2L^2 < -4$ requires $\phi_2 = 0$, which is equivalent to $a_4 = 0$. A vanishing mass m_Φ is equivalent to a marginal operator and negative (positive) mass squared corresponds to a(n) (ir)relevant operator in the boundary theory, where one exploits the relations $\lim_{\Phi\to 0} V(\Phi) = -12/L^2 + m_\Phi^2\Phi^2/2 + \ldots$ and $m_\Phi^2L^2 = \Delta(\Delta-4)$ to characterise the dual operator to Φ by its conformal dimension Δ.
Integrating the field equations (9, 10) at $f = 1$ with the side condition (4) means mapping $U_0(z)$ (or a particular hadron spectrum) on $V(\Phi)$. The corresponding profiles $A(z)$ and $\Phi(z)$

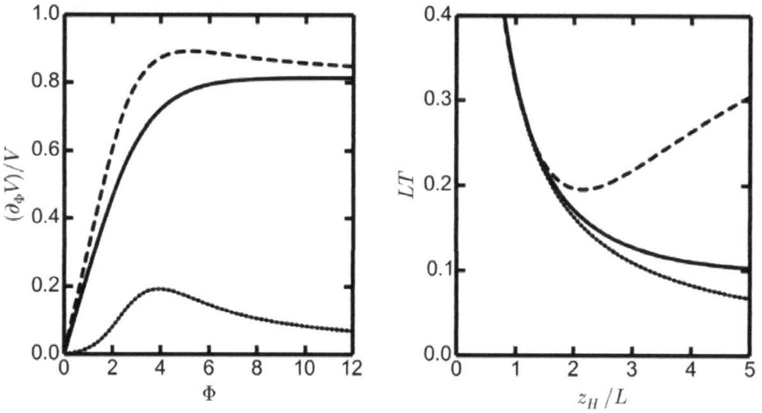

Figure 3. $(\partial_\Phi V)/V$ as a function of Φ (left panel) for $L^2V(\Phi,p) = -12\,{}_0F_1^2\left(\frac{1}{2p},\frac{\Phi^2}{24}\right) + \frac{p^2}{2}\Phi^2\,{}_0F_1^2\left(\frac{2p+1}{2p},\frac{\Phi^2}{24}\right)$ and the resulting temperature T as a function of the horizon position z_H (right panel). Dashed curves: $p = 1.5$, $T(z_H)$ has a global minimum (which induces a first-order phase transition; approximate Regge trajectories within U_0 at $T = 0$ but not within U_T); dotted curves: $p = 0.5$, $T(z_H)$ has an inflection point (which points to a cross over; at $T = 0$, the potentials U_0 and U_T do not allow for meson states at all); solid curves: $p = 1$, delineation case, where $T(z_H)$ asymptotes to a constant value at large values of z_H. The temperature independent potentials $V(\Phi,p)$ and U_0 follow here from (9, 10) at $f = 1$ by the ansatz $\Phi = (z/L)^p$.

are exhibited in Fig. 2 (left and middle panels) as functions of z/L. For a comparison, $A^{\mathrm{sw}} = -2\ln(z/L)$ and $\Phi^{\mathrm{sw}} = z^2/L^2$, which are elements of the soft wall model [5], are displayed too by dashed curves. In the displayed range, the soft wall profiles may serve as leading-order proxies of the solutions of the Einstein equations such to accommodate the Regge trajectory-determining potential U_0, albeit at $b = 0$ and for the vacuum, $f = 1$. This justifies a posteriori the ansätze in [14, 15] which explore certain deformations of the soft wall profiles with the goal to accommodate relevant QCD thermodynamics features.
The right panel in Fig. 2 exhibits $(\partial_\Phi V)/V$ as a function of Φ. With the aid of the adiabatic approximation [12], one can infer from $v_s^2 = 1-\frac{2}{3}[(\partial_\Phi V)/V]^2+\ldots$ that for $(\partial_\Phi V)/V > \sqrt{2/3}$ the velocity of sound, v_s, becomes imaginary, signalling the onset of a first-order phase transition, see also [24]. In fact, solving the set of field equations (8-10) for $f(z,z_H) \leq 1$ and $f(z = z_H, z_H) = 0$, i.e. admitting a horizon at z_H, with the same potential $V(\Phi)$ which is found for the above prescribed potential U_0, one sees that the temperature, $T(z_H) = -\partial_z f(z,z_H)\,|_{z=z_H}/(4\pi)$ has a global minimum at $z_H^{\mathrm{min}} = 1.2/L$ with $T(z_H^{\mathrm{min}}) = 116$ MeV for $1/L = 687$ MeV. As extensively discussed in [3, 4] this refers to a first-order phase transition with transition temperature $T_c = 1.02\,T(z_H^{\mathrm{min}})$.
It seems conceivable that a pure gravity-one–component dilaton model has a too restricted field content to mimic QCD thermodynamics and vacuum hadron (at least, vector meson) spectra by one unique potential $V(\Phi)$. For instance, considering a special ansatz for $V(\Phi)$ we find that either proper Regge trajectories and a first-order phase transition or QCD thermodynamics with a cross over and none discrete hadron states are realised in such a scenario, see Fig. 3. (This supplements the assessment in [16] for the glue sector based on (7): a gapped, discrete spectrum at $T = 0$ facilitates a first-order transition at $T > 0$.) A further issue is that holographic models face often a too low temperature of hadron melting [17, 18, 19, 20]. The above value of T_c is

a reminiscence of that insanity, as it is significantly below the QCD cross-over temperature of $\mathcal{O}(150\ \text{MeV})$.

4. Conclusions

The constraints from the mass spectra of radial ρ meson excitations forming Regge trajectories can be used to construct the dilaton potential $V(\Phi)$ within an Einstein-dilaton background model by solving the field equations at zero temperature (vacuum). However, the same $V(\Phi)$ generates a first-order (Hawking-Page type) phase transition in the case of non-zero temperatures. That kind of phase structure stands in conflict with 2+1 flavour QCD with physical quark masses [21, 22], however, it is in line with expectations of 2+1 flavour QCD near the chiral limit, where firm quantitative results are still lacking [23]. Hence, the strategy of our following work will be seeking other avenues to the gravity background and dilaton potential such to accommodate the vacuum hadron spectrum and QCD thermodynamics at the same time, as envisaged in the ansatz trials [14, 15, 25], as a prerequisite to uncover the T-μ plane (cf. [26, 27] for thermodynamics aspects and [28, 29] for the general layout) in the FAIR relevant region by holographic means.

Acknowledgments

Useful discussions with Profs. Mei Huang and Jorge Noronha are gratefully acknowledged. The work is supported by Studienstiftung des deutschen Volkes.

[1] Galatyuk T et al., *Eur. Phys. J.* A **52**, 131 (2016).
[2] Ammon M, Erdmenger J, *Gauge/Gravity duality*, Cambridge University Press (2015).
[3] Gürsoy U, Kiritsis E, *JHEP* 0802, 032 (2008).
[4] Gürsoy U, Kiritsis E, Nitti F, *JHEP* 0802, 019 (2008).
[5] Karch A, Katz E, Son D T, Stephanov M A, *Phys. Rev.* D **740**, 015005 (2006).
[6] Colangelo P, Giannuzzi F, Nicotri S, *Phys. Rev.* D **80**, 094019 (2009).
[7] Patrignani C et al. (Particle Data Group), *Chin. Phys.* C **40**, 100001 (2016).
[8] Ebert D, Faustov R N, Galkin V O, *Phys. Rev.* D **79**, 114029 (2009).
[9] Bartz S P, Kapusta J I, *Phys. Rev.* D **90**, 074034 (2014).
[10] Masjuan P S, Arriola E, Broniowski W, *Phys. Rev.* D **85**, 094006 (2012).
[11] Afonin S S, *Phys. Lett.* B **719**, 399 (2013).
[12] Gubser S S, Nellore A, *Phys. Rev.* D **78**, 086007 (2008).
[13] Li D, Huang M, *JHEP* 1311, 088 (2013).
[14] Zöllner R, Kämpfer B, *Phys. Rev.* C **94**, 045205 (2016).
[15] Zöllner R, Kämpfer B, *Eur. Phys. J.* A **53**, 139 (2017).
[16] Gürsoy U, Kiritsis E, Mazzanti L, Nitti F, *JHEP* 05, 033 (2009).
[17] Fujita M, Kikuchi T, Fukushima K, Misumi T, Murata M, *Phys. Rev.* D **81**, 065024 (2010).
[18] Ishii T, Kinoshita S, Murata K, Tanahashi N, *JHEP* 1404, 099 (2014).
[19] Braga N R F, Martin Contreras M A, Diles S, *Eur. Phys. J.* C **76**, 598 (2016).
[20] Bartz S P, Jacobson T, *Phys. Rev.* D **94**, 075022 (2016).
[21] Borsányi S et al., *Phys. Lett.* B **370**, 99 (2014).
[22] Bazavov A et al., *Phys. Lett.* B **737**, 210 (2014).
[23] Bazavov A et al., *Phys. Rev.* D **95**, 074505 (2017).
[24] Yaresko R, Kämpfer B, *Phys. Lett.* B **747**, 36 (2015).
[25] Zöllner R, Wunderlich F, Kämpfer B, *Preprint* arXiv:1611.04124 [hep-th] (2016).
[26] Knaute J, Yaresko R, Kämpfer B, *Preprint* arXiv:1702.06731 [hep-th] (2017).
[27] Critelli R et al., *Preprint* arXiv:1706.00455 [hep-th] (2017).
[28] Iatrakis I, Kiritsis E, Paredes A, *JHEP* 1011, 123 (2010).
[29] De Wolfe O, Gubser S S, Rosen C, *Phys. Rev.* D **83**, 086005 (2011).

IOP Conf. Series: Journal of Physics: Conf. Series **1024** (2018) 012004 doi:10.1088/1742-6596/1024/1/012004

Investigation into the charge collection efficiency of prototype microstrip sensors for the CBM Silicon Tracking System

I. Momot[1,2,3], M. Teklishyn[2,4], A. Lymanets[3], O. Bertini[3] and J. Heuser[3]

[1] Goethe University, Frankfurt am Main, Germany
[2] Kyiv Institute for Nuclear Research, Kyiv, Ukraine
[3] GSI Helmholtzzentrum für Schwerionenforschung GmbH, Darmstadt, Germany
[4] FAIR Facility for Antiproton and Ion Research, Darmstadt, Germany

E-mail: i.momot@gsi.de

Abstract.
Double-sided silicon microstrip sensors are known to be sensitive to the radiation dose received during operation. The lifetime fluence for the Silicon Tracking System of the CBM experiment is estimated as 10^{14} 1 MeV n_{eq} cm^{-2}. In order to maintain the signal-to-noise ratio sufficiently high during all the time of operation we study the newest sensor prototypes irradiated to the single and double lifetime doses. The results of the tests are addressed in this paper.

1. Introduction

The Compressed Barionic Matter (CBM) [1] experiment is planned as a part of the future FAIR facility in Darmstadt, Germany [2]. The aim of CBM is to explore the phase diagram of strongly interacting matter at high net baryon densities and moderate temperatures. The Silicon Tracking System (STS) [3] is the core tracking detector of the CBM experiment. STS will be placed in the 1 Tm dipole magnet 30 cm downstream of the target. It will provide track reconstruction and momentum determination of the charged particles. Signals registered by the charge sensitive electronics will be used to determine interaction points of the charged particles with the silicon sensors. During the operation of CBM, STS will cope with up to 10^3 charged particles per central Au-Au collision. As a result of the high interaction rates of up to 10 MHz, sensors are expected to be sufficiently radiation hard to survive fluences up to 10^{14} 1 MeV n_{eq} cm^{-2}.

STS will employ double-sided double-metal n-type silicon microstrip sensors. They are planned to be produced in different sizes: 2.2×6.2, 4.2×6.2, 6.2×6.2 and 12.2×6.2 cm^2 outer dimensions. Prototype sensors are shown in Fig. 1. On each sensor, there are 1024 strips per side with 58 μm pitch. To maintain a good position resolution in the bending plane of the magnet and to avoid excessive fake interaction points, strips on the *p*-side are oriented by 7.5° with respect to the sensor edge. Corner strips on both edges are interconnected via second metal layer so that sensor read-out routing lines can be achieved from one sensor edge only, matching the detector module concept.

IOP Conf. Series: Journal of Physics: Conf. Series **1024** (2018) 012004 doi:10.1088/1742-6596/1024/1/012004

2. Sensors characterization before and after irradiation

In order to evaluate the sensor performance the following procedures have to be accomplished before and after irradiation [4]:

- the study of the electrical characteristics of a sensor by measuring the leakage current as a function of the reverse bias voltage (IV) and the bulk capacitance (CV curve);
- the determination of the signal-to-noise ratio by measuring the signal amplitude from the radioactive source and the baseline width in the attached read-out electronics.

In particular the signal-to-noise ratio is the figure of merit for the detector performance. In this ratio the signal is defined as a mode of the particle energy distribution and the noise is estimated as the root-mean-square of the baseline variation in a particular channel.

2.1. Experimental setup

For the test purposes the prototype sensors are installed into printed circuit boards (PCB), which are populated with pitch-adapters in order to adjust the step of the traces from 58 μm to 130 μm. In-built R-C-L filters are mounted on the PCB after the high voltage input. In order to protect the wire-bonds, non-conductive glue was applied. A particular PCB configuration is shown in Fig. 2, the group of 128 central strips per side is ready for the charge collection measurements.

The measurements were performed in a custom made light-tight thermal enclosure with the temperature and humidity control. The top view of the setup is depicted in Fig. 3. Radiators with fans are placed aside, scintillator and photomultiplier are under the sensor. Right above of the sensor there is a collimator with the radiation source. In order to suppress the leakage current during data taking and to avoid annealing, sensors were constantly kept at the temperature of -10°C and at a relative humidity of 30 – 50%.

Figure 1. Silicon microstrip sensors in four different sizes: 2.2×6.2, 4.2×6.2, 6.2×6.2 and 12.2×6.2 cm^2.

Figure 2. Sensor installed into a printed circuit board (PCB).

Figure 3. Top view of the radiation source setup.

The detection of minimum ionizing particles (MIPs) is an ultimate task for the silicon sensors. To mimic MIPs, a ^{90}Sr β-source was used. It produces electrons with a continuous energy spectrum and E_{max}=2.28 MeV. A plastic scintillator (2.5 cm thick) with an attached photomultiplier was used to produce external trigger signals. The most energetic particles were selected by requiring a sufficiently high signal amplitude in the scintillator. In order to keep the interaction region of the same size as the sensitive area, a system of collimators was implemented in the setup.

A dedicated read-out chip was designed for the STS. STS-XYTER is suitable for the signal detection from double sided silicon microstrip sensors [5]. This chip is currently under evaluation

IOP Conf. Series: Journal of Physics: Conf. Series **1024** (2018) 012004 doi:10.1088/1742-6596/1024/1/012004

thus we can not use it for charge collection studies yet. For the purpose of the laboratory tests the Alibava system [6] was involved. This system uses the Beetle ASIC [7] which was applied for the LHCb Vertex Detector, Silicon Tracker and RICH [8]. Front-end boards of the Alibava were customized according to our needs: the fan-out with two 68-pin ERNI connectors was wire-bonded to the pitch adaptor of the Beetle chip.

2.2. Electrical and signal measurements

As it was mentioned above, the IV and CV measurement must be performed before and after irradiation. A particular example of electrical characteristics is shown in Fig. 4. The leakage current is strongly dependent on the temperature, it has been lately normalised to 20°C and to the sensor size. As one can see on the plots, the bulk capacitance does not change with the irradiation, but the leakage current increases by a factor of 1000, as expected.

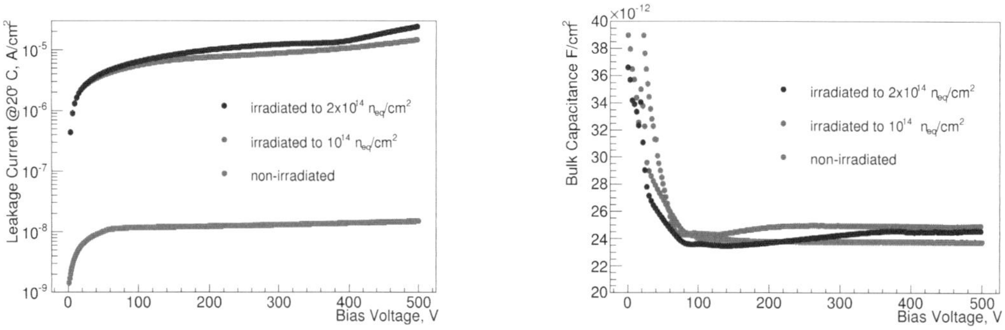

Figure 4. Comparison of the leakage current (left) and the bulk capacitance (right) as a function of the applied voltage for non-irradiated and irradiated sensors.

The noise level was estimated as the baseline width for each channel. All edge and noisy channels are removed from the analysis. The events with a single channel amplitude higher than 5σ of the noise level fill the signal distribution histogram. Geometrical considerations show that for the small crossing angles (i.e. close to perpendicular incidence) the total charge should not spread by more than 3 neighbouring strips. The search algorithm for the charge clusters starts from the strip with the highest signal-to-noise ratio and then moves to the left and to the right, stopping whenever the signal falls below the threshold.

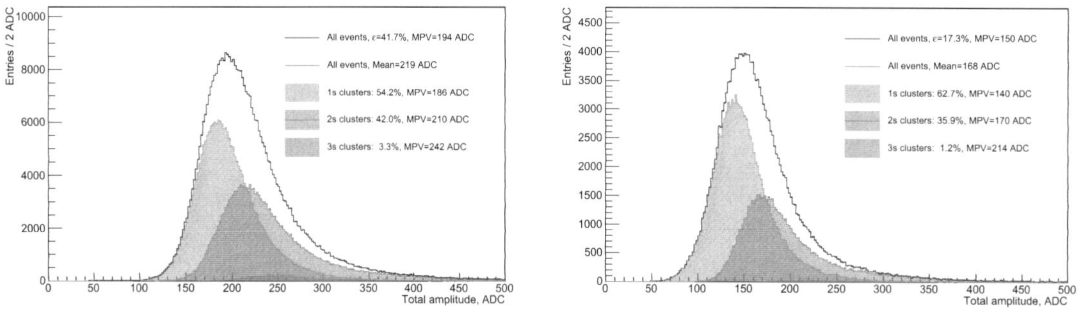

Figure 5. Spectra amplitude of non-irradiated sensor (left) and irradiated up to $2{\times}10^{14}$ 1 MeV n_{eq} cm^{-2} with spectra (right).

The most probable value obtained from the charge distribution is interpreted as the collected charge. Sample spectra of from a ^{90}Sr source are presented in Fig. 5.

The signal-to-noise ratio for one-strip clusters of non-irradiated 2.2×6.2 cm^2 sensors are shown in Fig. 6. *P*- and *n*-side read-out are denoted with p and n, the sensor name is composed of the vendor coding (c for CiS and h for Hamamatsu). Beyond the full depletion of the sensor, the signal-to-noise ratio remains constant. The systematic uncertainty is found to be much larger then statistical one. The error bar on the next figures represent uncertainty correlated between voltage scan points.

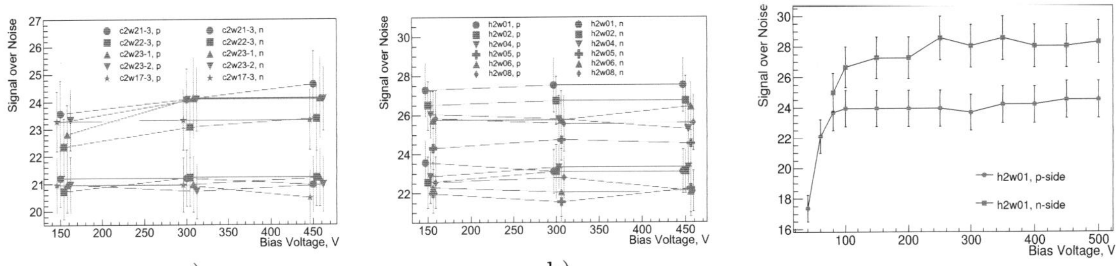

a) b) c)

Figure 6. Signal-to-noise ratio for non-irradiated 2.2×6.2 cm^2 sensors from different vendors at three applied voltages a) CiS and b) Hamamatsu. c) Dependence of the charge collection on the applied voltage in terms of signal-to-noise ratio. For the sake of illustration points were shifted on the X-axis by 3 V.

2.3. Charge collection efficiency after irradiation

The sensors were irradiated at the KIT Irradiation Center (Karlsruhe, Germany) with protons of E_{kin} = 22.9 MeV produced by the Compact Cyclotron [9]. There, sensors are scanned in 1 mm spaced rows by the 7 mm wide beam. In order to dissipate the heat and to minimize annealing effects, nitrogen flow at the temperature of -40°C was supplied during irradiation.

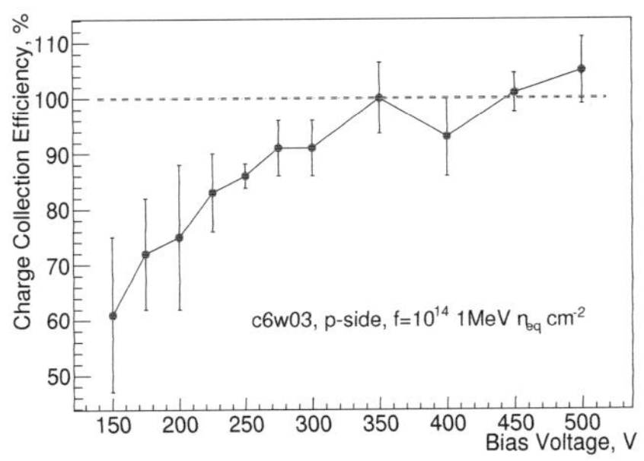

Figure 7. Dependence of the charge collection efficiency of the irradiated sensor on the applied voltage.

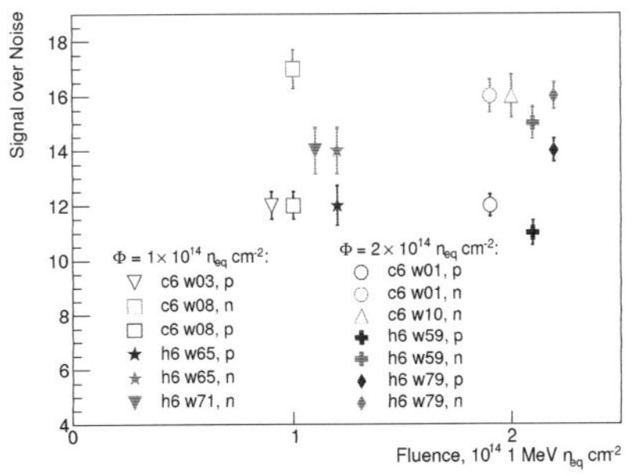

Figure 8. Signal-to-noise ratio for one strip cluster events of irradiated sensors as a function of fluence.

The radiation causes defects in the silicon lattice which act as traps for the charge carriers. To compensate this damage, one should apply higher bias voltage in order to speed up the charge collection in the sensor medium. In Fig. 7 one can see how the charge collection efficiency saturates to 100% with sufficient bias voltage applied. The 100% level of charge collection efficiency was defined as the charge collected by the non-irradiated sensor from the same vendor of given thickness. Fig. 8 depicts the signal-to-noise ratios obtained with different sensors.

Nevertheless, with an expected radiation environment of the CBM experiment it is still possible to collect almost 100% of the signal with increasing of the bias voltage. On Fig. 9 one can see performance of the sensors irradiated up to twice the lifetime fluence.

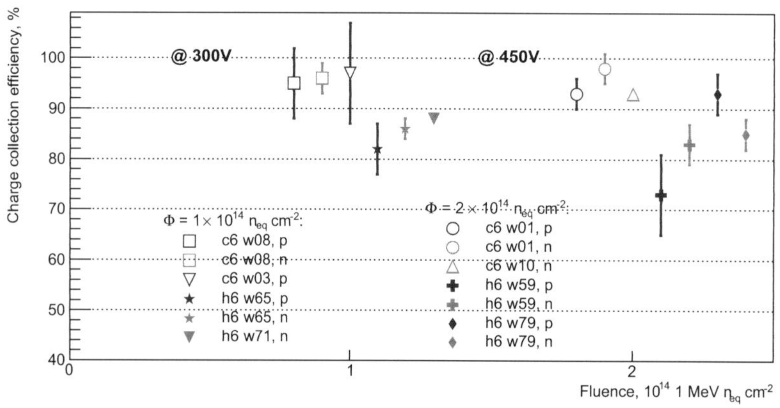

Figure 9. Charge collection efficiency for the *p*- and *n*-side of irradiated sensors.

3. Conclusions

The latest prototypes of the silicon microstrip sensors for the STS are found to be sufficiently radiation hard to survive the lifetime particle flux without critical performance degradation. At radiation levels reaching once and twice the expected nominal life-times of the STS detector, i.e. neutron equivalent fluences of $1{\times}10^{14}$ and $2{\times}10^{14}$ 1 MeV n_{eq} cm^{-2}, the signal-to-noise ratio after irradiation drops down from about 23 to 12. The sensor prototypes from two vendors show a reduction of the charge collection efficiency by 10% to 20% after being irradiated to the double lifetime non-ionizing dose.

Acknowledgments

Supported by HGS-HIRe.

References

[1] Friman B *et al.* 2011 *The CBM Physics Book: Compressed Baryonic Matter in Laboratory Experiments* vol. 814 (Lecture Notes in Physics, Springer)
[2] `http://www.fair-center.org/`
[3] Heuser J *et al.* 2013-4 *Technical Design Report for the CBM Silicon Tracking System, GSI Report 2013-4*
[4] Bertini O *et al. Specifications for Quality Assurance of Microstrip Sensors in the CBM Silicon Tracking System* CBM-STS Note 2016-1
[5] Kasinski K *et al.* 2014 *Proc. NSS/MIC IEEE* (Seattle) (IEEE)
[6] `http://www.alibavasystems.com/`
[7] Löhner S and Schmelling M LHCb-2005-105 CERN-LHCb-2005-105
[8] `http://www.kip.uni-heidelberg.de/lhcb/`
[9] `http://www.etp.kit.edu/english/irradiation_center.php`

IOP Conf. Series: Journal of Physics: Conf. Series **1024** (2018) 012005 doi:10.1088/1742-6596/1024/1/012005

Recent theoretical advances regarding α-spectroscopy

A Dumitrescu[1,2] **and DS Delion**[1,2,3]

[1] "Horia Hulubei" National Institute for R&D in Physics and Nuclear Engineering, 407 Atomiştilor, POB-MG6, Bucharest-Măgurele, RO-077125, Romania
[2] Academy of Romanian Scientists, Splaiul Independenţei, Bucharest, RO-013724, Romania
[3] Bioterra University, Gârlei str., Bucharest, RO-013724, Romania

E-mail: alexandru.dumitrescu@theory.nipne.ro

Abstract. We present an overview of a unified description of the structure and α-emission properties of even-even nuclei. The low-energy spectrum relevant for α-emission is described within the framework of the Coherent State Model (CSM). The treatment of the α-emission process is based on an α-daughter interaction containing a monopole component, calculated through a double folding procedure with a M3Y interaction plus a repulsive core simulating the Pauli principle, and a quadrupole-quadrupole (QQ) interaction. The decaying states are identified with the lowest narrow outgoing resonances obtained through the coupled channels method. The α-branching ratio to the first excited state is reproduced by means of the QQ strength. Simultaneously, a reasonable agreement is obtained for the α-branching ratios to the second and third excited states.

1. Introduction

The field of α-spectroscopy in an extension of traditional nuclear spectroscopy, where the probe is no longer of electromagnetic, but of strong origin. Thus, in addition to the usual γ-ray analysis, one may investigate nuclear structure through α-emission. This work summarizes recent results regarding the fine structure of the α-decay spectrum in even-even nuclei. For a comprehensive study of the presented structure properties and α-transitions, as well as a review of the literature, see Refs. [1, 2]. For a systematics of experimental data regarding decay intensities and partial half-lives, consult Ref. [3]. The topic of α-decay in odd-mass nuclei is expanded upon in Ref. [4], for the case of favored α-transitions. Remarks concerning the even-odd staggering of the α-particle spectroscopic factor and its relation to α-clustering can be found in Ref. [5].

2. The CSM Model

The main idea behind this approach consists in treating the surface vibrations of a deformed even-even nucleus as a superposition of quadrupole phonons [6, 7]. Following this idea, the model was greatly expanded towards the treatment of three interacting collective bands [8, 9, 10], as well as in terms of the particle-core interaction in odd-mass nuclei [11].

In its simplest realization, the model is based on the assumption that the intrinsic state of an even-even nucleus having an arbitrary axial deformation is given by a coherent superposition of quadrupole phonons

$$|\psi_g\rangle = e^{d\left(b_{20}^{\dagger}-b_{20}\right)}|0\rangle \ , \tag{1}$$

IOP Conf. Series: Journal of Physics: Conf. Series **1024** (2018) 012005 doi:10.1088/1742-6596/1024/1/012005

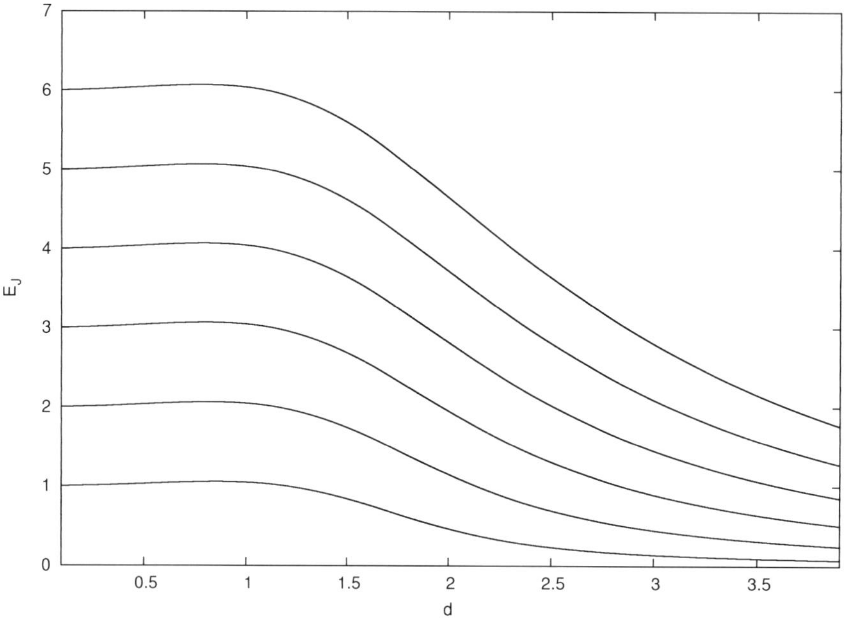

Figure 1. Normalized energy levels as function of CSM deformation for an even-even core.

where $\left(b^\dagger_{2\mu}, b_{2\mu}\right)$ are the phonon operators with projection $\mu = 0$ and d is the deformation parameter, proportional to the usual quadrupole deformation [8].

The ground band states are obtained by projecting out the angular momentum components of the intrinsic function

$$|\varphi^{(g)}_J\rangle = \mathcal{N}^{(g)}_J \hat{P}^J_{M0}|\psi_g\rangle \ , \tag{2}$$

by using the projection operator $\hat{P}^J_{M0}$ and normalizing with the factor $\mathcal{N}^{(g)}_J$. The corresponding energy levels E_J follow as the expectation values in this basis of the Hamiltonian

$$\hat{H} = A_1 b^\dagger_2 \cdot b_2, \tag{3}$$

where A_1 is a strength factor and the dot designates a scalar product. The resulting level structure normalized to A_1 as function of deformation is shown in Fig. 1. One observes that the spectrum is harmonic for vanishing deformation, goes through a transitional regime as the deformation increases and then forms a rotational band. Detailed applications of this method to the study of even-even α-emitters and its extension to the odd-mass case can be consulted in Refs. [2, 4].

3. The coupled-channels method

The general theory of resonant states is expanded upon in Refs. [12, 13]. Here, we will give a concise overview of the coupled-channels equation and its application to α-decay in the case of even-even nuclei.

The α-transition process can be represented as

$$P \to D(J) + \alpha(J) \ , \tag{4}$$

where, J is the spin/parity of the daughter nucleus and emitted α-particle, coupling to the 0 angular momentum of the parent nucleus.

The wave function that is under consideration has the form of a clustered α-daughter ansatz [13] with the total spin of the initial state

$$\Psi(b_2^\dagger, \mathbf{R}) = \sum_J \frac{f_J(R)}{R} \mathcal{Y}^{(J)}(b_2^\dagger, \hat{R}) \ . \tag{5}$$

The function $f_J(R)$ describes the α-daughter radial motion in the channel J. We introduce the core-angular harmonic

$$\mathcal{Y}^{(J)}(b_2^\dagger, \hat{R}) = \left[\varphi_J(b_2^\dagger) \otimes Y_J(\hat{R})\right]_0 \ , \tag{6}$$

where $\varphi_J(b_2^\dagger)$ designates the daughter internal wave function, with $b_2^\dagger$ the daughter degrees of freedom, while $Y_J(\hat{R})$ is the usual spherical harmonic that describes the relative angular motion of the α-daughter system. These core-angular harmonics are orthonormal.

The α-daughter dynamics is treated using a stationary Schrödinger equation, i.e.

$$H\Psi(b_2^\dagger, \mathbf{R}) = Q_\alpha \Psi(b_2^\dagger, \mathbf{R}) \ , \tag{7}$$

where Q_α is the Q-value of the decay process. Due to the fact that all measured decay widths are by many orders of magnitude smaller than the corresponding Q-values, the stationarity approximation is a very good assumption. Hence, an α-decaying state is identified with a narrow resonant solution that contains only outgoing components.

The Hamiltonian

$$H = -\frac{\hbar^2}{2\mu}\nabla_R^2 + H_D(b_2^\dagger) + V(b_2^\dagger, \mathbf{R}) \ , \tag{8}$$

contains the kinetic operator depending on the reduced mass, a term describing the dynamics of the daughter nucleus $H_D(b_2^\dagger)$ and the α-core interaction $V(b_2^\dagger, \mathbf{R})$, which is generally split into spherical and deformed parts. The spherical component follows from the matching of a harmonic oscillator with the double folding potential integrated from the M3Y nucleon-nucleon plus Coulomb force (see Ref. [13] for computational details, along with references contained therein), while for the deformed component it is enough to retain the QQ coupling.

This expansion of the interaction, together with the expansion of the wavefunction in the core-angular harmonic basis of Eq. (6), leads to a system of coupled differential equations for the radial functions

$$\frac{d^2 f_J(R)}{d\rho_J^2} = \sum_{J'} A_{JJ'}(R) f_{J'}(R) \ , \tag{9}$$

where ρ_J is the reduced radius. The coupling matrix $A_{JJ'}$ has off-diagonal components given by the matrix elements of the deformed part of the interaction. When evaluated in the CSM model, the QQ component of the interaction term present in Eq. (8) can be adjusted by a coupling strength dependent on the CSM deformation [2]. These matrix elements vanish at large distances, where the field is Coulombian, and there the asymptotic expression of the first resonance of this potential is an outgoing Coulomb-Hankel wave. One can show [13] that the partial decay widths that constitute the fine structure of the α-emission spectrum can be supplied by the asymptotic form of these functions

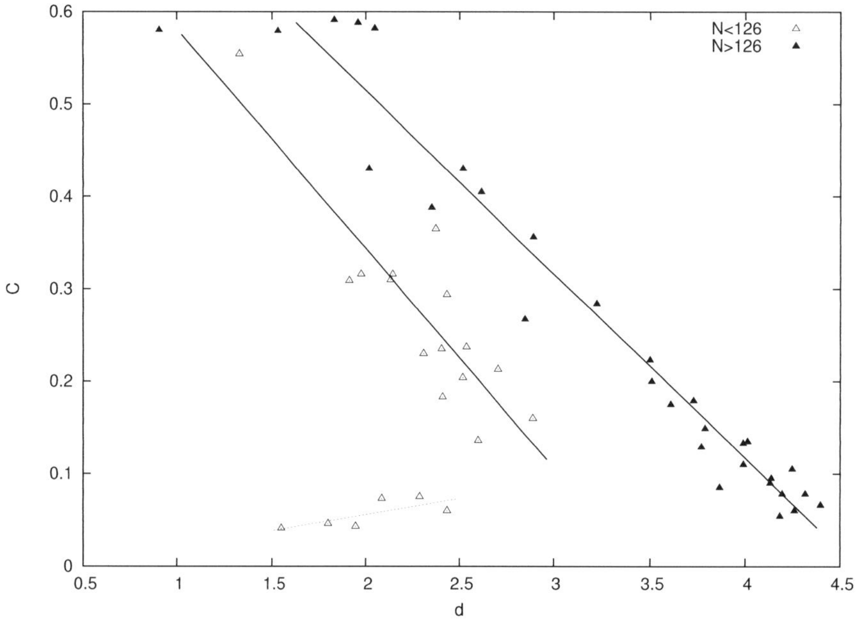

Figure 2. QQ coupling strength C versus CSM deformation d for even-even actinide emitters.

$$\begin{aligned}\Gamma &= \sum_J \Gamma_J = \sum_J {}^-hy \lim_{R\to\infty} |f_J(R)|^2 \\ &= \sum_J {}^-hy |N_J|^2 \,,\end{aligned} \tag{10}$$

in terms of the scattering amplitudes N_J and the center of mass velocity at infinity for each α-daughter channel J.

The goal is to give an accurate description of the decay intensities

$$\Upsilon_J = \log_{10}\frac{\Gamma_0}{\Gamma_J}, \tag{11}$$

which is easily checked by means of the so-called suppression factors

$$\log_{10} SF_J = \Upsilon_J^{\text{exp}} - \Upsilon_J^{\text{theor}}, \tag{12}$$

where the two values stand for experimental and theoretical quantities, respectively. Clearly, small suppression factors will indicate an accurate theory.

4. Application

Solving the coupled channels system of Eq. (9) in a manner that reproduces the decay intensities to 2^+ states leads to a linear dependence of the QQ coupling strength, denoted by C, on the CSM deformation d. This is shown in Fig. 2. One notices different slopes around the neutron number $N = 126$. For heavy nuclei, where several transitions are observed in the decay spectra, this dependence can be used to predict their structure. For example, in the case of even-even actinides, one can obtain a simultaneous description of the transitions to 2^+ and 4^+ states up to

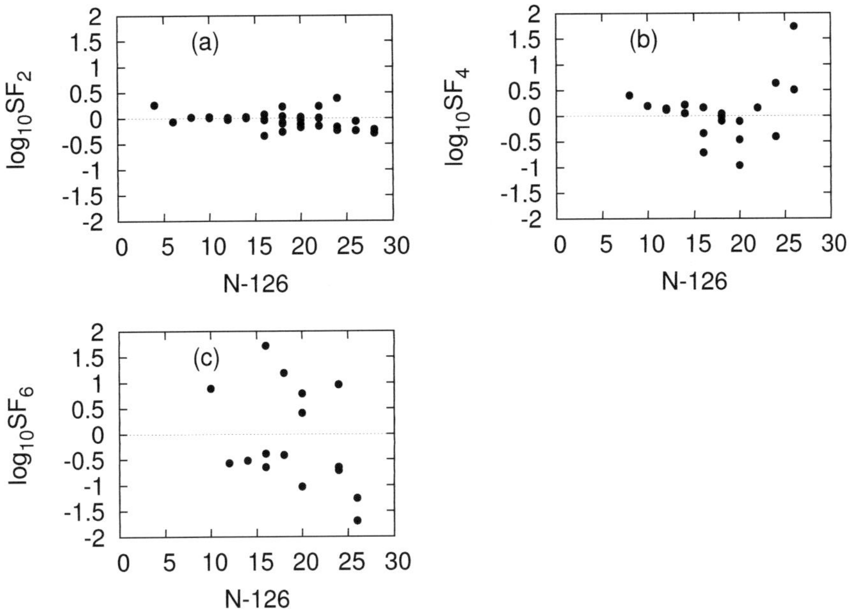

Figure 3. Suppression factors of α-transitions in even-even actinides as function of the neutron number above the closed shell at N=126, for decays from the ground state of the parent to the daughter states of angular momenta $J = 2, 4, 6$ in panels (a), (b) and (c) respectively.

a factor of 3, while the transitions to 6^+ states are usually described with an accuracy of about an order of magnitude, as shown in Fig. 3.

In conclusion, the CSM model offers a unified approach to the study of vibrational, transitional and well-deformed nuclei. Inserting the parametrization of the deformation that describes the nuclear structure in the coupled channels equation supplies good estimates for the fine structure of the α-emission spectrum.

Acknowledgments

In carrying out this work, the support from the Academy of Romanian Scientists through the project "*Approximations and other applications of mathematics in physics and engineering*" is gratefully acknowledged.

References

[1] D.S. Delion and A. Dumitrescu, Phys. Rev. C **87**, 044314 (2013).
[2] D.S. Delion and A. Dumitrescu, At. Data Nucl. Data Tables **100**, 1 (2015).
[3] D.S. Delion, and A. Dumitrescu, Phys. Rev. C **92** (2), 021303(R) (2015).
[4] A. Dumitrescu and D.S. Delion, Phys. Rev. C **93**, 024313 (2016).
[5] D.S. Delion, A. Dumitrescu and V.V. Baran, Phys. Rev. C **93**, 044321 (2016).
[6] P.O. Lipas and J. Savolainen, Nucl. Phys. A **130**, 77 (1969).
[7] P.O. Lipas, P. Haapakoski and T. Honkaranta, Phys. Scripta **13**, 339 (1976).
[8] A.A. Raduta, V. Ceausescu, A. Gheorghe, and R.M. Dreizler, Phys. Lett. B **99**, 444 (1981).
[9] A.A. Raduta, V. Ceausescu, A. Gheorghe, and R.M. Dreizler, Nucl. Phys. A **381**, 253 (1982).
[10] A.A. Raduta, R. Budaca, and Amand Faessler, Ann. Phys. (NY) **327**, 671 (2012).
[11] A. A. Raduta, D. S. Delion and N. Lo Iudice, Nucl. Phys. **A551** (1993).
[12] A.M. Lane and R.G. Thomas, Rev. Mod. Phys. **30**, 257 (1958).
[13] D.S. Delion, *Theory of particle and cluster emission* (Springer-Verlag, Berlin, 2010).

Measurements and understanding of fundamental properties of hot and dense nuclear matter with HADES

Georgy Kornakov for the HADES Collaboration

IKP, Technische Universität Darmstadt, Schlossgartenstr. 9, 64298 Darmstadt, Germany

E-mail: g.kornakov@gsi.de

Abstract. Heavy ion collisions at few GeV per nucleon energy range according to microscopic transport model calculations, create matter with densities several times larger that the ground density and heated to temperatures of 90 MeV. At such conditions the fundamental properties of the particles might be modified. Promising observables are the rare (multi-) strange hadrons and dileptons. The High Acceptance DiElectron Spectrometer was designed to measure rare and penetrating probes. It is a fixed target experiment located at SIS18 accelerator at GSI, Darmstadt, Germany. Recent results on dileptons and strangeness production in Au+Au collisions at center of mass energy $\sqrt{s_{NN}} = 2.42$ GeV are presented. Moreover, future experiments and upgrades of the apparatus are discussed.

1. Introduction

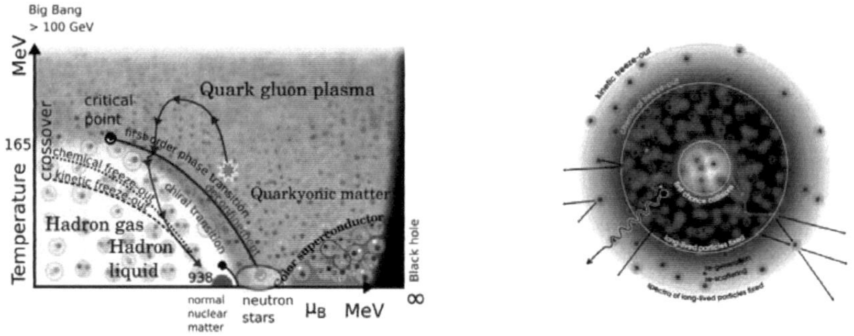

Figure 1. Cartoon of a phase diagram of QCD matter (left) and the time evolution of a heavy-ion collision. Red circles: inelastic collisions, blue: elastic (right).

Heavy-ion collisions at relativistic energies provide a laboratory in which hot and dense strongly interacting matter is created. According to a qualitative picture of the phase diagram of QCD matter as a function of temperature and baryochemical potential μ_B, shown in Figure 1, different phases might exist[1]. The transition boundary between confined matter into hadrons and the deconfined quark-gluon plasma remains a question, especially its location and width[2]. It is thought that hadronic degrees of freedom at high enough densities will disappear and another phase of chirally restored matter emerge. Also, theoretical and experimental efforts are done in addressing the possibility of existence of more exotic quark grouping structures as well as inhomogeneous or mixed phases[1][3]. At bombarding energies of few GeV the so called resonance matter is formed, composed of nucleons and their excited states with mesons, mostly pions. During the collision, different regions of the phase diagram can be accessed. In the beginning, the first nucleon-nucleon interactions take place, resulting in an increase of temperature and baryochemical potential by increasing the density of the system, then the formed matter cools down until inelastic and elastic collisions terminate[4] as shown in the left panel of Figure 1. This matter seems to have features of a thermalized system from measured abundances of hadrons in the final state[5][6]. The spectral function of the ρ meson, when produced within a hot and dense hadronic state is modified[7]. This feature is essential to describe the low-mass dilepton yields[8]. The production of virtual photons, strange and double-strange hadrons, proton number fluctuations, azimuthal flow are the main observables in HADES. Currently, HADES is the only experiment exploring the high baryochemical potential region.

2. The HADES

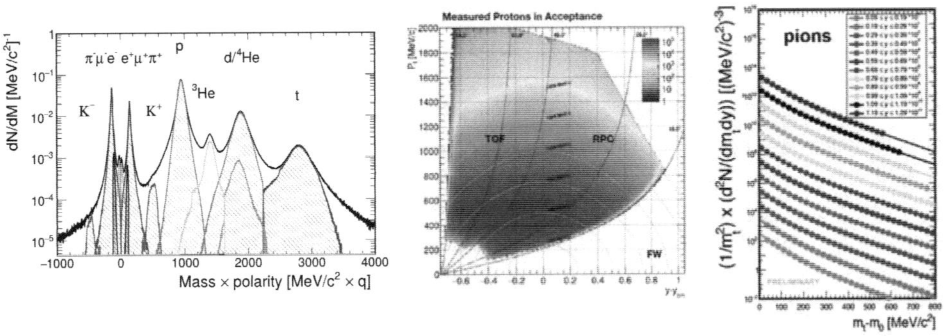

Figure 2. HADES performance: the mass times polarity spectrum (left), the measured protons in the P_T-Y plane (middle) and the differential distribution of negative pions (right).

The High Acceptance DiElectron Spectrometer is a fixed target, low mass, multipurpose detector optimized to measure rare and penetrating probes in elementary and heavy ion collisions. It is installed at SIS18 synchrotron at GSI, Darmstadt. The main particle identification method is the correlation between the velocity and the momentum of reconstructed traces. The time-of-flight measurement is performed by the combination of times from a diamond START, located in front of the target, and two STOP detectors located after the tracking stations based on fast plastic scintillator (150 ps) and resistive plate chamber technology (81 ps). Momentum is reconstructed from the bending of the particle trajectory in a strong magnetic field measured by four stations of low-mass Mini Drift Chambers with accuracy of 150 μm which corresponds with a mass resolution for dileptons at the ω pole mass of 16 MeV/c^2. Specific energy loss measured by the tracking and the scintillator detectors provides additional separation capability. Electron identification is enhanced by two detectors: a hadron blind Ring-Imaging Cherenkov and Pre-Shower detectors. For more details see [9].

3. Results from Au+Au at 2.42 GeV

The 2012 experiment was dedicated to measure Au+Au collisions at 2.42 GeV. A gold beam, with 2×10^6 ions/s was focused on a 15-fold gold target. The multiplicity trigger selected the 47% most central collisions[10]. In total 7.3×10^9 events where recorded. The reconstructed mass spectrum from velocity and momentum is shown in the left panel of Figure 2. A clear hierarchy in production is already observed at raw level. The large acceptance coverage and

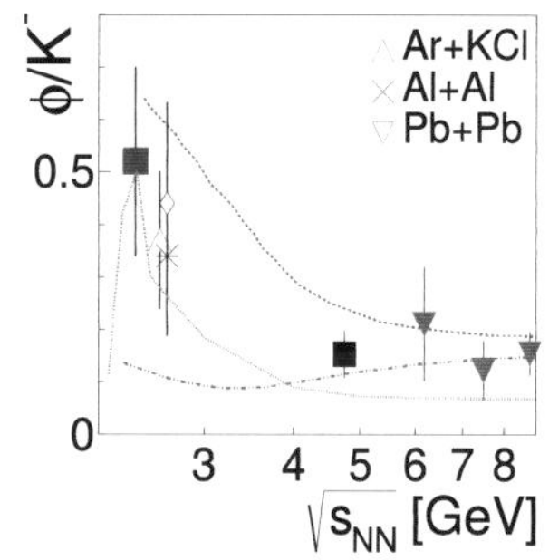

Figure 3. ϕ over K$^-$ ratio as a function of center of mass energy for different systems. Square symbols stand for Au+Au reactions[11].

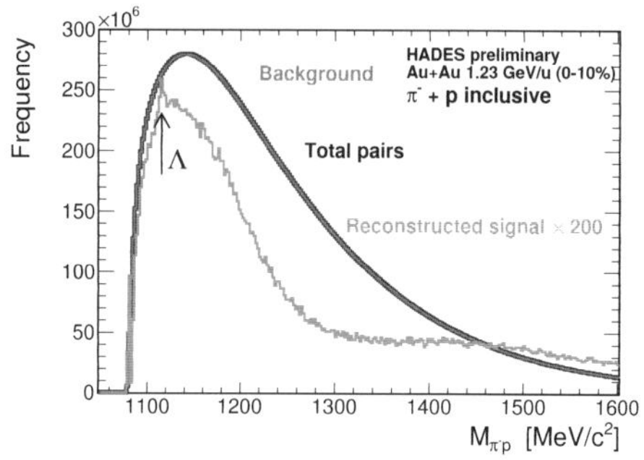

Figure 4. π^-p invariant mass in 10 % most central Au+Au collisions at 2.42 GeV (blue) shown together with the total pairs (black) and combinatorial background obtained with the iterative technique (red).

good efficiency allows differential analysis as shown for protons and pions in the middle and right panels of Figure 2.

Electron identification was improved by introducing multidimensional multivariate analysis combining different observables into an artificial neural network for identification of single electrons. Further selections were applied in order to suppress the combinatorial background produced by conversion photons, mainly from π^0 and η mesons and from partially reconstructed electrons. Then in each event all the possible pairs of electrons are created, being many of them background. The like-sign method is used to calculate the combinatorial background contribution. The good signal-to-background ratio and the high statistics will allow to investigate the differential production distribution of virtual photons and contrast measurement with models[12]. Moreover, the previously measured at same energy reference spectra by HADES of p+p and p+n reactions and the freeze-out contributions that can be obtained from hadron analyses allow to access medium radiation[13][14].

The excellent identification capabilities allowed to reconstruct a complete set of strange hadrons. The large ϕ yield accounts for 25 % of all K^-. The ϕ to K^- ratio grows with decreasing collision energy[11] as shown in the Figure 3. This fact implies that the K^- transverse spectra is modified by the kinematics of the K$^-$ from ϕ decay. The observation of a different slope for the strange and anti-strange kaons, interpreted as an effect of a sequential freeze-out, was revised. Once corrections of the K$^-$ spectrum are included, both slopes of the transverse momentum distributions of K^+ and K^- are compatible within uncertainties and the former interpretation becomes doubtful[11].

In order to characterize the produced matter the Collaboration currently focuses on the following observables: strangeness production[15], net-proton number fluctuations[16], flow of hadrons[10] and virtual photons[13][14], production of pions[17] and eta mesons[18] and production of baryonic resonances. An example of an uncorrected reconstructed signal of π^-p pairs is shown in Figure 4. In order to achieve an accurate combinatorial background, a special iterative technique has been developed[19]. The goal is to provide a complete set of observables to allow detailed comparisons to models.

4. Detector upgrade and the physics program for FAIR Phase-0

A new experimental campaign with HADES is planned in 2018-2019. Several detectors upgrades are ongoing to extend capabilities of the spectrometer. Photon reconstruction allows to measure

Figure 5. ECAL frame (left), the PaDiWa-AMPs read-out board (middle) and lead glass module resolution as a function of the incident photon energy (right).

Figure 6. Constituents of the upgraded RICH detector (left), one module for 6 MAPMTs (middle) and the time resolution of the FEE as a function of the pulse amplitude (right).

new reaction channels like neutral mesons, strange baryons $\Sigma(1385)$ and $\Lambda(1405)$ or direct photon production in both elementary and heavy ion reactions. For that reason, the HADES is replacing the Pre-Shower detector by an electromagnetic calorimeter (ECAL) which is going to cover forward angles from 16° to 45° and almost full azimuth[20]. The Cherenkov light is produced in 987 lead glass modules, recovered from the OPAL experiment, and collected by 1.5" EMI 9903KB and 3" R6091 Hamamatsu photomultipliers. The readout is based on the PaDiWa-AMPs boards with TDC and charge-to-width measurement connected to the TRB3[21]. The expected energy resolution is better than 6% for a 1 GeV photon. The recently installed frame, the read-out board and the module performance are shown in the Figure 5.

Quantum efficiency of detection of UV-photons has been significantly improved by recent developments in Bialkali photo-cathodes present in the Hamamatsu H12700 series flat panel multianode photomultipliers (MAPMT). These detectors where chosen to replace the CsI photon converters and the Multi-Wire Proportional Chambers of the RICH detector as shown in Figure 6. The same model will be used by the CBM RICH detector and both Collaborations are developing the required read-out electronics together. The large effective area together with a large packing due to a square geometry and a low cross-talk probability will provide an excellent single electron identification efficiency. In total, 432 MAPMTs with 27.6k readout channels will cover a sensitive area of about 1.3 m^2. The readout is segmented into modules of 10x15 cm^2, each of them houses 2x3 MAPMTs with 384 individual 5.8x5.8 cm^2 channels. The MAPMT read-out including all the stages between the analogue pre-amplification to the digital output data stream are implemented in the DiRich board based on off-the-shelf components. The core of the system is

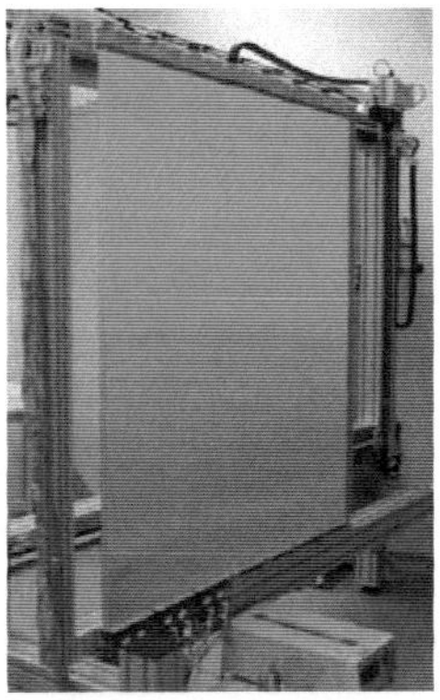

Figure 7. PANDA tracker plane[22].

an FPGA which works as a TDC, discriminator, threshold generator and DAQ network stack[23].

The forward acceptance at low polar angles between 0.5^o and 6.5^o is going to be significantly increased by the installation of a Forward Detector System. It comprises two low material budged tracking stations of 2% X_0, developed for the PANDA experiment[22], placed 3.1 and 4.6 m downstream of the target based on self-supporting straw tubes with 10 mm diameter filled with 90/10 Ar/CO_2 mixture arranged in four planes per station with different orientations. Simulations show that it will have 0.5 mrad angular resolution for a 2 GeV/c protons. The particle identification bases on the specific energy loss and the time of flight provided by a fast Resistive Plate Chamber detector located behind the two tracking modules as it operates in a magnetic-free region.

The main program for the upcoming experiments is to continue the study of baryon-rich matter in medium-heavy systems like Ag+Ag at $1.65A$ GeV, where abundant strangeness and vector meson are produced. Also the existence of the secondary pion beam [24] opens the door to measure the electromagnetic structure of baryons and the role of the ρ meson in their decays into di-leptons[25]. Differential cross-sections of hadronic final states can be analyzed in the Partial Wave Approach in order to access the baryon-meson coupling and to improve substantially the world data base for PWA[26].

The collaboration gratefully acknowledges the support by PTDC/FIS/113339/2009 LIP Coimbra, NCN grant 2013/10/M/ST2/00042 SIP JUC Cracow, Helmholtz Alliance HA216/EMMI GSI Darmstadt, VH-NG-823, Helmholtz Alliance HA216/EMMI TU Darmstadt, 283286, 05P12CRGHE HZDR Dresden, Helmholtz Alliance HA216/EMMI, HIC for FAIR (LOEWE), GSI F&E Goethe-Universität, Frankfurt VH-NG-330, BMBF 06MT7180 TU München, Garching BMBF:05P12RGGHM JLU Giessen, Giessen UCY/3411-23100, University Cyprus CNRS/IN2P3, IPN Orsay, Orsay MSMT LG 12007, AS CR M100481202, GACR 13-06759S NPI AS CR, Rez, EU Contract No. HP3-283286.

References

[1] Fukushima K and Sasaki C 2013 Prog. Part. Nucl. Phys. **72** 99
[2] Baym G, Hatsuda T, Kojo T, Powell P D, Song Y and Takatsuka T 2017 RIKEN-QHP-316 (*Preprint* astro-ph.HE/1707.04966)
[3] Neumann F, Buballa M and Oertel M 2003 Nucl. Phys. A **714** 481
[4] Galatyuk T, Hohler P M, Rapp R, Seck F and Stroth J 2016 Eur. Phys. J. A **52** no.5, 131
[5] Andronic A *et al* 2010 Nucl. Phys. A **837** 65
[6] Agakishiev G *et al* [HADES Collaboration] 2016 Eur. Phys. J. A **52** no.6 178
[7] Rapp R, Chanfray G and Wambach J 1996 Phys. Rev. Lett. **76** 368
[8] Rapp R and van Hees H 2016 Phys. Lett. B **753** 586
[9] Agakishiev G *et al* [HADES Collaboration] 2009 Eur. Phys. J. A **41** 243
[10] Kardan B 2016 J. Phys. Conf. Ser. **742** no.1 012008.
[11] Adamczewski-Musch J *et al* [HADES Collaboration] 2017 *Preprint* nucl-ex/1703.08418
[12] Seck F same proceedings.
[13] Harabasz S 2017 Ph.D. Thesis
[14] Sellheim P 2017 Ph.D. Thesis
[15] Lorenz M *et al* [HADES Collaboration] 2016 J. Phys. Conf. Ser. **668** no.1, 012022
[16] Szala M same proceedings.
[17] Gumberidze M 2016 EPJ Web Conf. **130** 05007
[18] Behnke C [HADES Collaboration] 2015 J. Phys. Conf. Ser. **599** no.1 012029
[19] Kornakov G [HADES Collaboration] 2016 EPJ Web Conf. **130** 07011
[20] Svoboda O *et al* [HADES Collaboration] 2014 JINST **9** C05002
[21] Rost A *et al* [TRB and HADES and CBM Collaborations] 2017 JINST **12** no.02 C02047
[22] Smyrski J *et al* [PANDA Collaboration] 2017 JINST **12** no.06 C06032
[23] Michel J *et al* [TRB Collaboration] 2017 JINST **12** no.01 C01072
[24] Salabura P *et al* [HADES Collaboration] 2015 Nucl. Phys. News **25** no.2 22
[25] Scozzi F [HADES Collaboration] 2017 EPJ Web Conf. **137** 05023
[26] Przygoda W [HADES Collaboration] 2016 EPJ Web Conf. **130** 01021

AstroBox2E: a detection system for very low energy β-delayed proton decay

I.C. Stefanescu[a], A. Spiridon[a], L. Trache[a], E. Pollacco[b], A. Saastamoinen[c], B. Roeder[c]

[a] Horia Hulubei National Institute for R&D in Physics and Nuclear Engineering, Bucharest-Magurele Ro-077125, Romania
[b] IRFU, CEA, Saclay, Gif-sur-Yvette, France
[c] Cyclotron Institute, Texas A&M University, College Station, TX 77843, USA

ionut.stefanescu@nipne.ro

Abstract. To study very low-energy β-delayed proton decay, which are of interest for astrophysics, it is essential to achieve an efficient suppression of the β-background. ASTROBOX is such a device: energetic precursor nuclei are produced, identified and then stopped in the gas volume of the detector. The resulting β or β-proton decay traces are ionizing paths in the gas. The electrons produced through ionization are drifted in an electric field and are amplified with a Micro Pattern Gas Amplifier Detector (MPGAD). High gain and high signal to noise ratio are expected to be obtained. The two predecessors of this detection system which is under construction now in our group are AstroBox and AstroBox2, which were built and commissioned at Texas A&M University. The goal of this project is to build and use AstroBox2E at European facilities.

1. Introduction

Resonant reactions are important in many explosive hydrogen-burning scenarios. The most significant parameters in understanding the astrophysical reaction rates are the energies and decay widths of the associated nuclear states along the reaction paths [1]. The relevant energies for proton capture reactions are located in the Gamow window (stars are "cold", reactions occurring at low energies, tens or hundreds of keV), just above the associated proton separation threshold. The properties of these states can be investigated by using indirect methods, which include β-decay studies.

The use of decay spectroscopy tools is an important method to find resonances relevant in various radiative proton capture reactions of significance in nuclear astrophysics. Over the past decade, a program of measurements was started at the Cyclotron Institute, Texas A&M University, using the excellent conditions given by the primary beams from the superconducting cyclotron K500 and the MARS recoil separator [2]. One of the main methods used was the spectroscopy of states populated by β-delayed proton emission.

2. *β*-delayed proton emission

This method is useful for cases where the proton capture is dominated by low-energy resonances. The resonant capture of protons is a two-step process, where the proton incident on a nucleus populates first a metastable state in the compound nucleus (1st step) that then de-excites (2nd step) by gamma-ray emission. The corresponding astrophysical reaction rates are given by the properties of the narrow, isolated resonances only: spin and parity, energy and resonant strength [3]. To study these resonances at astrophysical energies through direct measurements is not always easy or even possible. An alternative is to populate the same metastable states and determine their spectroscopic properties by other means. One method is focused on the decay spectroscopy: choose an exotic nucleus that will beta-decay to these same states. Important conditions should be verified:

i. the energy of the excited state of daughter nucleus (after its parent went through beta decay) has to be larger than the proton separation energy, and
ii. the spin and parity selection rules must allow the population of the states that represent the most relevant resonances to examine.

The relation between these decay spectroscopy studies and the proton capture reactions that occur through narrow, isolated resonances in stellar environments is presented in Figure 1.

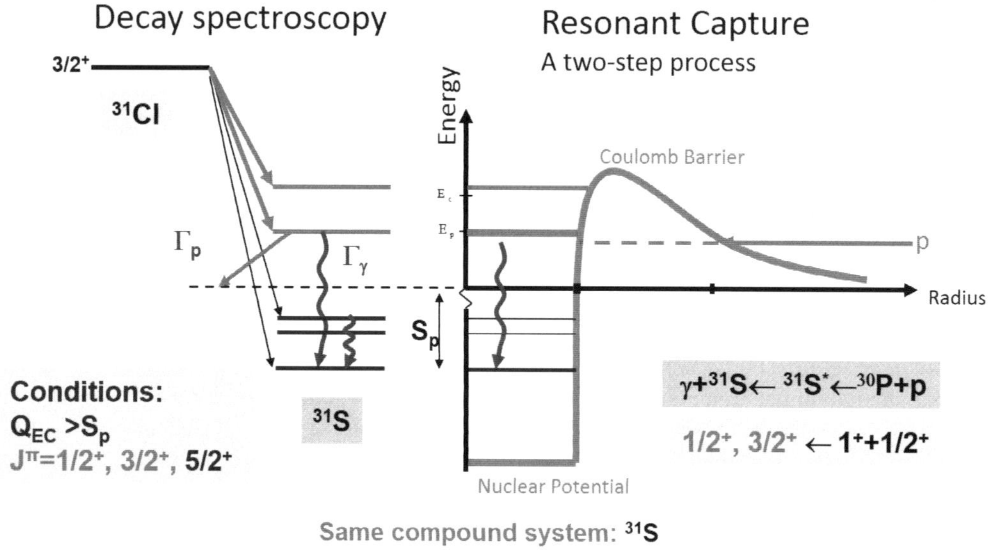

Figure 1. Schematic presentation of the relation allowing the use of beta-delayed proton decay in resonant proton capture (S_p - proton separation energy from the ^{31}S nucleus; Q_{EC} – beta decay energy; J^π – spin and parity of the excited state; Γ – transition width) [4].

3. The AstroBox detection system

A number of decay studies involving nuclei that undergo β-delayed proton emission have been done in the past at Cyclotron Institute, Texas A&M University [5-8]. The experimental procedure involved the separation of short-lived proton-rich radioactive nuclei, implanting them in a detector and measuring the gamma-rays and protons emitted after beta-decay. This type of set-up consisted of very thin Silicon(Si) strip detectors, which proved to be successful in the detection of protons, with energies as low as 200-400 keV, but with a large background from positrons. To lower the β-background and improve the resolution at low-energies, a new type of detector was developed.

3.1 AstroBox1

AstroBox1 was the first gas detector developed. The gas was used as a medium to stop particles and the MICROMEGAS (MICROMEsh GAseous Structure) were used for signal collection. A schematic drawing can be seen in Fig. 2. The whole system consisted of a stainless steel chamber using the

IOP Conf. Series: Journal of Physics: Conf. Series **1024** (2018) 012007 doi:10.1088/1742-6596/1024/1/012007

MICROMEGAS structure, the grid, the cathode, the equipotential rings and the gas [9] (see Fig. 3). MICROMEGAS is a detector consisting of a two-stage parallel-plate avalanche chamber with a drift region and a small amplification gap, providing large gas gain and very good energy resolution [10].

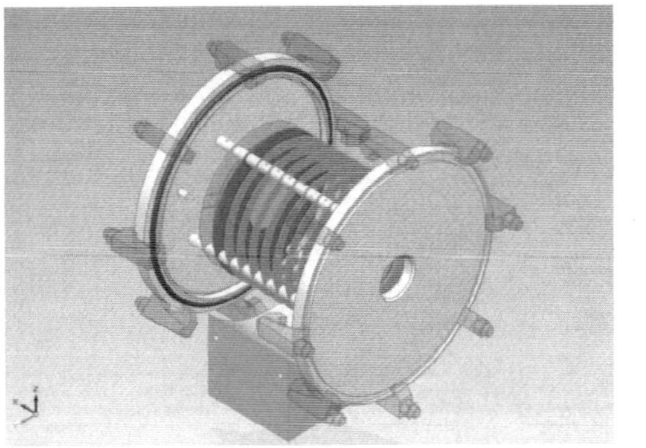

Figure 2. Schematic drawing of AstroBox.

Figure 3. Picture of AstroBox from a pre-experiment setup. Beam enters perpendicular to cylinder axis.

3.2 Astrobox2

The detector described above proved to be an efficient tool to investigate the beta-delayed proton decay. In the fall of 2014, a new type of detector was commissioned at Texas A&M University. The main difference of this detector compared to its predecessor is the more appropriate geometry, regarding to the stopping distribution of the beam in the gas of the detector. Another difference is that it has 29 separate pads and correspondingly 29 signals, compared with only 5 for AstroBox1. The 29 rectangular anode pads are arranged into a symmetric geometry along the beam axis. This new configuration allows better control of the heavy-ion implantation and also gives more possibilities for decay studies.

4. Description of the AstroBox2 detection system

The general operating principle of the detector and associated components is similar to the description given in Ref. [9]. The detector, shown in Fig. 4, is a rectangular-shaped gas detector, and is operated in a mode that allows the relevant ions to be stopped inside the gas volume and the decay be observed. The electrons created by the decay radiation that ionizes the gas drift towards a gas amplifier based on MICROMEGAS technology [10].

The MICROMEGAS detector element has an active area of 100 mm x 145 mm, and is divided into 29 rectangular anode pads of various sizes arranged symmetrically along the beam axis. At the moment of commissioning, only a detector with a 128 μm amplification gap had been tested. The pad layout is shown in Fig. 5 and Fig. 6. The PCB (Printed Circuit Board) doubles as a sub-flange that allows the signals to be transmitted directly to the readout electronics, reducing the electronic noise that is induced by the cables connecting the detector to the pre-amplifiers.

The field cage equipotential rings with spacing of 16.5 mm are mounted on the detector PCB. The gating grid is the equipotential ring closest to the MICROMEGAS and contains two small PCBs with a set of 50 μm thick wires between them. This is used to limit the transmission of the drifting electrons during implantation to protect the detector.

Since the fall of 2015, the detection system has been tested, in-beam and off-beam, and was also used in various experiments at Texas A&M University, involving nuclear astrophysics. The next section presents a short summary about the tests and one of the experiments that involved the beta-delayed proton decay of ^{31}Cl.

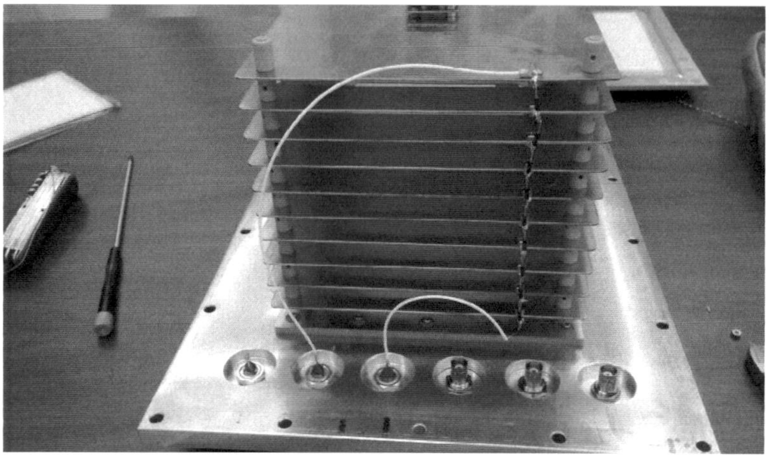

Figure 4. Photograph of the AstroBox2 detector during off-line testing showing the anode flange with the PCB attached and the copper equipotential rings.

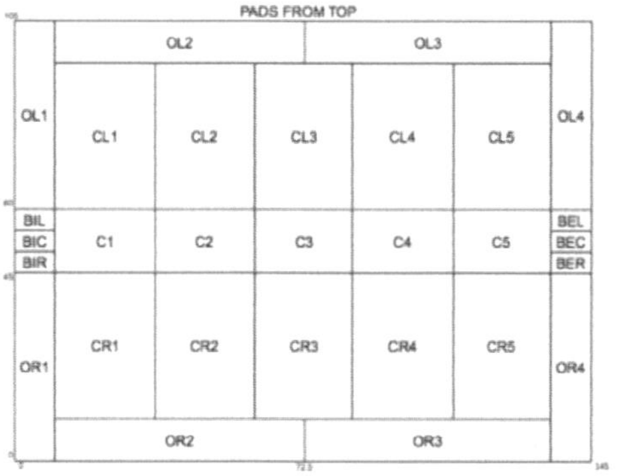

Figure 5. Schematic of the MICROMEGAS anode. BI-beam-in pads; C-center pads; CL-center left pads; CR-center right pads; OL-outer left pads; OR-outer right pads; BE-beam exit pads. Beam enters from the left.

Figure 6. Photograph of the MICROMEGAS anode attached as a sub-flange to the feed-through flange of the detector chamber.

5. Tests and beta-delayed proton decay of ^{31}Cl

All tests were carried out using a P5 gas mixture (5% CH4 + 95% Ar) at 800 torr pressure. The detector response for α sources was tested with a mixed α source containing ^{148}Gd, ^{239}Pu, ^{241}Am and ^{244}Cm. A typical resolution of 3% was achieved for the 5.486 MeV ^{241}Am α decay. Also, a set of tests was carried out with a ^{55}Fe X-ray source. The resolution achieved for the 5.9 keV X-ray was 14-16% across the whole detector.

The astrophysical experiment focused on the beta-delayed proton decay of ^{31}Cl was done at Cyclotron Institute. The MARS recoil spectrometer [2] was used to produce a high purity and high intensity ^{31}Cl secondary beam which was implanted into the AstroBox2 detector. Data was taken by pulsing the beam, alternating implantation and decay measurement cycles. Using the versatility of the anode pad geometry it was possible to reduce the β-background in the software to 100 keV. As a consequence of this level of beta-suppression, it was possible to observe clearly the known proton states around 800-1000 keV, and also detect a potentially new proton group in the energy region 200-600 keV.

IOP Conf. Series: Journal of Physics: Conf. Series **1024** (2018) 012007 doi:10.1088/1742-6596/1024/1/012007

6. Conclusions and future perspectives

In the last decade many experiments that involved unstable nuclei which undergo beta-delayed proton decay were performed at Texas A&M University. After the extensive study with various Si detectors reached its limits in what concerns lowering of the β-background, a new type of detection system was developed. AstroBox1, and a few years later, AstroBox2, proved to be an improved tool to perform decay studies. In particular, AstroBox2 has very good characteristics in what concerns the resolution, the particle identification, the geometry between the radioactive beam and the whole detection system. An additional advance of this detector is its limited cost, due to decades of developed MICROMEGAS technology, making the effort building this detector financially feasible.

At the time of submitting this paper, the Nuclear Astrophysics Group (NAG) at IFIN-HH has in development and construction an European version of AstroBox2E, to be used at European facilities which can provide radioactive ion beams. Broadly, the geometry of the detection system will not be changed. We will continue to use MICROMEGAS detectors and already have acquired two of them, one with an amplification gap of 128 μm, and the other with an amplification gap of 64 μm. Also, the gas handling system is under construction, as well as the stainless steel housing for the MICROMEGAS detectors.

References

[1] Saastamoinen A *et al.*, *Nucl. Instr. Meth. Phys. Res.* B **376** (2016) 357-360
[2] Tribble R E, Burch R H and Gagliardi C A, *Nucl. Instr. Meth. Phys. Res.* A **285** (1989) 441-446
[3] Rolfs C and Rodney W S, *Cauldrons in the Cosmos*, University of Chicago Press, 1988
[4] Trache L, *10th Russbach Workshop on Nuclear Astrophysics, March 10-16, 2013*
[5] McCleskey M *et al.*, *Nucl. Instr. And Meth.* A **700** (2013) 124
[6] Saastamoinen A, Trache L, Banu A *et al.*, *Phys. Rev.* C **83**, 045808 (2011)
[7] Trache L *et al.*, *Proc. 10th Symposium in Nuclei in the Cosmos, July 27-Aug 1, 2008, Mackinac Island, MI*, PoS (NIC X) 163
[8] Trache L *et al.*, *PROCON2011*, AIP Conf. Proc. **1409**, 76 (2011)
[9] Pollacco E *et al.*, *Nucl. Instr. Meth. Phys. Res.* A **723** (2013) 102-108
[10] Giomataris Y *et al.*, *Nucl. Instr. und Meth. in Phys. Res.* A **376** 29 (1996)
[11] Spiridon A *et al.*, *Exotic Nuclei and Nuclear/Particle Astrophysics (IV). From Nuclei to Stars*, AIP Conf. Proc. **1498**, 369-375 (2012)

IOP Conf. Series: Journal of Physics: Conf. Series **1024** (2018) 012008 doi:10.1088/1742-6596/1024/1/012008

Higgs $\rightarrow \tau\tau$ Branching Ratio Measurement at CEPC

Dan YU

E-mail: yudan@ihep.ac.cn

Abstract.
The CEPC (Circular Electron Positron Collider) and ILC (International Linear Collider) detector design provides a high efficiency vertex and Arbor Particle Flow Algorithm[1] uses a new charged particle identification with efficiency higher than 98%. Taking these advantages, the Higgs decaying to $\tau\tau$ channels have been studied in full simulation samples, and the precision of the production cross section times the branching ratio have been improved to 1% level.

1. Introduction

Taus are extremely intriguing physics objects. As the heaviest lepton in the Standard Model (SM), a significant fraction of the SM Higgs boson decays into di-τ final states. As a result, g(H$\rightarrow$ $\tau\tau$) could be measured to an accuracy better than 1% at the CEPC. Measuring the polarization of Tau at the Z pole leads to a precise determination of forward-backward asymmetry $A_{fb}(\tau)$ and therefore $\sin^2\theta$ and could be used to check the lepton universality. The reconstruction of τ functional spectral is of key importance to the CEPC Electro-Weak program, as it grants the access to a precise reconstruction of α_S and EW parameters. In this note, the branching ratio of H$\rightarrow$ $\tau\tau$ is measured in full simulated ZH samples and SM backgrounds.

Most of the tau decaying modes[2] are with one or three charged particles in the final state and an odd number of photons, which is the main idea of the τ tagging we plan to perform. From the decay modes, the topology of tau's is simpler than jets, allowing a clear disentanglement of tau events from the others.

2. Monte-Carlo Samples

The CEPC luminosity is planed to be 5000 fb^{-1}, the cross section for different Z decay modes are summarized in [3]. All the samples mentioned in this note are generated by the MC generator Whizard, version 1.95[4]. The detector used in the simulation is the CEPC_v1 detector, containing a tracking system with excellent momentum resolution and a high granular calorimeter system (ECAL and HCAL)[5].

3. Event categories

A successful reconstruction of the τ lepton is not a trivial task, since it could be generated with various event topologies, and it decays in different final states. In the CEPC collision environment, the τ events are divided into two categories according to their topology. The reconstruction algorithm and the spectrometer performances have been optimized following this selection criterium.

Due to the limited computing resource, the inclusive ZH events, and SM categories background events are filtered by some preselection using MC truth information to simplify the samples. The excellent performance of PFA ensures that this preselection would not lose information.

3.1. llH channel

The first category is the leptonic one, where no physics objects, or only lepton / photon / missing energy is generated together with the τ candidates.[1] These events include, for example:

- ZH, Z$\to l^+l^-$ / $\nu\nu$, $H \to\tau\tau$ events; golden channel for $g(H\tau\tau)$ measurements
- ZZ, l^+l^- / $\nu\nu$ / $\tau\tau$ events
- WW events with $l\nu\tau\nu$ final states.
- $Z \to \tau\tau$ events at Z pole operation.

In these events, the global multiplicity is limited while the additional physics objects, if they exists, are easy to identify[6]. A successful identification of these events relies highly on the reconstruction of photons and charged hadrons. In the following section, the physics performances of τ reconstruction at $\mu\mu H$ and $\nu\nu H$ channel are shown as well as their Br($H \to\tau\tau$) measurement.

The steps for di-τ events tagging are:

- For $\mu\mu$H, veto the μs decayed from Z by choosing the μ pair with invariant mass closest to Z mass
- Find the leading track among the remaining particles and collect the tracks and photons close to this track (¡ 1 rad, to be grouped in region A), and their numbers are noted as NTrkA and NPhA.
- Collect the rest tracks and photons and group them in region B with their numbers noted as NTrkB and NPhB.
- Get the angle between the leading tracks in region A or B and the furthest track in this region, noted as Cone$_{T-T}$(A/B).
- Cone$_{T-P}$(A/B) is the angle between the leading tracks in region A or B and the furthest photon in this region.
- Cone$_{P-P}$(A/B), the angle between the leading photon in a region and the furthest photon in this region.

The inclusive SM background could be efficiently subtracted by requesting the proper number of leptons, if exist, and limit the number of photon and charged hadrons. For $\mu\mu$H events, the event selection could be further enhanced by request on the recoil mass against the di-lepton system.

The pre-selections applied to select $\mu\mu$H and $\nu\nu$H suppress the SM background by 4 orders of magnitude and the remaining backgrounds are dominated by ZH events. The cut chain of the event selection for $\mu\mu$H and $\nu\nu$H are shown in Table 1 and Table 2.

3.2. Hadronic

The second category is the hadronic one, where the τ lepton(s) are generated with jets. For instance, we have:

- ZH, Z$\to$qq, H$\to\tau\tau$

[1] The charge is ignored for event classifications.

IOP Conf. Series: Journal of Physics: Conf. Series **1024** (2018) 012008 doi:10.1088/1742-6596/1024/1/012008

Table 1. $\mu\mu$H cut flow

	$\mu\mu H\tau\tau$	$\mu\mu H$ inclusive bkg	ZZ	WW	single$W$	single$Z$	$2f$
total generated	2292	33557	5711445	44180832	15361538	7809747	418595861
after preselection	2246	32894	122674	223691	0	86568	1075886
$N_{Trk}(A/B) < 6$ & $N_{Ph}(A/B) < 7$	2219	1039	2559	352	0	9397	25583
BDT>0.78	2135	885	484	24	0	157	161
efficiency	93.15%	2.63%	<0.01%	<0.01%	<0.01%	<0.01%	<0.01%

Table 2. nnH cut flow

	$\nu\nu$H$\tau\tau$	$\nu\nu$H inclusive bkg	ZZ	WW	singleW	single Z	$2f$
total generated	15497	231670	5711445	44180832	17361538	7809747	418595861
after preselection	9434	214830	1239457	7463105	3327803	956694	12826280
$N_{Trk}(A/B) < 6$ & $N_{Ph}(A/B) < 7$	9260	8858	24760	1354852	17389	676185	1535029
BDT > 0.78	8836	6587	15450	89729	1355	10739	11243
efficiency	57.02%	2.84%	0.27%	0.20%	<0.01%	0.14%	<0.01%

- ZZ→qq$\tau\tau$
- WW→qqlτ
- ZH, Z→qq, H→WW→lvτv

To find the τ lepton in the hadronic categories is much more challenge than that in the leptonic environments. The identification algorithm would always be a compromise between the signal efficiency and purity. Taking the Br(H→$\tau\tau$) measurement at qqH events for example, the τ candidate finding steps are:

- Find tracks with energy higher than a defined E_{min} as the seed
- Collect tracks and photons within a small angle
- Calculate invariant mass with these particles
- Calculate the energy in a larger cone ConeB around the seed.

The parameters for cut of τ tagging is:

- Number of tracks/photons
- Energy proportion in the smaller cone

Here the cuts of parameters are optimized to the value $\epsilon \cdot p$, where ϵ is the efficiency of finding an opposite charged τ pair in qq$\tau\tau$ events and p is the probability of tagging a opposite charged τ pair in the backgrounds.

After these cuts, the remaining τs in an event is collected and the two leading energetic ones with opposite charge are chosen to calculate the invariant mass of the di-τ, The recoil mass and invariant mass of qq system are also applied for the background suppression. The total efficiencies for the signal and the dominating background, ZZ and WW, are 49.97%, 0.41% and 0.04%, respectively. The cut chain is summarized in Table 3.

IOP Conf. Series: Journal of Physics: Conf. Series **1024** (2018) 012008 doi:10.1088/1742-6596/1024/1/012008

Table 3. Cut Flow of MC sample for qqH→ $\tau\tau$ selection on signal and inclusive SM backgrounds

	$qqH\tau\tau$	qqH inclusive bkg	ZH inclusive bkg	ZZ	WW	single$W$	single$Z$	$2f$
total generated (scaled to 5 ab^{-1})	45597	678158	357249	5711445	44180832	17361538	7809747	418595861
1st preselection	45465	677854	310245	5039286	42425195	1267564	1398362	148401031
2nd preselection	45145	174650	226059	293306	12452091	125735	117306	547402
$N_{\tau^+} > 0, N_{\tau^-} > 0$	24674	7342	33721	93955	723989	33887	54386	103642
$20GeV < M_{\tau^+\tau^-} < 20GeV$	24284	6290	32344	88245	597480	24927	36039	56615
$70GeV < M_{qq} < 110GeV$	22937	2103	4887	65625	21718	738	1893	556
$100GeV < M_{qq}^{Rec} < 170GeV$	22703	2045	4524	23789	13154	315	306	193
efficiency	49.97%	0.31%	1.26%	0.41%	0.04%	<0.01%	<0.01%	< 0.01%

4. Results

To conclude, the τ reconstruction at the CEPC is currently categorized into leptonic and hadronic events and reconstructed using different strategies and τ finding algorithms. In the leptonic events, where the τ lepton is generated only in association with leptons, photons or missing energy, the τ events identification relies strongly on a successful reconstruction of the photons and charged hadrons.

In the hadronic events, it is more difficult to suppress the background, for further study, the correlation with other channels might be applied.

The total signal yields are extracted from a fit to a variable defined by the impact parameters on the leading tracks from the two tau objects, and the fit results are shown in Figure. 1. In summary, with 5 ab^{-1} of CEPC data and all three channels combined, the Branching Ratio of $H \to \tau\tau$ measurement is $6.28 \pm 0.07\%$, and the fractional error on the $\sigma \times BR$ can reach 0.81%.

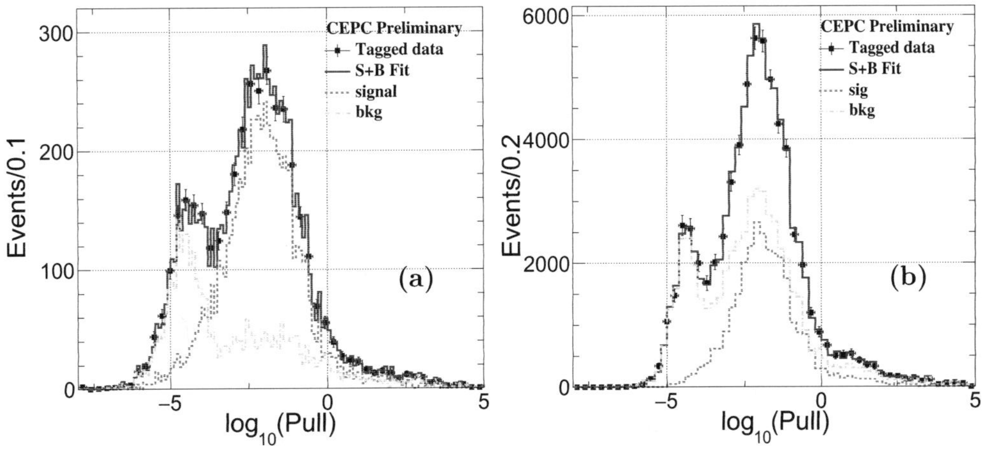

Figure 1. Post-fit distributions of the impact parameter variable based on the leading tracks from the two taus in the $Z \to \ell\ell$ (a) and $Z \to qq$ (b) channel, expected for 5 ab^{-1} of CEPC ZH data.

With these channels analyzed and fitted, the cross section of Higgs decaying to $\tau\tau$ can be summarized as in Table 4

Table 4. Combined cross section

	BR ($H \to \tau\tau$)	δ ($\sigma\times$ BR)/($\sigma\times$ BR)
$\mu\mu$H	6.40 ± 0.18	2.26%
eeH(extrapolated)	6.37 ± 0.18	2.72%
$\nu\nu$H	6.19 ± 0.17	4.29%
qqH	6.23 ± 0.04	0.93%
combined	6.28 ± 0.07	0.81%

In both cases, a precise reconstruction of the impact parameter is essential for the τ events identification, as shown in the figures, the statistics can be fitted only if the position resolution is good enough to distinguish the two peaks for τs and backgrounds.

5. Extrapolating in ILC

Comparing these cross sections for three polarization scenarios in ILC at 250GeV with the cross section at CEPC as shown in previous section, a simple extrapolation can be done as in Table 5. The assumption here is that the efficiency for each signal and background stays the same for ILC and CEPC.

Table 5. Extrapolated accuracy δ ($\sigma\times$ BR)/($\sigma\times$ BR) in ILC 250GeV (2000 fb^{-1})

Type	(-80%,30%)	(80%,-30%)	non-pol
$\mu\mu$H	2.98	3.63	3.57
$\nu\nu$H	5.48	4.34	6.78
qqH	1.26	1.36	1.47
combined	1.13	1.22	1.33

6. Discussion

In this paper, different channels with Higgs decaying into $\tau\tau$ at CEPC have been studied and the combined accuracy is reaching 1% level. This result is also extrapolated to ILC and also gives the reasonable accuracy.

This accuracy has still space to be improved. One choice is to use the collinear approximation to recover the momentum of neutrino(s) from τ. This method needs to assume that τ decay products almost fly back-to-back. The collinear approximation will help to reconstruct the invariant mass of tau pair system and its comparison to the Higgs mass could be a powerful variable to suppress ZZ/WW backgrounds with τ final states.

Another method is to fully reconstruct hadronically decaying τ momenta by making use of the interaction point position, the impact parameters of the τ decay products, and the transverse momentum of the Z boson recoiling against the $\tau\tau$ system. Since more than 60% of τs decays into hadrons, this method will help to improve the performance of these channels.

Besides, a jet clustering algorithm can be applied in the qqH channel in order to suppress the 2f backgrounds with jets.

However, this study here is based on a perfect vertex detector, the resolution is not taken into account. Since the result was obtained from the impact parameter, the influence of the vertex detector design to the performance should be studied in the future.

IOP Conf. Series: Journal of Physics: Conf. Series **1024** (2018) 012008 doi:10.1088/1742-6596/1024/1/012008

7. Bibliography

[1] M. Ruan, *Arbor, a new approach of the Particle Flow Algorithm.* arXiv:1403.4784 (2014).
[2] C. Patrignani et al. Particle Data Group, *Chin. Phys. C*, **40**, 100001 (2016)
[3] X. Mo, G. Li. Generated Sample Stauts for CEPC Simulation Studies. *CEPC NOTE 2015*
[4] Kilian W, Ohl T, Reuter J. WHIZARDsimulating multi-particle processes at LHC and ILC[J]. *The European Physical Journal C*, 2011, **71(9)**: 1742.
[5] M. Ahmad et al (The CEPC-SPPC Study Group), CEPC-SppC Preliminary Conceptual Design Report: Physics and Detector, *http://cepc.ihep.ac.cn/preCDR/main preCDR.pdf*, retrieved 4th May 2015
[6] D. YU, M. Ruan, V. Boudry, H. Videau. Lepton identification at particle flow oriented detector for the future e^+e^- Higgs factories. *Eur. Phys. J. C* (2017) **77**: 591.

IOP Conf. Series: Journal of Physics: Conf. Series **1024** (2018) 012009 doi:10.1088/1742-6596/1024/1/012009

Calculating hadron properties from dynamical hadronization in the Functional Renormalisation Group

J. París-López[1], R. Alkofer[1], A. Maas [1], W. Mian[1], M. Mitter[1,2], J. Pawlowski[3,4], H. Sanchis-Alepuz[1]

[1] Institut für Physik, Karl-Franzens-Universität Graz, NAWI Graz, 8010 Graz, Austria
[2] Brookhaven National Laboratory (BNL) - Department of Physics, P.O. Box 5000, Upton, NY 11973-5000, USA
[3] Institut für Theoretische Physik, Universität Heidelberg, Philosophenweg 16, 69120 Heidelberg, Germany
[4]ExtreMe Matter Institute EMMI, GSI, Planckstr. 1, 64291 Darmstadt, Germany

Abstract. We report on a functional renormalisation group approach to bound state properties similar to the Dyson-Schwinger–Bethe-Salpether approach. The current approach is set-up for the two-flavour quark-meson model for an illustration of the basic properties. This allows us to access the pion and sigma poles and decay properties. First numerical results are presented and evaluated. We also discuss the next steps.

1. Introduction

Hadron properties such as masses or decay widths are, at least in principle, determined by the underlying dynamics of quarks and gluons. Due to their strong interactions at small scales and the non-perturbative nature of bound states, non-perturbative methods as well as real-time computations are requested for calculating such observables. Functional methods, mainly the Dyson-Schwinger and Bethe-Salpeter equations (DSE-BSE) approach, have been successfully applied to access hadron resonances, for a recent review on the successes and open problems within this approach see [1].

In the present work we report on first steps towards a similar programme within the functional renormalisation group (FRG) approach. This programme is based on two cornerstones: the dynamical hadronisation procedure, [2, 3, 4], as formulated for QCD in [5, 6, 7], and the recent development of real-time approaches to the FRG in the context of Minkowski space-time, see e.g. [8, 9, 10, 11]. Dynamical hadronisation allows to reparameterise resonant hadronic channels in multi-quark interactions in terms of composite operators, the effective hadronic degrees of freedom. For an access to bound state properties one has to evaluate the correlation functions of these composite operators for real time frequencies. In such an approach the pole masses of the hadronic resonances are encoded in the analytic properties of the propagators of the effective hadronic degrees of freedom. Decay widths are encoded in the vertex functions that are directly related to Bethe-Salpether wave functions. In summary this provides a uniform approach only based on the effective action of the theory. More details on this approach will be presented elsewhere.

2. The Functional Renormalisation Group

The FRG approach in its modern form is based on the effective action $\Gamma_k[\Phi]$ of the theory, where the superfield Φ includes all fields in the theory at hand. In the present work we consider the two-flavour quark-meson model with $\Phi = (q, \bar{q}, \phi)$ with $\phi = (\sigma, \vec{\pi})$. After introducing an infrared cutoff k the scale-dependent effective action Γ_k satisfies the Wetterich equation [12],

$$\partial_t \Gamma_k = \frac{1}{2} \mathrm{Tr} \, \frac{1}{\Gamma_k^{(2)} + R_k} \partial_t R_k \,, \tag{1}$$

where k is the RG scale, $\partial_t = k\partial_k$, and $\Gamma_k^{(n)}$ is the nth field derivative of the effective action. The trace sums over species of fields, internal indices and four-momenta, and $R_k(p^2)$ is a regulator function that suppresses the propagation of the infrared modes of all fields Φ. The present work is based on and uses the same notation as the QCD-works [5, 6, 7], for more details and literature we refer to these works. Finally, we use analytic regulators that have the full Euclidean $O(4)$ space-time symmetry, and hence also preserves $O(1,3)$ Minkowski space-time symmetry. In our opinion keeping the full space-time symmetry in the regularisation is key to an access to the higher resonances. Also, it allows for a fully numerical access to correlation functions at real-time also at finite temperature and density, see [10, 11].

3. The Quark-Meson model with dynamical hadronisation

Two-flavour QCD with dynamical hadronisation for the scalar–pseudo-scalar multiplet has been studied in [5, 6, 7] in the Landau gauge. There it has been shown, that the decoupling of gluons at very low energies leads to a low-energy effective theory in terms of quarks $q, \bar{q}$ and $\phi = (\sigma, \vec{\pi})$ below some cutoff scale k_{EFT} that serves as an ultraviolet (UV) cutoff for the effective theory. Accordingly, the two-flavour quark-meson model serves as an ideal testing ground for the techniques and mechanisms at work that will be applied to QCD later.

3.1. QCD-induced quark-meson model

Let us briefly describe how the quark-meson model emerges as a low energy effective theory from QCD with dynamical hadronisation: the QCD-flows in [5, 6, 7] are initated at a high inital cutoff scale $k \approx 10^2\,\text{GeV}$. There, perturbation theory works and the effective action at this initial scale is well approximated by the classical action. When flowing down to non-perturbative scales $k \approx 1\,\text{GeV}$ multi-quark interactions to all order are generated by (1). In particular scalar-pseudo-scalar four-quark interactions are generated. The η and $\vec{a}$ fields have masses that are far larger than the UV cutoff scale k_{EFT} of the low-energy effective theory, and hence do not contribute to the dynamics below this scale. Note however, that they are present in the effective action and cannot be neglected in the evaluation of physical processes. In the present structural work we simply drop them. Concentrating on the local part of the remaining scalar-pseudoscalar interactions we are led to

$$\Gamma_k^{(4-int)} = -\lambda_k \int_x \left(\frac{(\bar{q}q)^2}{4} + (\bar{q} i\gamma_5 \tau^a q)^2 \right) , \tag{2}$$

where the $\tau^a = \sigma^a/2$ with the Pauli matrices σ^a and $\int_x = \int d^4x$. The scale-dependent coupling λ_k changes via quark-gluon diagrams as well as pure quark or gluon diagrams. In the vicinity of the chiral symmetry breaking scale the four-fermi interaction λ_k grows large and even diverges in the chiral limit. It is convenient to describe such resonant channels in terms of effective degrees of freedom. To that end we rewrite (2) as

$$\Gamma_k^{(4-int)} = \int_x \left[\frac{1}{2}(\sigma^2 + \vec{\pi}^2)\,(p^2 + m_k^2) + h_k\, \bar{q} \left(\frac{\sigma}{2} + i\gamma_5 \tau^a \pi^a \right) q \right]_{\text{EoM}(\sigma,\vec{\pi})} \tag{3}$$

by means of a scale-dependent Hubbard-Stratonovich (HS) transformation: with the identification $\lambda_k = h_k^2/(2m_\pi^2)$ the right hand side of (3) reduces to (2) on the equations of motion for $\sigma, \vec{\pi}$. This procedure introduces effective mesonic σ and $\vec{\pi}$ degrees of freedom.

3.2. Dynamical hadronisation

If the above procedure is performed at one cutoff scale k, the four-fermi interaction (2) is regenerated in QCD in the next RG-step via quark-gluon diagrams as well as pure quark or gluon diagrams. Even if switching off the remaining QCD interactions the hadronised interaction (3) itself regenerates the four fermi-interaction. A suitable scale-dependent HS transformation will reshuffle this remnant four-fermi interaction again into the Yukawa interaction which in turn induces a four-fermi interaction in the next RG step. This successive interplay of generating the four-quark interaction, or more generally multi-quark interactions, and then nullifying them, is called Dynamical Hadronization [2, 3, 4]. Here we have described it for the local (momentum-independent) part of a four-fermi interaction, but it can be applied to a full momentum-dependent channel, see [5, 7] for the application to full QCD. By construction the hadronised theory still is full QCD, no information is lost and no double-counting problem arises as typically in the case of a fixed-scale HS transformation. For a more detailed discussion see [5, 7].

While simply being an identity transformation, dynamical hadronisation provides a significant practical advantage: A resonant channel of the four-quark interaction is rewritten in the form of Yukawa terms, and therefore one channel of the four-point correlation function is represented by a meson exchange, which facilitates calculations. The scale-adaptive nature of the dynamical hadronisation implies scale-dependent mesonic fields, $\Phi = (\sigma, \vec{\pi}) \to \Phi_k = (\sigma_k, \vec{\pi}_k)$ leading to

$$\partial_t \Gamma_k \to \partial_t \Gamma_k + \int \frac{\delta \Gamma_k}{\delta \Phi_k} \partial_t \Phi_k \,, \tag{4}$$

on the left hand side of the flow equation (1). Even in the case discussed above, where we remove the local four-quark interaction, $\lambda_k \to 0$ we still have a reparameterisation freedom of the mesonic fields due to $\lambda_k = h_k^2/(2m_\pi^2)$: neither h_k nor m_π^2 is fixed. This is an important example as it entails that m_π is not necessarily the physical mass of the pion. Here we choose

$$\partial_t \sigma_k = \frac{1}{2} \int \partial_t A_k \, \bar{q} q \,, \qquad \partial_t \pi_k^a = \int \partial_t A_k \, \bar{q} i \gamma_5 \tau^a q \,, \tag{5}$$

such that the requirement $\partial_t \lambda_k = \text{flow}^{(4)} - h_k \partial_t A_t \overset{!}{=} 0$ (as suggested in refs. [2, 3, 4, 5, 6, 7]) determines the quantity A_k in every RG step. Here $\text{flow}^{(4)}$ stands for the diagrams that are derived from the right hand side of (1) for the flow of the local part of the scalar-pseudoscalar part of the four-quark interaction.

3.3. Current Approximation to the quark-meson model

Above we have outlined how the quark-meson model emerges as a low-energy effective theory from full QCD flows with dynamical hadronisation. Apart from the quark-meson interactions discussed in the previous section further interactions as well as the kinetic terms are generated. At the UV scale of the effective field theory, k_{EFT} we can read off the UV action of the quark-meson model as the result of QCD flows from the initial scale $k \approx 10^2$ GeV to k_{EFT}.

Here we concentrate on the extraction of real-time quantities and take the effective field theory point of view: we fix the UV couplings of the model at k_{EFT} by adjusting observables at $k = 0$. It is left to discuss the approximation used in the present work. Finally we want to use and extend the approximation to the quark-meson sector used in the QCD flows [5, 7], for more details see the discussion there. Here we shall use a much reduced version of this

IOP Conf. Series: Journal of Physics: Conf. Series **1024** (2018) 012009 doi:10.1088/1742-6596/1024/1/012009

approximation which still far surpasses those used before in quark-meson model computations in terms of momentum and frequency dependence. This approximation will allow us to study the qualitative feature of the real-time properties of the quark-meson sector of QCD. Moreover, it is embedded into the QCD flows and can be successively extended towards full QCD flows. Space limitation do not permit to detail the approximation, schematically it is described by

$$\Gamma_k = \int_x \left[Z_{k,\psi}\bar{q}\partial\!\!\!/ q + \frac{1}{2}\left[Z_{k,\sigma}(\partial\sigma)^2 + Z_{k,\pi}(\partial\vec{\pi})^2 \right] + h_k\bar{q}\left(\frac{\sigma}{2} + i\gamma_5\tau^a\pi^a\right) q + V_k[\rho] \right], \tag{6}$$

with $\rho = 1/2(\sigma^2 + \vec{\pi}^2)$ and four-momentum dependent wave function renormalisations $Z_{k,\Phi} = Z_{k,\Phi}(p)$ as well as a momentum-dependent Yukawa coupling $h_k = h_k(p^2/2 + q^2/2)$ with quark and anti-quark momentum p and q. The full effective potential $V_k(\rho)$ is expanded about a fixed value ρ_0 of the unrenormalised field which leads to rapid convergence, see [13],

$$V_k[\rho] = \sum_{n=0}^{\infty} \frac{V_k^{(n)}(\rho_0)}{n!}(\rho - \rho_0)^n \,. \tag{7}$$

ρ_0 is chosen such that it is close to the physical expansion point $\rho_{\text{phys}} \propto f_\pi^2$. Note that the latter is not at the minimum of V_k in the case of explicit chiral symmetry breaking, for a discussion see e.g. [13]. At the minimum the first and second expansion coefficients of V_k define the curvature masses of the mesons, $m^2_{\pi,\text{cur}} = V_k^{(1)}(\rho_{\text{phys}})$ and $m^2_{\sigma,\text{cur}} = V_k^{(1)}(\rho_{\text{phys}}) + \sqrt{2\rho_{\text{phys}}}V_k^{(2)}(\rho_{\text{phys}})$.

In principle, one may consider all orders of the potential. However, it has been checked that a polynomial approximation about a fixed expansion point ρ_0 up to the order ρ^6 provides a robust and quantitatively accurate representation of the potential, [5, 6, 7, 13].

4. Observables and analytical continuation

At vanishing cutoff scale $k = 0$ we are now in the position to determine observables such as the pole masses. To that end we need to perform an analytical continuation of the Euclidean results. In future work this is done by a direct computation along the lines of [10, 11].

However, here we will exploit two different and less demanding methods of analytical continuation leaving the direct calculation for future work. The first one is using a Padé approximant and determining the respective poles, e.g. [14], while the second method is a modified Padé approach based on the Schlessinger point method, e.g. [15].

The following results are obtained by tuning to certain infrared values. In particular we require the curvature masses $m_{k\to 0}$ to be $m_{\text{IR},\pi} = 138$ MeV, $m_{\text{IR},\sigma} = 536$ MeV, and $m_\psi = 297$ MeV. Furthermore, $f_\pi = 93$ MeV is used as input. Starting the RG flow from the UV scale to be chosen to be 0.95 GeV we obtained the results displayed in figs. 1 and 2.

Using the analytical continuation methods described above, one obtains pole masses that can be compared with the curvature masses, both for $k \to 0$. Hereby the uncertainties for the Padé pole masses are derived from the convergence pattern of different orders of the approximant. The results agree very well with the curvature mass for the pion, in accordance with [14], but not with the same precision for the sigma meson. Further results will be published elsewhere.

Table 1. Pole masses vs. curvature masses, all in MeV.

Particle	Curvature Mass (Input)	Pole Mass Padé	Pole Mass Schlessinger
Pion	138.22	136.2 ± 0.1	136.4 ± 1.0
Sigma meson	536.37	472 ± 7	479 ± 10

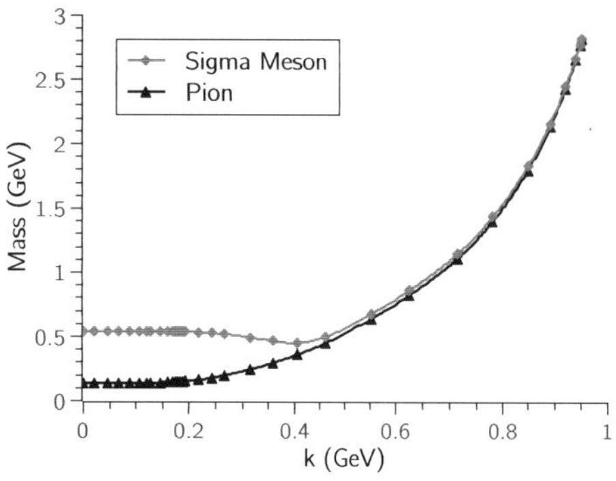

Figure 1. Curvature mass for mesons as a function of the RG scale k.

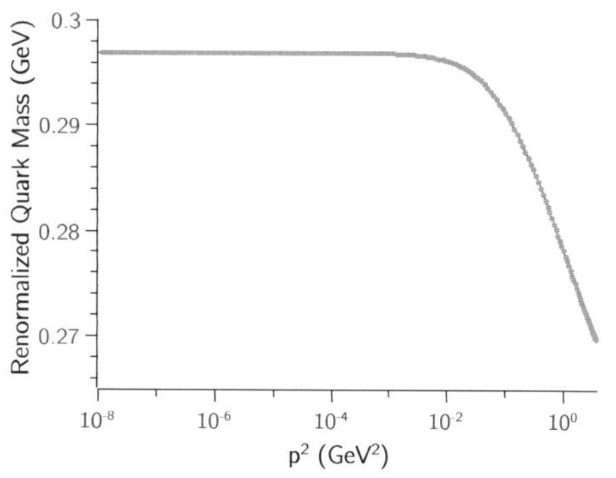

Figure 2. Momentum-dependent renormalized quark mass at $k = 10^{-9}$ GeV.

5. Summary and Outlook

In this preliminary study we have demonstrated that the FRG is a functional method suitable for calculating observables related to bound states. Within the Quark-Meson model we determined the propagators of the sigma and the pion fields, and from them the respective pole masses via Padé approximants. In a next step we will compute real time correlation functions with the methods deviced in [10, 11] as well as improving the approximation towards full QCD along the lines of [5, 6, 7] .

Acknowledgments

J. P.-L. and W. A. M. are supported by the FWF Doctoral Program W1203 "Hadrons in vacuum, nuclei and stars" and by the mobility program of the Science Faculty of the University of Graz. M. M. is supported by the FWF grant J3507-N27, the DFG grant MI 2240/1-1 and the U.S. Department of Energy under contract de-sc0012704. H. S. A. is supported by the FWF grant P29216-N36. This work is also supported by EMMI, the BMBF grant 05P12VHCTG, and is part of and supported by the DFG Collaborative Research Centre SFB 1225 (ISO-QUANT).

References

[1] Eichmann G, Sanchis-Alepuz H, Williams R, Alkofer R and Fischer C, *Prog. Part. Nucl. Phys.* **91** (2016) 1 (*Preprint* 1606.09602)
[2] Gies H and Wetterich C, *Phys. Rev. D* **65** (2002) 065001 (*Preprint* hep-th/0107221)
[3] Pawlowski J, *Annals Phys.* **322** (2007) 2831 (*Preprint* hep-th/0512261).
[4] Floerchinger S and Wetterich C, *Phys. Lett. B* **680** (2009) 371 (*Preprint* 0905.0915)
[5] Mitter M, Pawlowski J and Strodthoff N, *Phys. Rev. D* **91** (2015) 054035 (*Preprint* 1411.7978)
[6] Braun J, Fister L, Pawlowski J and Rennecke F, *Phys. Rev. D* **94** (2016) no.3, 034016 (*Preprint* 1412.1045)
[7] Cyrol A, Mitter M, Pawlowski J and Strodthoff N, *Preprint* 1706.06326
[8] Floerchinger S, *JHEP* **1205** (2012) 021 (*Preprint* 1112.4374)
[9] Tripolt R, Strodthoff N, von Smekal L and Wambach J, *Phys. Rev. D* **89** (2014) no.3, 034010 (*Preprint* 1311.0630)
[10] Pawlowski J and Strodthoff N, *Phys. Rev. D* **92** (2015) no.9, 094009 (*Preprint* 1508.01160)
[11] Strodthoff N, *Phys. Rev. D* **95** (2017) no.7, 076002 (*Preprint* 1611.05036)
[12] Wetterich C, *Phys. Lett. B* **301** (1993) 90.
[13] Pawlowski J and Rennecke F, *Phys. Rev. D* **90** (2014) no.7, 076002 (*Preprint* 1403.1179)
[14] Helmboldt A, Pawlowski J and Strodthoff N, *Phys. Rev. D* **91** (2015) no.5, 054010 (*Preprint* 1409.8414)
[15] Tripolt R, Haritan I, Wambach J and Moiseyev N, *Preprint* 1610.03252

IOP Conf. Series: Journal of Physics: Conf. Series **1024** (2018) 012010 doi:10.1088/1742-6596/1024/1/012010

S- and *D*-wave vector charmonia

A Krassnigg[1,2] **and T Hilger**[1]

[1]Institute of Physics, University of Graz, NAWI Graz, A-8010 Graz, Austria

[2]Institute of High Energy Physics, Austrian Academy of Sciences, A-1050 Vienna, Austria

E-mail: `andreas.krassnigg@uni-graz.at`

Abstract. We revisit earlier calculations of leptonic decay constants of vector charmonia and present and illustrate our decomposition of the corresponding covariant Bethe-Salpeter amplitudes in terms of orbital angular momentum as interpreted in the meson's rest frame. Our results confirm our previous conclusions drawn from the magnitudes of vector-meson leptonic decay constants, identifying the $\Psi(3770)$ and $\Psi(4160)$ as *D*-wave states in our setup.

1. Fundamental approaches to hadron spectroscopy

Hadrons have been around for a long time and, apparently, also will be [1]. They have their prominent place in particle physics for a number of reasons. One of these reasons is presented by reference to the strong interaction, which, in the face of our understanding of the atomic nucleus as a composite object made of only electrically positively charged and neutral nucleons, was necessary to provide a mechanism for the stability of matter as we know it. Thus, it is important to keep in mind that *what a hadron is* is defined by the strong interaction and not by the hadron's inner workings.

As a consequence, it is a choice what is meant by a fundamental approach to hadron spectroscopy, simply because we can choose or invent different mechanisms to underly the hadron spectrum. Many scientists also make many clever choices of this kind. A delectably large number of them were presented at least in some respect at the FAIRNESS2017 workshop, which should be understood as a merit of the organizing committee. The landscape of such approaches has been impacted significantly by the quark concept and the advent of quantum chromodynamics (QCD) several decades ago. Some people are tempted to rely on the quark-gluon picture of hadrons too easily, e. g., when it comes to the distinction between mesons and baryons, which is actually and simply made by the value of the hadron's baryon number. The fact that composing hadrons out of quarks and gluons reproduces the correct pattern is necessary as a feature of a successful explanation, but not acceptable as a definition. Still, one can get the impression that, cf. [2]

> All approaches to hadron spectroscopy are fundamental, but some approaches are more fundamental than others ...

However, this notion, often referring to the relevant scales in different descriptions of hadrons, is sometimes unjustly used to imply truth or importance. Instead, we understand the term *fundamental approach* to hadron spectroscopy as one that explains the hadron spectrum by a well-defined set of rules and assumptions. We refer the reader to the other corresponding contributions to these proceedings for more details.

2. DSBSE approach

Our approach to hadron spectroscopy is the Dyson-Schwinger-Bethe-Salpeter-equation (DSBSE) incarnation of QCD. It is fundamental at the level of quarks and gluons and starts at the QCD Lagrangian. This framework of infinitely many coupled nonlinear integral equations is nonperturbative, manifestly covariant, and uses continuum quantum field theory [3]. In numerical studies, truncations are employed in order to render the computation of any given hadronic observable feasible. Investigation of schemes of truncations, e.g. [4–9] and references therein, have pointed to the rainbow-ladder (RL) truncation [10] as a meaningful first step for any numerical treatment, in particular, if the set of results is to be comprehensive [11–14]. In addition, in the DSBSE approach, pioneering phenomenological studies always used the RL truncation, be it for the properties of mesons [15–29] or baryons [30–39].

It rapidly becomes clear to any practitioner of theoretical hadron spectroscopy that dynamical chiral symmetry breaking (DχSB), an apparent phenomenon in QCD, is a necessary ingredient in a successful description of the light-hadron spectrum. For example, DχSB provides a mass scale that, starting from current quarks with a mass of a few MeV each, results in the nucleons' mass of about 1 GeV. While constituent-quark masses can explain a lot in terms of the masses of hadrons simply by adding them appropriately and thinking in terms of binding energies, the limits of this concept are illustrated best by investigation of the lightest hadron, the pion [40].

A better explanation is offered by symmetry arguments. The chiral symmetry of QCD with massless quarks, together with its dynamical breaking results in a massless Goldstone boson. The explicit breaking of QCD's chiral symmetry by the current-quark masses makes the pion the massive but very light realization of this (pseudo-)Goldstone boson. As a consequence, pion properties are an important focus of modern hadron theory.

3. DSBSE Model Setup

In the DSBSE approach, chiral symmetry as well as its dynamical breaking are described by the axial-vector Ward-Takahashi identity (AVWTI) [42]. In any given truncation, satisfaction of the AVWTI leads to correct anchoring of results both in the absence [43] and presence of DχSB [44]. DχSB is already visible at the quark level by inspection of the solution of the quark Dyson-Schwinger equation (DSE), where the momentum dependent quark mass function is qualitatively different for the chirally symmetric and broken phases [41, 45]. At the meson level, one finds that, if a truncation satisfies the AVWTI, one automatically has a massless pion in the chiral limit, if chiral symmetry is dynamically broken, together with well-known features such as a generalized Gell-Mann-Oakes-Renner relation valid for all isovector pseudoscalar mesons irrespective of quark-mass content and level of excitation [44, 46]. In RL truncation, the quark-antiquark interaction is modeled by parameterizing an effective dressed one-gluon exchange as

Table 1: Calculated pseudoscalar- and vector-meson decay constants f_{Calc} in MeV compared to experimental data for selected charmonia, together with our orbital-angular-momentum assignment L_{Calc} and contributions in percent.

State	J^{PC}	f_{Exp} [1]	f_{Calc} [41]	L_{Calc}	S	S-P	P	P-D	D	S-D
η_c	0^{-+}	338(14)	378	S	99	1	0	–	–	–
J/Ψ	1^{--}	416(6)	411	S	99	1	0	0	0	0
$\Psi(2S)$	1^{--}	295(5)	155	S	100	0	0	0	0	0
$\Psi(3770)$	1^{--}	100(4)	45	D	0	0	0	2	98	0
$\Psi(4040)$	1^{--}	187(17)	188	S	100	0	0	0	0	0
$\Psi(4160)$	1^{--}	143(32)	1	D	0	0	0	1	99	0

IOP Conf. Series: Journal of Physics: Conf. Series **1024** (2018) 012010 doi:10.1088/1742-6596/1024/1/012010

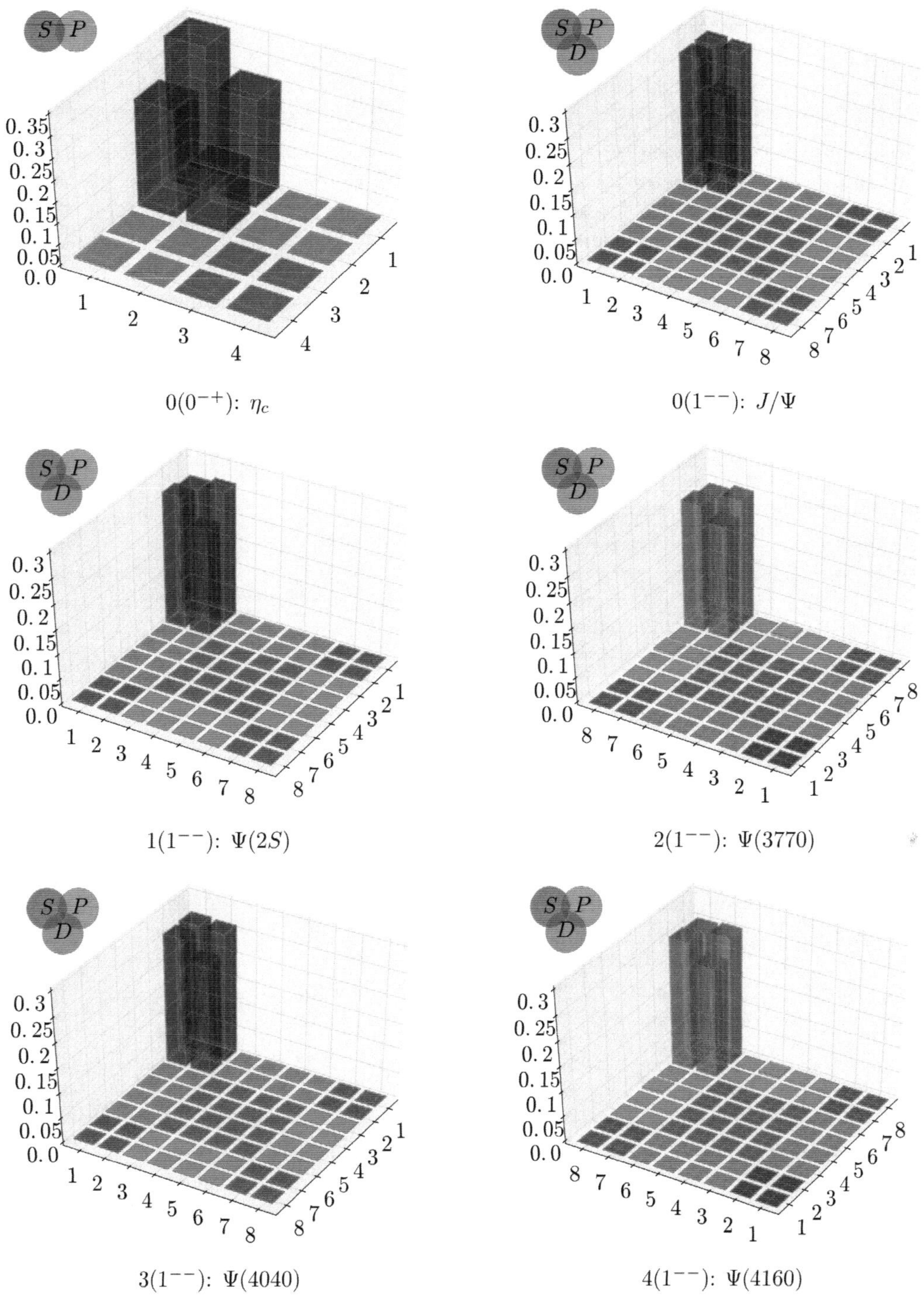

Figure 1: Orbital angular momentum decompositions of charmonia. The set of quantum numbers underneath each subfigure denotes $n(J^{PC})$, where n is the excitation quantum number, and J, P, and C are the meson's total spin, parity, and charge-conjugation parity.

IOP Conf. Series: Journal of Physics: Conf. Series **1024** (2018) 012010 doi:10.1088/1742-6596/1024/1/012010

a function of the gluon momentum squared. Several functional forms have been used for this purpose, and we chose the one in [15], whose intermediate-momentum domain is characterized by two parameters, an inverse effective range ω and an overall strength D. For more details and an illustration of this effective interaction, see Fig. 1 of [47] and the related discussion there.

4. Orbital Angular Momentum

The solution of the Bethe-Salpeter-equation (BSE) is the Bethe-Salpeter amplitude (BSA), which we obtain numerically, see [48–52] for computational details. It is the covariant analog of a wave function and contains all information about the state [53–56]. While being covariant, it also contains information that can be interpreted as the quark-antiquark orbital angular momentum in a meson by virtue of its covariants T_i, $i = 1, \ldots, N$, where $N = 4$ for spin-0 mesons or $N = 8$ otherwise, and the index i refers to the numbering on the horizontal axes in Fig. 1. The details of the orbital angular momentum decomposition of a meson can be found in App. A of [47] and for baryons in [57]. What is important to note here is that spin-0 allows for S-wave, P-wave and S-P mixed components [58], while in a spin-1 meson BSA one can have S-wave, P-wave, D-wave, and S-P, P-D, S-D mixed components [47]. This situation as well as the quantification of these components via the canonical norm of the BSA [59] are explained and illustrated in Fig. 6 of [47] and the corresponding discussion there.

5. Results and Discussion

In previous studies we have focussed on heavy-quarkonium masses [13, 60] and leptonic decay constants [41]. In the latter work, we provided a likely assignment of S- and D-wave to vector quarkonia based on the calculated values for their leptonic decay constants, which we present again herein as f_{Calc} in Tab. 1. These numbers, as well as the orbital-angular-momentum assignment L_{Calc} presented in the same table, were calculated with $\omega = 0.3$ GeV and $D = 1.3$ GeV^2 in the interaction of [15]. In this table, we compare them to the updated experimental values from [1] denoted by f_{Exp}. In addition, we have computed the contributions of orbital angular momentum to the BSA for the states listed in Tab. 1, where the percentages for each component fill the right part of the table. The corresponding illustrations can be found in Fig. 1.

It is apparent that in all cases presented here the assignment of an overall or dominating quark-antiquark orbital angular momentum L_{Calc} is unambiguous. Our assignments confirm the statements made previously by means of f_{Calc} alone [41]. Our leptonic-decay-constant and orbital-angular-momentum results are of interest in comparison to other recent approaches that have a similar or even more direct handle on such contributions in their respective meson amplitudes, e. g., [61–68].

Acknowledgments

A. K. would like to thank the organizers of the FAIRNESS2017 workshop for their kind invitation, their hospitality in Sitges, and the once again stimulating and productive atmosphere. This work was supported by the Austrian Science Fund (FWF) under project no. P25121-N27.

References

[1] Patrignani C *et al.* (Particle Data Group) 2016 *Chin. Phys. C* **40** 100001 and 2017 update
[2] Orwell G 1945 *Animal Farm: A Fairy Story* (Secker and Warburg, London, England)
[3] Bashir A *et al.* 2012 *Commun. Theor. Phys.* **58** 79–134
[4] Bhagwat M S, Höll A, Krassnigg A, Roberts C D and Tandy P C 2004 *Phys. Rev.* C **70** 035205
[5] Höll A, Krassnigg A and Roberts C D 2005 *Nucl. Phys. Proc. Suppl.* **141** 47–52
[6] Gomez-Rocha M, Hilger T and Krassnigg A 2015 *Few Body Syst.* **56** 475–480
[7] Sanchis-Alepuz H and Williams R 2015 *J. Phys. Conf. Ser.* **631** 012064
[8] Gomez-Rocha M, Hilger T and Krassnigg A 2015 *Phys. Rev.* D **92** 054030
[9] Gomez-Rocha M, Hilger T and Krassnigg A 2016 *Phys. Rev.* D **93** 074010

[10] Eichmann G, Alkofer R, Cloët I C, Krassnigg A and Roberts C D 2008 *Phys. Rev.* C **77** 042202(R)
[11] Krassnigg A 2009 *Phys. Rev.* D **80** 114010
[12] Popovici C, Hilger T, Gomez-Rocha M and Krassnigg A 2015 *Few Body Syst.* **56** 481–487
[13] Hilger T, Popovici C, Gomez-Rocha M and Krassnigg A 2015 *Phys. Rev.* D **91** 034013
[14] Hilger T, Gómez-Rocha M, Krassnigg A and Lucha W 2017 Eur. Phys. J. A, in press (*Preprint* 1702.06262)
[15] Maris P and Tandy P C 1999 *Phys. Rev.* C **60** 055214
[16] Krassnigg A and Roberts C D 2004 *Fizika* B **13** 143–152
[17] Krassnigg A and Roberts C D 2004 *Nucl. Phys.* A **737** 7–15
[18] Krassnigg A and Maris P 2005 *J. Phys. Conf. Ser.* **9** 153–160
[19] Höll A, Krassnigg A, Roberts C D and Wright S V 2005 *Int. J. Mod. Phys.* A **20** 1778–1784
[20] Alkofer R, Kloker M, Krassnigg A and Wagenbrunn R F 2006 *Phys. Rev. Lett.* **96** 022001
[21] Höll A, Krassnigg A, Maris P, Roberts C D and Wright S V 2005 *Phys. Rev.* C **71** 065204
[22] Cloët I C, Krassnigg A and Roberts C D 2007 *eConf* **C070910** 125
[23] Eichmann G, Alkofer R, Krassnigg A and Nicmorus D 2008 *PoS* **CONFINEMENT8** 077
[24] Dorkin S M, Hilger T, Kämpfer B and Kaptari L P 2010 (*Preprint* 1012.5372)
[25] Blank M, Krassnigg A and Maas A 2011 *Phys. Rev.* D **83** 034020
[26] Mader V, Eichmann G, Blank M and Krassnigg A 2011 *Phys. Rev.* D **84** 034012
[27] Raya K *et al.* 2016 *Phys. Rev.* D **93** 074017
[28] Hilger T and Krassnigg A 2017 *Eur. Phys. J. A* **53** 142
[29] Hilger T and Krassnigg A 2017 *EPJ Web Conf.* **137** 01010
[30] Bloch J C R, Krassnigg A and Roberts C D 2003 *Few-Body Syst.* **33** 219–232
[31] Alkofer R, Höll A, Kloker M, Krassnigg A and Roberts C D 2005 *Few-Body Syst.* **37** 1–31
[32] Höll A, Alkofer R, Kloker M, Krassnigg A, Roberts C D and Wright S V 2005 *Nucl. Phys.* A **755** 298–302
[33] Eichmann G, Krassnigg A, Schwinzerl M and Alkofer R 2008 *Ann. Phys.* **323** 2505–2553
[34] Nicmorus D, Eichmann G, Krassnigg A and Alkofer R 2009 *Phys. Rev.* D **80** 054028
[35] Nicmorus D, Eichmann G, Krassnigg A and Alkofer R 2008 *PoS* **CONFINEMENT8** 052
[36] Eichmann G, Cloët I C, Alkofer R, Krassnigg A and Roberts C D 2009 *Phys. Rev.* C **79** 012202(R)
[37] Eichmann G, Alkofer R, Krassnigg A and Nicmorus D 2010 *Phys. Rev. Lett.* **104** 201601
[38] Eichmann G, Alkofer R, Krassnigg A and Nicmorus D 2010 *EPJ Web of Conferences* **3** 03028
[39] Eichmann G *et al.* 2016 *Prog. Part. Nucl. Phys.* **91** 1–100
[40] Horn T and Roberts C D 2016 *J. Phys. G* **43** 073001
[41] Krassnigg A, Gomez-Rocha M and Hilger T 2016 *J. Phys. Conf. Ser.* **742** 012032
[42] Munczek H J 1995 *Phys. Rev. D* **52** 4736–4740
[43] Hilger T 2016 *Phys. Rev.* D **93** 054020
[44] Maris P and Roberts C D 1997 *Phys. Rev. C* **56** 3369–3383
[45] Blank M and Krassnigg A 2010 *Phys. Rev.* D **82** 034006
[46] Höll A, Krassnigg A and Roberts C D 2004 *Phys. Rev.* C **70** 042203(R)
[47] Hilger T, Gomez-Rocha M and Krassnigg A 2017 *Eur. Phys. J. C* **77** 625
[48] Bhagwat M S, Höll A, Krassnigg A, Roberts C D and Wright S V 2007 *Few-Body Syst.* **40** 209–235
[49] Krassnigg A 2008 *PoS* **CONFINEMENT8** 075
[50] Blank M and Krassnigg A 2011 *Comput. Phys. Commun.* **182** 1391
[51] Blank M and Krassnigg A 2011 *AIP Conf. Proc.* **1343** 349
[52] Dorkin S M, Kaptari L P, Hilger T and Kämpfer B 2014 *Phys. Rev.* C **89** 034005
[53] Krassnigg A and Blank M 2011 *Phys. Rev.* D **83** 096006
[54] Dorkin S M, Hilger T, Kaptari L P and Kämpfer B 2011 *Few Body Syst.* **49** 247–254
[55] Hilger T U 2012 *Medium Modifications of Mesons* Ph.D. thesis TU Dresden
[56] Hilger T, Gomez-Rocha M and Krassnigg A 2015 *Phys. Rev.* D **91** 114004
[57] Eichmann G 2009 Ph.D. thesis University of Graz (*Preprint* 0909.0703)
[58] Bhagwat M S, Krassnigg A, Maris P and Roberts C D 2007 *Eur. Phys. J.* A **31** 630–637
[59] Smith C H L 1969 *Nuovo Cim. A* **60** 348–351
[60] Blank M and Krassnigg A 2011 *Phys. Rev.* D **84** 096014
[61] Brambilla N *et al.* 2011 *Eur. Phys. J. C* **71** 1534
[62] Dudek J J 2011 *Phys. Rev. D* **84** 074023
[63] Li Y, Maris P, Zhao X and Vary J P 2016 *Phys. Lett. B* **758** 118–124
[64] Rohrhofer C, Pak M and Glozman L Ya 2016 (*Preprint* 1603.04665)
[65] Leitao S, Li Y, Maris P, Pena M T, Stadler A, Vary J P and Biernat E P 2017 (*Preprint* 1705.06178)
[66] Leitão S, Stadler A, Peña M T and Biernat E P 2017 *Few Body Syst.* **58** 91
[67] Piotrowska M, Reisinger C and Giacosa F 2017 *Phys. Rev. D* **96** 054033
[68] Leitao S, Stadler A, Pena M T and Biernat E P 2017 *Phys. Rev. D* **96** 074007

IOP Conf. Series: Journal of Physics: Conf. Series **1024** (2018) 012011 doi:10.1088/1742-6596/1024/1/012011

Thermal dileptons as QCD matter probes at SIS

F Seck[1], **T Galatyuk**[1,2], **R Rapp**[3] **and J Stroth**[2,4]

[1] IKP, Technische Universität Darmstadt, Schlossgartenstr. 9, 64298 Darmstadt, Germany
[2] GSI Helmholtzzentrum für Schwerionenforschung, Planckstr. 1, 64291 Darmstadt, Germany
[3] Cyclotron Institute, Texas A&M University, College Station (TX) 77843-3366, USA
[4] IKF, Goethe-Universität, Max-von-Laue-Str. 1, 60438 Frankfurt, Germany

E-mail: f.seck@gsi.de

Abstract. Electromagnetic radiation is emitted during the whole course of a heavy-ion collision and can escape from the collision zone without further interactions. This makes it an ideal tool to study the properties of hot and dense QCD matter. To model the space-time evolution of the collision at SIS energies a coarse-graining approach is used to convert transport simulations into meaningful temperatures and densities. These parameters serve as input for the determination of the pertinent radiation of thermal dileptons based on an in-medium ρ spectral function that describes available spectra at ultrarelativistic collision energies. The resulting excitation function of the thermal excess radiation provides a baseline for future measurements by the HADES and CBM experiments at GSI/FAIR, and experiments proposed at NICA and J-PARC.

1. Introduction

Heavy-ion collisions (HICs) provide insights into the states of matter under extreme densities and temperatures. While the conditions in the fireball produced in ultrarelativistic collisions resemble the early universe a few microseconds after the big bang, the matter created in collisions at lower bombarding energies might be similar to the conditions in the center of two merging neutron stars [1, 2]. Thus, changing the energy and the type of the nuclei which induce the reactions permits systematic investigations of the properties and the composition of matter across the QCD phase diagram.

For example, chiral symmetry which is spontaneously broken in the QCD vacuum due to a non-zero quark-antiquark condensate $\langle\bar{q}q\rangle$ gets restored in a hot medium [3, 4]. This induces modifications to the spectral distributions of chiral partners in the light hadron sector such as the ρ and a_1 mesons – ultimately leading to their degeneracy. A substantial melting of the chiral condensate is conjectured to be relevant already for HICs at low collision energies [5, 6].

Electromagnetic (EM) radiation represents an excellent probe for such investigations as it is emitted during the whole time evolution of a HIC and decouples from the strongly interacting collision zone once created. In contrast to real photons which are massless, virtual photons decaying into a pair of leptons (e^+e^- or $\mu^+\mu^-$) have an additional observable in terms of their invariant-mass. Thus, one can access the information about the properties of matter produced inside the hot and dense fireball which is irretrievable from the spectra of final-state hadrons due to rescattering.

In particular, the excess yield of low-mass dileptons above the contribution from the hadronic freeze-out cocktail was identified to be sensitive to the fireball lifetime, while the slope in the

IOP Conf. Series: Journal of Physics: Conf. Series **1024** (2018) 012011 doi:10.1088/1742-6596/1024/1/012011

intermediate-mass region of the dilepton invariant-mass spectrum can serve as a thermometer which is not distorted by the collective expansion of the medium [7, 8].

2. Coarse-grained transport approach for thermal dilepton rates

For a proper theoretical description of the contribution of in-medium signals to the dilepton invariant-mass spectrum realistic thermal dilepton emission rates need to be convoluted with an accurate modeling of the chemical potentials and temperatures reached during the space-time evolution of the fireball.

For HICs at ultrarelativistic energies (URHICs) a hydrodynamic description can be used to evaluate the thermodynamic properties of the medium and the thermal dilepton rates can directly be employed. The applicability of such an approach for the system evolution down to the SIS18 energy range of a few GeV is unclear. One issue is the justification of thermalization due to the long time it takes until a full overlap of the two incoming nuclei is reached. For this reason hadronic transport models like UrQMD are commonly used to describe the system evolution of HICs at SIS energies. The dilepton radiation is then obtained perturbatively by integrating for each created resonance the probability of a decay into a dilepton over the whole lifetime of this hadronic resonance. The incorporation of medium effects on broad resonances into these off-equilibrium approaches remains however challenging.

To bridge this gap between the microscopic transport and macroscopic hydrodynamic approaches a coarse-graining procedure was proposed [9]. It was recently applied to the SIS18 energy regime [10, 11, 12] where the HADES collaboration has measured the dielectron spectra in collisions of Ar+KCl at E_{lab}=1.76A GeV and Au+Au at E_{lab}=1.23A GeV (corresponding to $\sqrt{s_{NN}}$=2.6 and 2.4 GeV respectively) [13, 14]. By dividing the space-time evolution into 4-dimensional cells and averaging over an ensemble of many simulated transport events one obtains smooth particle distributions. Reasonable temperatures, baryon and pion densities as well as collective flow patterns can then be extracted from cells which fulfill certain criteria that are favorable for (the onset of) thermalization, *i.e.*, a minimum number of three collisions that the majority of nucleons in the cell need to have undergone so that their momentum distributions in all Cartesian directions acquire a Gaussian shape with comparable width as well as the transverse mass spectra of pions that can be described by an exponential shape. In this way the premise of a full hydro simulation is mitigated as the deviations from the vanishing mean-free path limit are kept in the evolution. Figure 1 shows how the distribution of nucleons in central Au+Au collisions at $\sqrt{s_{NN}}$=2.4 GeV which have experienced a given number of collisions is modified over time. During the first few time steps the distribution changes quite rapidly due to many interactions in the system, while it stays almost the same after 21 fm/c indicating kinetic freeze-out. The evolution of the extracted temperature, effective baryon density $\rho_{\mathit{eff}} = \rho_N + \rho_{\bar{N}} + \frac{1}{2}(\rho_R + \rho_{\bar{R}})$ (N refers to nucleons, R to baryonic resonances) for central collisions averaged over the inner cube of 7x7x7 cells (each of a volume of 1 fm^3) is shown in Fig. 2. In the center of the Au+Au collision system temperatures of up to 90 MeV and densities of up to 3 times normal nuclear matter density are reached.

These bulk properties of the cells are used as input parameters for the calculation of thermal dilepton radiation using the expression

$$\frac{d^8N}{d^4x\, d^4p} = \frac{\alpha_{EM}^2}{\pi^2 M^2} f_B(p_0, T)\, \varrho_{EM}(M, p; T, \rho_{\mathit{eff}}, \mu_\pi)\,, \qquad (1)$$

where $M = \sqrt{p_0^2 - p^2}$ is the invariant mass of the virtual photon, f_B denotes the thermal Bose distribution function, and ϱ_{EM} the EM spectral function of the QCD medium depending on the temperature, the effective baryon density, and an effective chemical potential of pions, μ_π. For ϱ_{EM} we employ a parametrization of the in-medium ρ spectral function of Ref. [15] which

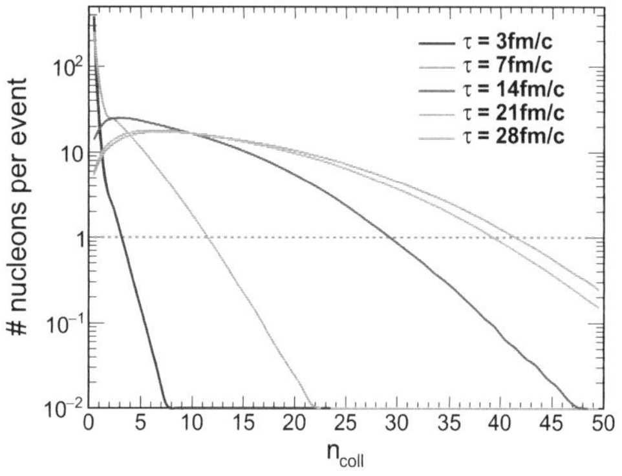

Figure 1. Distribution of the number of binary collisions of nucleons per event in central Au+Au collisions at $\sqrt{s_{NN}}$=2.4 GeV for different time steps in the evolution.

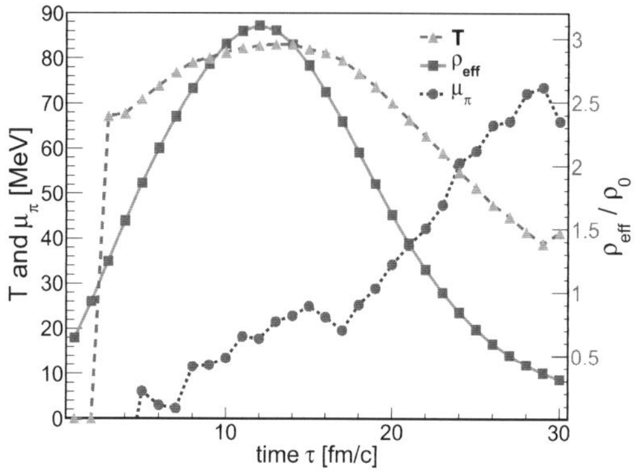

Figure 2. Time evolution of temperature, effective baryon density (right scale) and effective pion chemical potential averaged over an inner cube of 7x7x7 cells (1 fm^3 each) in central Au+Au collisions at $\sqrt{s_{NN}}$=2.4 GeV.

describes dilepton data in URHICs. The effect of the chemical potential of pions on the EM spectral function is encoded in an overall fugacity factor, $z^\kappa = \exp(\mu_\pi/T)^\kappa$, where κ specifies the average number of pions participating in the production of ρ mesons. Using UrQMD we estimated this number to be close to 1 at SIS energies [11] as the main production channels are baryon resonance decays rather than $\pi\pi$ annihilation.

3. Results

An experimental acceptance filter was applied to the full phase-space distribution in order to compare the resulting invariant-mass spectrum of thermal radiation to available data of the Ar+KCl system. Figure 3 shows a fair agreement between the calculated spectrum inside the HADES acceptance and the measured excess yield above the "cocktail" of long-lived EM decays at freeze-out plus a superposition of p+p and n+p reactions [13, 16] to model the contribution of first-chance NN collisions. The high baryon densities reached during the collisions lead to strong medium modifications of the spectral shape of the ρ meson resulting in an almost exponential excess spectrum of thermal radiation.

One remarkable finding is that the cumulative yield of thermal dileptons in the low-mass region, M_{ee}=0.3-0.7 GeV/c^2, reveals the same temporal development as the build-up of the collective radial flow [11, 12]. This underlines a close relation of these quantities as the same interactions which drive the collectivity in the system also excite resonances which in turn emanate dileptons. It therefore consolidates the use of the low-mass dilepton yield as a fireball chronometer. The interacting fireball lifetime amounts to approximately 13 fm/c in central Au+Au collisions at $\sqrt{s_{NN}}$=2.4 GeV and to about 8 fm/c for the lighter Ar+KCl system.

The quantitative link between the fireball lifetime and the thermal dilepton yield established for URHICs [7] normalizes the latter to the number of charged particles, $N_{\rm ch}$, at midrapidity. To generalize this finding also to lower collision energies one has to keep in mind that $N_{\rm ch}$ is no longer a good proxy for the thermal excitation energy in the system as the fireball is more and more dominated by the incoming nuclei. Thus, Fig. 4 shows the low-mass thermal dilepton yield normalized to the number of charged pions, $N_{\pi^\pm}$. The curves have been calculated using a fireball model with QGP and in-medium hadronic emission [7], while the diamonds and circles represent the yields and lifetime obtained within our coarse-graining approach. The proportionality of the

IOP Conf. Series: Journal of Physics: Conf. Series **1024** (2018) 012011 doi:10.1088/1742-6596/1024/1/012011

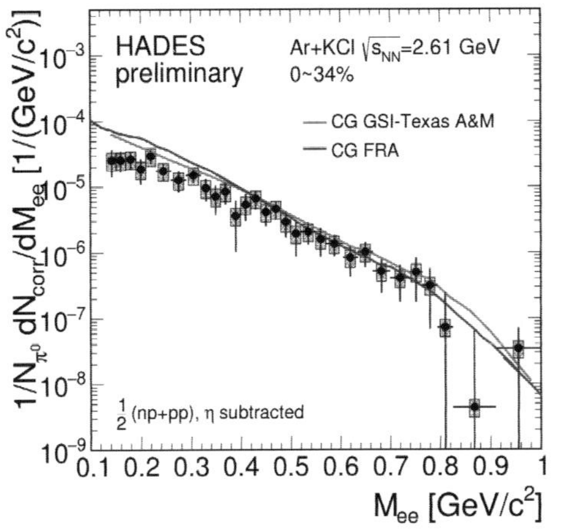

Figure 3. Comparison of dilepton excess spectra from two independent coarse-graining approaches [10, 11] and the experimentally extracted yield above the freeze-out cocktail in Ar+KCl at 1.76AGeV from HADES [13, 16].

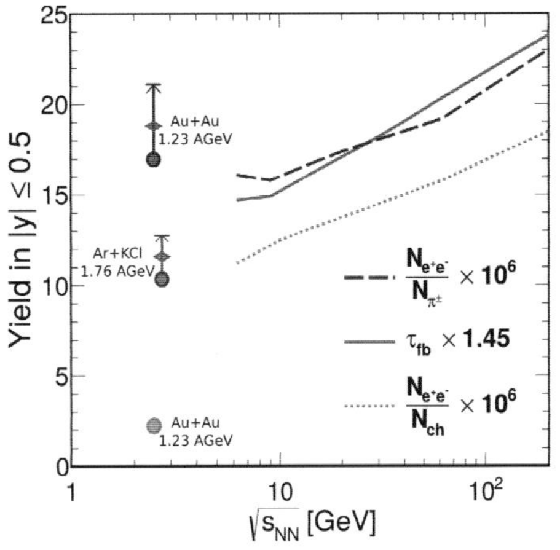

Figure 4. Excitation function of the thermal low-mass dilepton yield in central Au+Au collisions normalized to $N_{\pi^\pm}$ (blue dashed line, blue circle) and $N_{\rm ch}$ (green dotted line, green circle), together with the lifetime of the system (red solid line, red diamond).

yield to the lifetime for URHICs stays intact with a slightly larger normalization of 1.45 (which includes strong final-state decays in the dilepton yield) [11]

$$\left.\frac{N_{ll}}{N_{\pi^\pm}}\right|_{|y|\leq 0.5} \times 10^6 = 1.45\,\tau_{\rm fb}\,[{\rm fm}/c]. \tag{2}$$

The excitation function remains rather flat from top RHIC down to the SIS18 energies. This offers the opportunity to utilize the measurement of the thermal low-mass yield in the search for the phase transition in the so-far unexplored (with dileptons) FAIR energy regime. A critical slowing down of the evolution in the transition region could manifest itself by an increased lifetime of the fireball and in turn an enhanced dilepton yield.

Translating the extracted temperatures and baryon densities to the baryon chemical potential μ_B one can plot the trajectories in the QCD phase diagram of the central regions of Au+Au and Ar+KCl collisions at SIS energies together with the expectation value of the chiral condensate employing the Nambu-Jona-Lasinio (NJL) model [6], see Fig. 5. The trajectories are traversed in counter-clockwise direction in time steps of 1 fm/c. The highlighted regions show the time interval from which most (90%) of the thermal dileptons emanate. The chiral condensate during the emission window is depleted significantly. Note that the leading density approximation using the sigma term of nucleons suggest a 25-30% drop already at normal nuclear matter density which is not reflected in the model calculation. Shown is also the location of the chemical freeze-out point extracted from a statistical hadronization model (SHM) fit to the abundances of several hadron species in Ar+KCl collisions [17]. Shortly after this point on the trajectory the emission of thermal dileptons creases indicating thermal freeze-out.

4. Conclusions and outlook

The study of HICs at relativistic bombarding energies becomes increasingly important as results from several theoretical frameworks locate the QCD critical point in regions accessible to

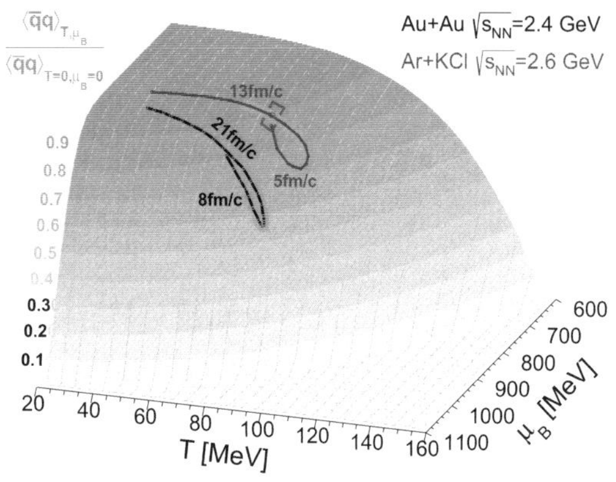

Figure 5. Trajectories of the central regions of Au+Au and Ar+KCl collisions in the QCD phase diagram. The orange contour lines indicate the reduction of the chiral condensate compared to the vacuum expectation value based on a NJL model calculation [6]. The green (cyan) highlighted part of the trajectory of Au+Au (Ar+KCl) shows the time period which radiates most of the thermal dilepton yield. The chemical freeze-out point of the Ar+KCl system [17] is depicted by the red square.

FAIR [18, 19]. Thermal dilepton emission in HICs at such energies can be obtained with the presented coarse-graining approach that couples the space-time evolution of hadronic transport simulations with in-medium dilepton rates. The procedure was applied to Au+Au and Ar+KCl collision at $\sqrt{s_{NN}}$=2.4 GeV or 2.6 GeV, respectively, and shows a fair agreement with the measured dilepton excess. Interesting insights emerge from the close correlation between the build-up of collectivity and the time window of thermal dilepton emission which also corroborate the possibility to track the lifetime of the system with the radiation yield in the mass window $0.3\,\mathrm{GeV/c^2} \leq M_{ll} \leq 0.7\,\mathrm{GeV/c^2}$. This opens the opportunity to apply this approach to the FAIR energy regime where up to now no dilepton measurements exist and to study the impact of different scenarios onto the invariant-mass spectrum of thermal lepton pairs.

Acknowledgments

We thank the HADES collaboration for providing the data points of the dilepton excess spectrum in Ar+KCl collisions and H. van Hees and S. Endres for their coarse-graining curve. This work was supported by the U.S. National Science Foundation under grant PHY-1614484, the A.-v.-Humboldt Foundation (Germany), the Helmholtz-YIG grant VH-NG-823 at GSI and TU Darmstadt (Germany), the Hessian Initiative for Excellence (LOEWE) through the Helmholtz International Center for FAIR (HIC for FAIR) and by the DFG through the grant CRC-TR 211.

References

[1] Kurkela A and Vuorinen A 2016 *Phys. Rev. Lett.* **117** 042501
[2] Hanauske M *et al.* 2017 *Phys. Rev.* D **96** 043004
[3] Wuppertal-Budapest coll., Borsanyi S *et al.* 2010 *J. High Energy Phys.* **09** 073
[4] Bazavov A *et al.* 2012 *Phys. Rev.* D **85** 054503
[5] Klimt S, Lutz M F M and Weise W 1990 *Phys. Lett.* B **249** 386-90
[6] Schaefer B J, Pawlowski J M and Wambach J 2007 *Phys. Rev.* D **76** 074023, Schaefer priv. comm.
[7] Rapp R and van Hees H 2016 *Phys. Lett.* B **753** 586-90
[8] Specht H J for the NA60 coll. 2010 *AIP Conf. Proc.* **1322** 1-10
[9] Huovinen P, Belkacem M, Ellis P J and Kapusta J I 2002 *Phys. Rev.* C **66** 014903
[10] Endres S, van Hees H, Weil J and Bleicher M 2015 *Phys. Rev.* C **92** 014911
[11] Galatyuk T, Hohler P M, Rapp R, Seck F and Stroth J 2016 *Eur. Phys. J.* A **52** 131
[12] Seck F, Galatyuk T, Rapp R and Stroth J 2017 *Acta Phys. Pol.* B *Proc. Suppl.* **10** 717
[13] HADES coll., Agakishiev G *et al.* 2011 *Phys. Rev.* C **84** 014902
[14] Kornakov G for the HADES coll. 2017 *FAIRNESS 2017 proceedings*
[15] Rapp R and Wambach J 1999 *Eur. Phys. J.* A **6** 415-20
[16] HADES coll., Agakishiev G *et al.* 2010 *Phys. Lett.* B **690** 118-22
[17] HADES coll., Agakishiev G *et al.* 2011 *Eur. Phys. J.* A **47** 21
[18] Fischer C S, Luecker J and Welzbacher C A 2014 *Phys. Rev.* D **90** 034022
[19] Critelli R *et al.* 2017 *Phys. Rev.* D **96** 096026

Effects of $\Lambda\Lambda - \Xi$N mixing in the decay of ${}^{6}_{\Lambda\Lambda}$He

J. Maneu, A. Parreño, A. Ramos

Departament de Física Quàntica i Astrofísica and Institut de Ciències del Cosmos, Universitat de Barcelona, Martí i Franquès, 1, E-08028 Barcelona, Spain

E-mail: maneu@fqa.ub.edu

Abstract. A one-meson exchange model including the ground state of the pseudoscalar octet is used to describe the weak two-body interactions responsible for the decay of ${}^{6}_{\Lambda\Lambda}$He. Strong interaction effects are taken into account by a microscopic study based on the solution of $G-$matrix and $T-$matrix equations for the initial and final interacting pairs respectively. Results for the decay induced by $\Lambda\Lambda \to \Lambda N(\Sigma N)$ transitions are given.

1. Introduction

This work focuses on the new decay modes associated to the presence of two Λ's in the ${}^{6}_{\Lambda\Lambda}$He hypernucleus. In particular, the weak decay process that may occur from a $\Lambda\Lambda$, ΞN or $\Sigma\Sigma$ state, which can be excited via the strong interaction from the initial $\Lambda\Lambda$ pair, while the strong interaction also determines the final hyperon-nucleon (YN) wave function, which may have transitioned from either a ΛN or a ΣN intermediate state. For the weak transition we employ a meson-exchange model, built upon the exchange of mesons belonging to the ground state of the pseudoscalar octet, thoroughly employed for single-Λ hypernuclei[1]. The tree-level values for the baryon-baryon-meson coupling constants are derived using SU(3) symmetry. The essential development with respect to previous calculations[2] is the consideration of new decay transitions that emerge when we allow for the intermediate states that can be coupled strongly to the initial $\Lambda\Lambda$ state. Solving a $G-$matrix equation for the initial state, with the input of realistic baryon-baryon potentials[3] gives us the wave functions for the coupled transitions $\Lambda\Lambda$-$\Lambda\Lambda$, $\Lambda\Lambda$-ΞN and $\Lambda\Lambda$-$\Sigma\Sigma$ (the latter, being much smaller than the other two components, will be disregarded in the present calculation). On the other hand, it is assumed that the two final particles are emitted with high enough momenta so that nuclear medium effects can be disregarded and the T-matrix formalism may then be used. The new $\Xi N \to \Lambda N(\Sigma N)$ weak transitions require additional baryon-baryon-meson coupling constants to be derived. These two new ingredients have allowed us to obtain an update on the decay rate for the $(\Lambda\Lambda - \Lambda\Lambda) \to (YN - Y'N)$ channel as well as new results for the $(\Lambda\Lambda - \Xi N) \to (YN - Y'N)$ channel, where $Y, Y' = \Lambda,\ \Sigma$.

2. Hypernuclear decay rate

The nonmesonic decay rate of a hypernucleus is given by

$$\Gamma_{nm} = \int \frac{d^3k_1}{(2\pi)^3} \int \frac{d^3k_2}{(2\pi)^3} \sum_{\substack{M_I\{R\} \\ \{1\}\{2\}}} (2\pi)\delta(M_H - E_R - E_1 - E_2) \times \frac{1}{2J+1} |\mathcal{M}_{fi}|^2, \tag{1}$$

where M_H, E_R, E_1 and E_2 correspond to the mass of the hypernucleus, the energy of the $(A-2)$-particle system, and the total asymptotic energies of the emitted baryons, respectively. The integration variables $\vec{k}_1$ and $\vec{k}_2$ stand for the momenta of the two particles in the final state, where the momentum of the residual nucleus has been integrated using momentum conservation. The sum, together with the factor $1/(2J+1)$, indicates an average over the initial hypernucleus spin projections, M_I, and a sum over all quantum numbers of the residual $(A-2)$-particle system, $\{R\}$, as well as the spin and isospin projection of the emitted final particles, $\{1\}$ and $\{2\}$. $\mathcal{M}_{fi}$ stands for the transition amplitude from the ${}^{6}_{\Lambda\Lambda}$He hypernucleus to a final state composed of a residual nucleus plus two outgoing baryons. We use a shell-model framework and a weak coupling scheme of the $\Lambda\Lambda$ pair to the nuclear core to write $\mathcal{M}_{fi}$ in terms of elementary two-body transitions. In the present case the ${}^{6}_{\Lambda\Lambda}$He wave function can be decoupled as:

$$|{}^{6}_{\Lambda\Lambda}\mathrm{He}\rangle = |\Lambda\Lambda\rangle^{J_{\Lambda\Lambda}=M_{\Lambda\Lambda}=0}_{T_{\Lambda\Lambda}=M_{T_{\Lambda\Lambda}}=0} \otimes |{}^{4}\mathrm{He}\rangle^{J_c=M_c=0}_{T_c=M_{T_c}=0}, \qquad (2)$$

where antisymmetry forces the two Λ hyperons to be in a ${}^1S_0$ state, since they are assumed to be in the lowest s-shell $(1s_{1/2})$ before the decay occurs.

The evaluation of the $\Lambda N \to NN$ transition rate requires to decompose the nonstrange nuclear core as one nucleon coupled to a three-particle system, while decoupling one of the two Λ particles, so that the initial ΛN pair can convert into a final NN pair, conveniently antisymmetrized with the residual system. The details and final expression for the hypernuclear decay amplitude in terms of two-body $\Lambda N \to NN$ ones can be found in Ref. [2]. As mentioned in the introduction, we focus on the $\Lambda\Lambda \to YN$ decay mode. In this case, the residual 4-particle system, which coincides with the ^{4}He nucleus, contains no strangeness, while the final two-particle state contains one hyperon that can be either a Λ ($|YT_Y\rangle = |00\rangle$), a Σ^- ($|Yt_Y\rangle = |1-1\rangle$) or a Σ^0 ($|Yt_Y\rangle = |10\rangle$). The $\Lambda\Lambda \to YN$ amplitude is then given by

$$\begin{aligned}\mathcal{M}_{\Lambda\Lambda\to YN} &= \langle \vec{k}_N s_N t_N, \vec{k}_Y s_Y t_Y; {}^4\mathrm{He}|\hat{O}|{}^{6}_{\Lambda\Lambda}\mathrm{He}\rangle \\ &= \sum_{S,M_S} \langle \tfrac{1}{2}s_N, \tfrac{1}{2}s_Y|SM_S\rangle\langle \tfrac{1}{2}t_N, Yt_Y|TM_T\rangle \\ &\times \langle \vec{K}|\Psi^{CM}_{\Lambda\Lambda}\rangle\langle \vec{k}, SM_S, TM_T|\hat{O}|\Psi^{rel}_{\Lambda\Lambda}, S_0M_{S_0}, T_0M_{T_0}\rangle \ , \end{aligned} \qquad (3)$$

where the initial $\Lambda\Lambda$ wave function has been written as a product of relative and center-of -mass wave-functions, $\Psi^{rel}_{\Lambda\Lambda}$ and $\Psi^{CM}_{\Lambda\Lambda}$, respectively, and $\vec{k}$ and $\vec{K}$ are the relative and total momentum of the emitted YN pair. The amplitude $\langle \vec{k}, SM_S, TM_T|\hat{O}|\Psi^{rel}_{\Lambda\Lambda}, S_0M_{S_0}, T_0M_{T_0}\rangle$ represents the two-body transition matrix element, calculated using a two-body interaction potential based on a meson-exchange picture and including initial and final correlated wave functions. Note that the isospin quantum numbers of the $\Lambda\Lambda$ should contain the coupling to an isospurion field, $|\frac{1}{2} - \frac{1}{2}\rangle_\Lambda$, introduced to account for the $\Delta I = 1/2$ rule in the transition[1].

In the case of the $\Lambda\Lambda \to YN$ matrix elements, and since the strong interaction allows for the conversion to other baryon-baryon channels, the weak interaction will take place not only from the initial $\Lambda\Lambda$ pair, but also from pairs containing other members of the baryon octet, $\Xi^- p$, $\Xi^0 n$ and $\Sigma^+\Sigma^-$, which will in turn decay weakly into either $\Sigma^0 n$, Λn or $\Sigma^- p$ states. Moreover, the strong interaction acting between the final baryons will produce additional $YN - Y'N$ transitions, with Y, Y' being Λ or Σ hyperons .

3. Final and initial state strong interactions

The formalism to include the strong interaction between baryons is well stablished and may be found in the literature[4]. This is known as the Brueckner-Goldstone theory, which considers the interactions of a pair of particles within the Fermi sea, with the collisions fulfilling the requirements of the Pauli principle. We solve this problem in infinite nuclear matter and use the same correcting factors (correlated wave functions over uncorrelated ones) to our wave functions for ${}_{\Lambda\Lambda}^{6}$He.

Working within the $\left(\frac{1}{2}\right)^+$ baryon octet, the strange $\Lambda\Lambda$, ΞN and $\Sigma\Sigma$ pairs can couple through the strong interaction to the initial $\Lambda\Lambda$ state. The correlated state, $|\Psi\rangle$, is defined through the relation $G|\Psi\rangle = V|\Phi\rangle$, where $|\Phi\rangle$ is the free-particle state and G is given in terms of the bare baryon-baryon potential, V. This yields an integral equation which includes a Pauli blocking operator, Q, which restricts the summation to unoccupied states above the Fermi level and the energy of the correlated two-body system, E. An equivalent formalism is used for the final state, but without the inclusion of the Pauli operator, *i.e.* $Q = 1$.

$$|\Psi\rangle = |\Phi\rangle + \frac{Q}{E - H_0 + i\epsilon} G|\Phi\rangle, \qquad \text{where} \qquad G = V + V\frac{Q}{E - H_0 + i\epsilon} G. \tag{4}$$

In Fig. 1 we represent the initial $\Lambda\Lambda$ ${}^1S_0$ wave function as a function of the relative distance between the two Λ particles and for a representative value of the relative momentum. The black line displays the uncorrelated harmonic oscillator wave function, while the green line displays the correlated wave function for the dominant diagonal $\Lambda\Lambda - \Lambda\Lambda$ component. The red and blue lines represent, respectively, the $\Lambda\Lambda - \Sigma\Lambda$ and the $\Lambda\Lambda - \Xi N$ components. It is clear that the dominant contribution to the $\Lambda\Lambda \to YN$ decay mode of ${}_{\Lambda\Lambda}^{6}$He will come from the diagonal $\Lambda\Lambda - \Lambda\Lambda$ component of the wave function, which at large distances behaves as an uncorrelated harmonic oscillator while at short distances gets reduced due to the short-distance repulsive behavior of the strangeness $S = -2$ baryon-baryon NSC97f interaction employed [3]. With regard to the non-diagonal components of the wave function one can see that, despite having a comparable size at the origin, the strength of the $\Lambda\Lambda - \Sigma\Sigma$ term is essentially located at distances under 0.5 fm and it will be strongly suppressed by the r^2 factor in the integrand of the two-body matrix element. On the other hand, the $\Lambda\Lambda - \Xi N$ component is still sizable around 1 fm and it is expected to contribute non-negligibly to the $\Lambda\Lambda \to YN$ decay mode of ${}_{\Lambda\Lambda}^{6}$He.

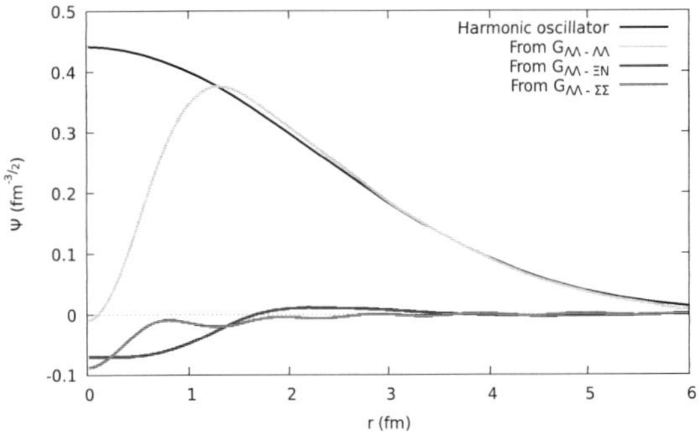

Figure 1. Different components of the relative $\Lambda\Lambda$ wave function, corresponding to the ${}^1S_0$ channel.

4. One-Meson-Exchange Potential

The evaluation of the two-body transition matrix elements of Eq. (3) requires the knowledge of the operator that triggers the weak $\Delta S = -1$ transition from an initial baryon pair to a final one. In the meson exchange description employed here these transitions are assumed to proceed via the exchange of virtual mesons belonging to the pseudoscalar and vector meson octets.

As we detail below, the general structure of the transition potential for pseudoscalar meson exchange reads:

$$V^{\phi}(\vec{q}) = \sum_k \left(A^Y_{k\phi} + \frac{B^Y_{k\phi}}{2\overline{M}} \vec{\sigma}_1 \vec{q} \right) \frac{\vec{\sigma}_2 \vec{q}}{\vec{q}^{\,2} + m^2_{\phi}} \hat{O}_{k\phi}, \tag{5}$$

where $\overline{M}$ is the average mass of the baryons involved in the weak vertex, m_ϕ is the mass of the exchanged pseudoscalar meson and the index k in the sum runs over the different isospin structures associated to each type of meson.

In order to build the isospin operators $\hat{O}_k$ one needs to know the isospin nature of the meson being exchanged (isoscalar for η, isodoublet for K, and isovector for π) and the specific baryons involved in the two-body weak transition. We focus on developing the isospin structure for the transitions $\Xi N \to YN$, with $Y = \Lambda, \Sigma$, which are the new contributions developed in the present work. Attending only to the isospin quantum numbers, the general structure of the $\Xi N \to YN$ matrix element is:

$$g_1\langle Y\,t_Y, \tfrac{1}{2}\,t_{Nf}|\hat{O}_1|0\,t_\Xi - \tfrac{1}{2}, \tfrac{1}{2}\,t_{Ni}\rangle\langle 0\,t_\Xi - \tfrac{1}{2}|\tfrac{1}{2}\,t_\Xi, \tfrac{1}{2}\,-\tfrac{1}{2}\rangle$$
$$+\langle Y\,t_Y, \tfrac{1}{2}\,t_{Nf}|\left(g_2\hat{O}_2 + g_3\hat{O}_3\right)|1\,t_\Xi - \tfrac{1}{2}, \tfrac{1}{2}\,t_{Ni}\rangle\langle 1\,t_\Xi - \tfrac{1}{2}|\tfrac{1}{2}\,t_\Xi, \tfrac{1}{2}\,-\tfrac{1}{2}\rangle. \quad (6)$$

where the isospurion has coupled to the isospin $\frac{1}{2}$ of the Ξ to give states with isospin $I = 0, 1$, which in turn couple to the initial nucleon isospin (with projection $t_{Ni} = \frac{1}{2}$ for a p and $-\frac{1}{2}$ for a n) to give the final YN state. We first examine those cases where the final state is of the ΛN type, i.e. $|Y t_Y\rangle = |0\,0\rangle$. The only possible operators entering Eq. (6) can be argued to be

$$\widehat{O}_1^\Lambda \equiv \mathbb{I}_1 \otimes \mathbb{I}_2, \qquad \widehat{O}_2^\Lambda \equiv \vec{\mathbb{T}}_{01} \otimes \vec{\tau}_2, \qquad \widehat{O}_3^\Lambda \equiv 0 \quad (7)$$

where $\vec{\tau}$ stands for the Pauli matrices and $\vec{\mathbb{T}}_{01}$ is an operator that allows the transition from a $I = 1$ state to a $I = 0$ one. Their spherical coordinates have the following matrix elements

$$\langle 00|\mathbb{T}_{01}^k|1m\rangle = (-1)^k\langle 0\,0, 1\,-k|1\,m\rangle = (-1)^k\delta_{m,-k}. \quad (8)$$

We note that, in the case of a ΛN final state, the operator $\widehat{O}_3$ is zero since there is no other scalar operator that can connect the initial $|1\,t_\Xi - \frac{1}{2}, \frac{1}{2}\,t_{Ni}\rangle$ pair with a the final $\langle 0\,0, \frac{1}{2}\,t_{Nf}|$ one. In the case of a ΣN final state, we have $|Y t_Y\rangle = |1 t_Y\rangle$ and the appropriate set of operators is:

$$\widehat{O}_1^\Sigma \equiv \vec{\mathbb{T}}_{10} \otimes \vec{\tau}_2, \qquad \widehat{O}_2^\Sigma \equiv \mathbb{I}_1 \otimes \mathbb{I}_2, \qquad \widehat{O}_3^\Sigma \equiv \vec{\mathbb{T}}_{11} \otimes \vec{\tau}_2, \quad (9)$$

where $\vec{\mathbb{T}}_{10}$ mediates transitions from isospin 0 to isospin 1, as can be inferred taking the adjoint in Eq. (8). Likewise, $\vec{\mathbb{T}}_{11}$ induces transitions from an initial $I = 1$ state to a final $I = 1$ one. Its matrix elements are given by:

$$\langle 1m|\mathbb{T}_{10}^k|00\rangle = \delta_{m,k}\ , \qquad \langle 1m'|\mathbb{T}_{11}^k|1m\rangle = \sqrt{2}\langle 1m'|1m1k\rangle\ . \quad (10)$$

5. Results and discussion

The results for the non mesonic, Λ-induced, $\Lambda\Lambda \to YN$ decay rate of ${}^{\ 6}_{\Lambda\Lambda}\mathrm{He}$ are displayed in Tables 1 and 2, and are given in terms of the free decay rate for the Λ hyperon, $\Gamma_\Lambda = 3.8\times10^9\ s^{-1}$. We also give the contribution of each individual meson separately, in order to asses its importance in a given transition, and we add up the contribution of the lightest pseudoscalar mesons sequentially for a better interpretation of our results.

The novelty of the present paper is the consideration of the strong non-diagonal $\Lambda\Lambda - \Xi N$ mixing of the $\Lambda\Lambda$ wave function. We observe that the effect of this mixing on each meson exchange contribution leading to a Λn state is different. By comparing both tables one sees that the final decay for this channel decreases by 50% when the ΞN channel is accounted for, from $\Gamma_{\Lambda n} = 1.41 \times 10^{-2}\Gamma_\Lambda$ to $6.53 \times 10^{-3}\Gamma_\Lambda$. The ΣN decay experiences a minor increase, from $\Gamma_{\Sigma N} = 6.29 \times 10^{-3}\Gamma_\Lambda$ to $7.24 \times 10^{-3}\Gamma_\Lambda$, with the inclusion of the $\Lambda\Lambda - \Xi N$ mixing. Altogether, the $\Lambda\Lambda - \Xi N$ component of the wave function brings the value of the $\Gamma_{\Lambda n}/(\Gamma_{\Sigma^0 n} + \Gamma_{\Sigma^- p})$ ratio to 0.90, a factor 2.5 times smaller than found when this contribution is neglected.

Adding the Λn, $\Sigma^0 n$ and $\Sigma^- p$ partial rates, the total contribution from the $\Lambda\Lambda \to YN$ decay mode amounts to $\Gamma_{YN} = 1.38 \times 10^{-2}\Gamma_\Lambda$, which represents an overall decrease of around 10% compared to the rate for the diagonal $\Lambda\Lambda - \Lambda\Lambda$ channel only.

Table 1. Meson-exchange contributions to the nonmesonic decay of ${}^{6}_{\Lambda\Lambda}\mathrm{He}$ considering final state interactions and only the $\Lambda\Lambda - \Lambda\Lambda$ component of the initial w.f., in units of $\Gamma_\Lambda = 3.8 \times 10^9\, s^{-1}$.

Meson	**Λn**	**ΣN**		
		$\mathbf{\Sigma^0 n}$	$\mathbf{\Sigma^- p}$	**ΣN**
π	1.35×10^{-4}	2.38×10^{-3}	4.76×10^{-3}	7.15×10^{-3}
K	2.22×10^{-2}	3.23×10^{-4}	6.46×10^{-4}	9.69×10^{-4}
η	8.95×10^{-4}	2.56×10^{-7}	5.11×10^{-7}	7.67×10^{-7}
$\pi + K$	2.17×10^{-2}	2.11×10^{-3}	4.23×10^{-3}	6.34×10^{-3}
$\pi + K + \eta$	1.41×10^{-2}	2.10×10^{-3}	4.20×10^{-3}	6.29×10^{-3}

Table 2. Meson-exchange contributions to the nonmesonic decay of ${}^{6}_{\Lambda\Lambda}\mathrm{He}$ considering final state interactions and both $\Lambda\Lambda - \Lambda\Lambda$ and $\Lambda\Lambda - \Xi N$ initial transition channels, in units of Γ_Λ.

Meson	**Λn**	**ΣN**		
		$\mathbf{\Sigma^0 n}$	$\mathbf{\Sigma^- p}$	**ΣN**
π	9.73×10^{-5}	1.96×10^{-3}	3.91×10^{-3}	5.87×10^{-3}
K	2.42×10^{-2}	1.61×10^{-5}	3.22×10^{-5}	4.83×10^{-5}
η	2.08×10^{-3}	5.90×10^{-5}	1.18×10^{-4}	1.77×10^{-4}
$\pi + K$	1.57×10^{-2}	2.24×10^{-3}	4.48×10^{-3}	6.73×10^{-3}
$\pi + K + \eta$	6.53×10^{-3}	2.41×10^{-3}	4.83×10^{-3}	7.24×10^{-3}

The complete non-mesonic decay rate Γ of ${}^{6}_{\Lambda\Lambda}\mathrm{He}$, contains also the processes induced by a ΛN pair and as such one may write $\Gamma = \Gamma_{\Lambda N \to NN} + \Gamma_{\Lambda\Lambda \to YN}$. The decay rate for $\Lambda N \to NN$ channel has been computed[1] to be $\Gamma_{\Lambda N \to NN} = 0.95\Gamma_\Lambda \approx 2\Gamma\left({}^{5}_{\Lambda}\mathrm{He}\right)$. Comparing this result to those of the $\Lambda\Lambda$ induced mode calculated in the present work, one can see that the decay rate for the $\Lambda\Lambda \to YN$ transition with $\Lambda\Lambda - \Lambda\Lambda$ diagonal correlations amounts to a 2.15% of the one-nucleon induced rate $\Gamma_{\Lambda N \to NN}$, while the inclusion of the $\Lambda\Lambda - \Xi N$ mixing reduces this percentage to 1.45%. A word of caution has to be said when using this number as reference in the current work, since the contribution of vector mesons has yet to be included[5].

Acknowledgments

This work is partly supported by the Spanish Ministerio de Economia y Competitividad under the project MDM-2014-0369 of ICCUB (Unidad de Excelencia 'María de Maeztu'), with additional European FEDER funds under the contract FIS2014-54762-P. Support has also been received from the Generalitat de Catalunya contract 2014SGR-401 and FI-DGR 2015.

References

[1] Parreno A, Ramos A and Bennhold C 2001 *Phys. Rev.* **C65**
[2] Parreno A, Ramos A and Bennhold C 2002 *Phys. Rev.* **C65** 015205 (*Preprint* `nucl-th/0106054`)
[3] Stoks V G J and Rijken T A 1999 *Phys. Rev.* **C59** 3009–3020 (*Preprint* `nucl-th/9901028`)
[4] Vidaña Haro I 2001 *Description of hyperonic matter and hypernuclei within the Brueckner-Hartree-Fock theory* Ph.D. thesis Barcelona U., ECM
[5] Maneu J, Parreño A and Ramos A *In preparation*

Performance of a prototype Straw Tube Tracker for the $\bar{\text{P}}$ANDA experiment

A Apostolou[1,2], J Messchendorp[1], N Kalantar-Nayestanaki[1], J Ritman[2], P Wintz[2]

[1] KVI-CART, University of Groningen, The Netherlands

[2] IKP, Forschungszentrum Jülich GmbH, Germany

E-mail: a.apostolou@rug.nl

Abstract. The $\bar{\text{P}}$ANDA experiment is one of the pillars of the future Facility for Antiproton and Ion Research (FAIR) in Darmstadt, Germany. The $\bar{\text{P}}$ANDA physics program is focused on answering fundamental questions related to Quantum Chromo Dynamics (QCD), using antiproton collisions on proton and nuclear targets. The central Straw Tube Tracker (STT) will be the main tracking detector which will reconstruct tracks induced by charged particles (with a transversal spatial resolution of $\simeq$ 150 μm) and it will measure the specific energy-losses for particle identification (PID) with an energy resolution better than 10%. The PID information is especially needed to separate protons, kaons and pions in the momentum region below 1 GeV/c. Results obtained with a prototype STT using a proton and a deuteron beam in the momentum range from 0.5 to 3.0 GeV/c are shown and discussed.

1. Introduction

FAIR is a new international accelerator facility which is under construction in Darmstadt. It will offer an accelerator complex, which will create the opportunity to perform experiments with antiprotons, ions and exotic nuclei thanks to the diversity of beams available. $\bar{\text{P}}$ANDA experiment is one of the pillars of the new FAIR facility. It will cover a diversified and broad hadron physics program. The design of the detector together with the unique high quality antiproton beam will give the opportunity for unprecedented research. The aim is to broaden our knowledge in the field of the strong interaction and hadron structure. $\bar{\text{P}}$ANDA will also contribute to the existing knowledge of the field by supplying very high statistics and improvements to the precision of the measurements. In order to achieve the physics objectives, a sophisticated detector system is needed. This system should have 4π acceptance, high resolution for tracking, calorimetry and particle identification, together with high rate capabilities of the detectors and their electronic readout systems. The $\bar{\text{P}}$ANDA detector (see figure 1) is about 13 meters long and has a diameter of about 5 meters.

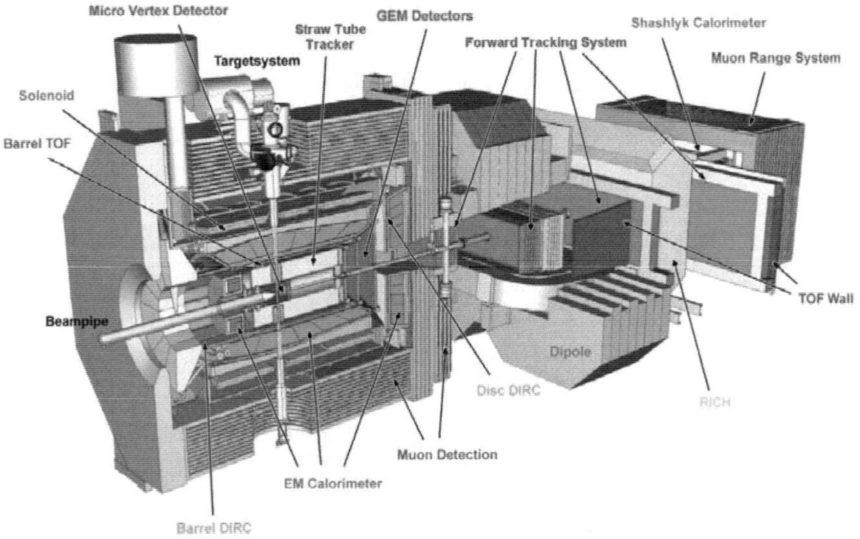

Figure 1: Schematic representation of the $\bar{\mathrm{P}}$ANDA detector [1].

2. Overview of the Straw Tube Tracker

The Straw Tube Tracker (STT) is the main tracking detector of $\bar{\mathrm{P}}$ANDA (see figure 2). It has two main tasks: the spatial reconstruction of the tracks of the charged particles (with a precision of 150 μm) for momenta of a few hundred MeV/c to about 8 GeV/c. The second task is the PID through specific energy-loss measurements (dE/dx). More specifically, the separation of kaons, protons and pions for momenta below 1 GeV/c is of crucial importance. The detector is using gas-filled drift tubes as the basic detection element. The STT has a cylindrical shape with an inner radius of 150 mm, an outer radius of 420 mm, a length of 1650 mm and an angle coverage from 10° to 140°. It consists of 4224 straw tubes with 27 layers in radial direction. The layers are positioned in such a way that 19 of them are in parallel with the beam direction and 8 of them are skewed. This will allow for a 3D reconstruction of the trajectories of the particles. The tubes (see figure 3) have a length of 1400 mm, a diameter of 10 mm and they are made of aluminized Mylar films with a total thickness of 27 μm as cathode. Inside the tubes, a gold-plated tungsten-rhenium anode wire with a diameter of 20 μm is used.

Figure 2: The STT detector [2].

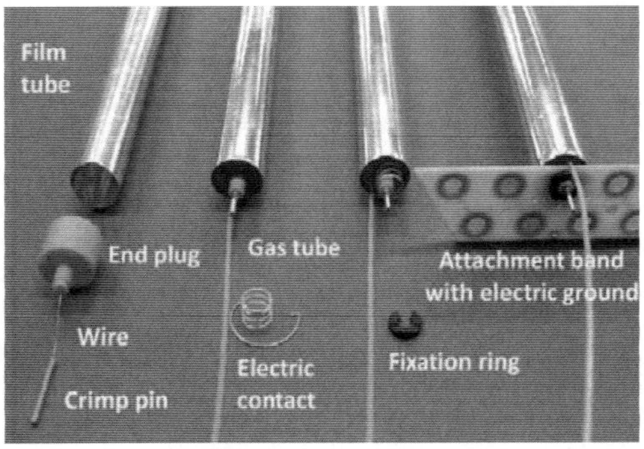

Figure 3: Components of each straw [2].

IOP Conf. Series: Journal of Physics: Conf. Series **1024** (2018) 012013 doi:10.1088/1742-6596/1024/1/012013

3. Experimental setup

Two beam tests took place in order to study the performance of a prototype STT detector in the COSY-TOF experimental hall at Forschungszentrum Jülich. The first test was performed with a proton beam while the second one with a deuteron beam. The momenta available for the proton beam were: 0.55, 0.75, 1.00, 2.95 GeV/c, while for the deuteron beam was 0.65, 0.75 and 1.5 GeV/c. The reason for using different particle species was to examine the PID performance of the prototype. For both beam tests, the same STT setup was used (see figure 4), which was of rectangular shape and a total of 144 tubes divided into 6 layers. Each layer had 24 tubes along the beam direction.

Figure 4: The prototype STT.

An Application Specific Integrated Circuit (ASIC) chip is used for the readout of the signals. This chip is using the Time Over Threshold (TOT) technique as an approach for the PID via measurements of the energy losses in the straw tubes. When a particle crosses the tube, the gas inside is ionized and pairs of ions and electrons are created. The electrons are collected to the anode. Two timing values are registered by the readout system, namely, the leading-edge time and the trailing-edge time. There is a common threshold for all the channels of the setup, and an individual baseline threshold for each of the channels. The leading-edge time is the first point of the signal that is above both thresholds. Similarly, the trailing-edge time is the last point of the signal over both thresholds. The TOT value, used for the PID, is simply the difference between the trailing time and the leading time. The leading time itself is used for the tracking reconstruction and for the calculation of the isochrone radius, which is the distance between the reconstructed track and the wire of each tube. The difference between the reconstructed track and the isochrone radius is called residual. This is the value used for the calculation of the spatial resolution.

4. Results

The PASSTRECv1 chip [3] can have several settings concerning gain and peaking time of the signal. The results shown in this section have a gain equal to 1.8 mV/fC and a peaking time equal to 35 ns. The threshold was set to 20 mV and the voltage applied was 1800 mV, which will be the operational value for $\bar{\text{P}}$ANDA. In table 1 the spatial resolution values obtained for all momenta are given. The values are obtained from a Gaussian fit to the distribution of the residuals. During the proton beam, several problems with the setup and the electronics were present and thus, the results are worse than those of the deuteron beam.

Table 1: Spatial resolution values obtained from both beam tests.

Momentum (GeV/c)	Proton beam (μm)	Deuteron beam (μm)
0.55	185	—
0.65	—	170
0.75	202	172
1.00	217	—
1.50	—	198
2.95	223	—

To examine the PID performance, the mean TOT/dx value is used, where dx is the total distance the particles travelled through the whole setup. A truncation of 30% is used on the higher values of TOT/dx in order to improve the resolution of our energy loss measurements. In figure 5 the results obtained are shown for all momenta, while in table 2 the resolutions of the mean TOT/dx values are given which are obtained from a Gaussian fit to the distributions.

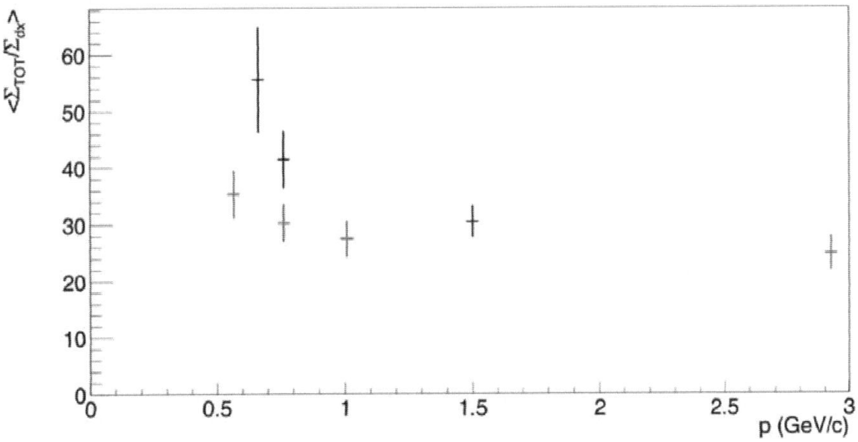

Figure 5: Mean TOT/dx values after 30% truncation for all momenta and beam types. Crosses with a black colour show the proton beam results while the red crosses depict the results from the deuteron beam. The errors shown are the sigma values from the Gaussian fit applied.

IOP Conf. Series: Journal of Physics: Conf. Series **1024** (2018) 012013 doi:10.1088/1742-6596/1024/1/012013

Table 2: Mean TOT/dx resolutions obtained from both beam tests.

Momentum (GeV/c)	Proton beam (%)	Deuteron beam (%)
0.55	5.1	—
0.65	—	7.1
0.75	4.6	5.2
1.00	6.7	—
1.50	—	4.1
2.95	5.3	—

5. Conclusions

The performance of a prototype STT detector was studied extensively during two beam tests. The first goal of the tests was to calculate the spatial resolution that can be obtained and compare it with the value calculated from simulations. The second goal was to examine the PID capability of the prototype, i.e how well it can separate particles of different species. The results for the spatial resolution and the PID capability through the TOT/dx method are presented which show that both goals have been achieved to a great extent. It can be easily seen that the results of the deuteron beam are much better than those from the proton beam. The reason for this is the presence of several problems during the first beam test. One can also conclude that the ASIC readout is quite sensitive to the first electrons reaching the anode wire. This is one of the main reasons that the spatial resolution becomes worse towards the minimum ionizing region. Concerning the PID capability of the prototype, the fact that we have two different particle species, helps us to draw safe conclusions about the performance. One can study the separation for different momenta within the same species, i.e only investigate protons or deuterons for different momenta. However, since there is a common momentum for both, it is more safe to investigate the separation between the different particle species. One can see from figure 5 that there is a clear separation between protons and deuterons at 0.75 GeV/c. The resolution obtained for the PID, is better than the expected value ($\sim$ 10%). Improvements on the prototype STT, the readout and the analysis methods, can only further improve the existing results.

References

[1] Erni W *et al* 2009 *Physics Performance Report for: $\bar{\text{P}}$ANDA Strong Interaction Studies with Antiprotons* Tech. rep. $\bar{\text{P}}$ANDA collaboration
[2] Erni W *et al* 2013 *Technical Design Report for the: $\bar{\text{P}}$ANDA Straw Tube Tracker* Tech. rep. $\bar{\text{P}}$ANDA collaboration
[3] D Przyborowski *et al* 2013 *Journal of Instrumentation* **11** 08009

IOP Conf. Series: Journal of Physics: Conf. Series **1024** (2018) 012014 doi:10.1088/1742-6596/1024/1/012014

Inclusive Prompt Photons from the Color Glass Condensate at NLO

Sanjin Benić[1], Kenji Fukushima[2], Oscar Garcia-Montero[3,*], Raju Venugopalan[4]

[1]Physics Department, Faculty of Science, University of Zagreb, Zagreb 10000, Croatia
[2] Department of Physics, The University of Tokyo, 7-3-1 Hongo, Bunkyo-ku, Tokyo 113-0033, Japan
[3] Institut für Theoretische Physik, Universität Heidelberg, Philosophenweg 16, 69120 Heidelberg, Germany
[4] Physics Department, Brookhaven National Laboratory, Bldg. 510A, Upton, NY 11973, USA

*Speaker at FAIRNESS2017. Email: garcia@thphys.uni-heidelberg.de

Abstract. The cross-section for photons radiated by quarks in proton-nucleus collisions at collider energies was obtained using the Color Glass Condensate framework, in the dense-dilute kinematics regime. We observe that the inclusive photon cross-section is proportional to all-twist Wilson line correlators in the nucleus. These correlators also appear in quark-pair production; unlike the latter, photon production is insensitive to hadronization uncertainties and therefore more sensitive to multi-parton correlations in the gluon saturation regime of QCD.

1. Introduction

In hadron-hadron collision experiments, photons free-stream to the detectors without interacting strongly with the produced debris. This renders them as excellent probes for the study of gluon saturation at the very high energies of the RHIC and LHC colliders. In these regimes, prompt photon observables can be calculated pertubatively using the Color Glass Condensate (CGC) effective theory [1–5]. In the CGC, there is an emergent scale, $Q_S(x_0)$, which separates partons with a momentum fraction $x > x_0$ (fast partons) from those with $x < x_0$ (slow partons), where x_0 stands for an arbitrary energy measurement scale which can be evolved using the JIMWLK equation. The fast partons, highly dilated, become static sources, while the slower ones can be treated as classical Yang-Mills fields. The interplay between these two can be found by solving the classical Yang-Mills equations. Unfortunately, in this framework, a fully analytical solution for a collision of two nuclear projectiles is out of reach. However, for the case of a p+A collision in the kinematic region available to LHC, one can approximate the system to the so-called *dilute-dense limit* in which the static sources of the projectile are much smaller than its characteristic transverse momentum, $\rho_p/\boldsymbol{k}_{1\perp}^2 \ll 1$, while for the target $\rho_A/\boldsymbol{k}_{2\perp}^2 \sim 1$. In this limit, the analytical solution is known, and was derived in ref. [6].

From the CGC framework, the parametric leading order (LO) process [7] for prompt photon production is given by the emission of a photon from a valence quark from the projectile after multiple-scattering off the nucleus (see LO in fig. 1) . Since the target is in the high occupancy regime ($gA_\mu \sim \mathcal{O}(1)$) , the $q\bar{q}g$ vertex is of $\mathcal{O}(1)$, which makes the inclusive cross-section at LO

IOP Conf. Series: Journal of Physics: Conf. Series **1024** (2018) 012014 doi:10.1088/1742-6596/1024/1/012014

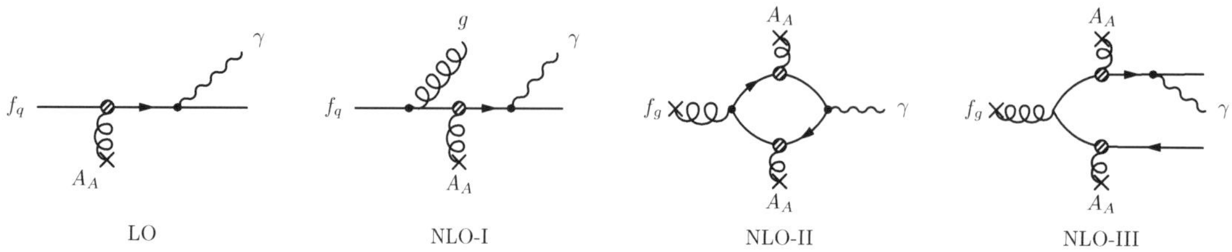

Figure 1. Leading and next-to-leading order processes for prompt photon production in proton-nucleus collisions. The LO diagram describes the bremsstrahlung of a photon from a valence quark after multiple scattering off the classical gluon field in the nucleus.

of $\mathcal{O}(\alpha_e)$. The next parametric order, $\mathcal{O}(\alpha_e\alpha_S)$, can be found by the inclusion of an incoming gluon from the proton, or strong bremsstrahlung from a quark line. The NLO processes can then be categorized in three classes, listed in fig. 1. They consist of (I) the bremsstrahlung of a photon and a gluon from a valence quark line. The remaining two NLO process consist of $q\bar{q}$ creation from a gluon of the proton, and (II) subsequent annihilation into a photon (see ref. [8]) or (III) photon emission.

It is important to note that, as stated in refs. [8, 9], while at large x the LO should be a good description of the cross-section, at smaller x, the valence quark distribution functions, f_q, decrease, while the gluon distribution, f_g, rises. In the kinematic window of $10^{-2} > x > 10^{-3}$, which is available to RHIC and LHC, the quark and gluon distributions will satisfy $f_q < \alpha_S f_g$, making the NLO dominant. Because of this, one has to include the NLO effects in the cross-section for a complete description of prompt photon emission in p+A collision. In this short report, we present the analytical results previously derived in [9] for diagram NLO-III in fig. 1.

2. Setting up the gluon field

To calculate the amplitude of this process in the dilute-dense limit, we used the Lorentz gauge, $\partial_\mu A^\mu = 0$. Within the CGC effective theory, the gluon fields for a projectile moving independently in the positive z direction and a nucleus moving in the opposing direction are

$$A_p^\mu(x) = -g\delta^{\mu+}\delta(x^-)\frac{1}{\nabla_\perp^2}\rho_p(\boldsymbol{x}_\perp) \qquad \text{and} \qquad A_A^\mu(x) = -g\delta^{\mu-}\delta(x^+)\frac{1}{\nabla_\perp^2}\rho_A(\boldsymbol{x}_\perp)\,, \tag{1}$$

where ρ_p and ρ_A are the static (and stochastic) color charge densities of partons at large x. Using these results as construction blocks, one can calculate the result for both sources in the dilute-dense limit together using the Yang-Mills equations. For the complete derivation see ref. [6]. The complete result is given then by[1].

$$A^\mu(q) = A_p^\mu(q) + \frac{\mathrm{i}g}{q^2+\mathrm{i}q^+\epsilon}\int_{\boldsymbol{k}_{1\perp}}\int_{\boldsymbol{x}_\perp} \mathrm{e}^{\mathrm{i}(\boldsymbol{q}_\perp-\boldsymbol{k}_{1\perp})\cdot\boldsymbol{x}_\perp}\Big\{C_U^\mu(q,\boldsymbol{k}_{1\perp})[U(\boldsymbol{x}_\perp)-1]+C_V^\mu(q)[V(\boldsymbol{x}_\perp)-1]\Big\}\frac{\rho_p(\boldsymbol{k}_{1\perp})}{\boldsymbol{k}_{1\perp}^2}\,, \tag{2}$$

where $A_\mu(q) = A_\mu^a(q)T^a$ and T^a are the generators of SU(N_c) in the adjoint representation.

We can dissect eq. (2) into its components. $A_\mu(q)$ corresponds to the field of a projectile travelling alone. The vectors C_U^μ and C_V^μ appear as factors for the multiple scattering of the gluon from the proton off the nucleus. $\boldsymbol{k}_{1\perp}$ here is the momentum exchanged from the proton,

[1] Here we use a shorthand notation, $\int_{\boldsymbol{k}_\perp} \equiv \int \frac{\mathrm{d}^2\boldsymbol{k}_\perp}{(2\pi)^2}$ and $\int_{\boldsymbol{x}_\perp} \equiv \int \mathrm{d}^2\boldsymbol{x}_\perp$

while $\boldsymbol{k}_{2\perp} \equiv \boldsymbol{q}_\perp - \boldsymbol{k}_{1\perp}$ is exchanged from the nucleus. These vectors are explicitly given by

$$\begin{aligned} &C_U^+(q,\boldsymbol{k}_{1\perp}) \equiv -\frac{\boldsymbol{k}_{1\perp}^2}{q^- + \mathrm{i}\epsilon}, \quad C_U^-(q,\boldsymbol{k}_{1\perp}) \equiv \frac{(\boldsymbol{q}_\perp - \boldsymbol{k}_{1\perp})^2 - \boldsymbol{q}_\perp^2}{q^+}, \quad \boldsymbol{C}_U^\perp(q,\boldsymbol{k}_{1\perp}) \equiv -2\boldsymbol{k}_{1\perp}, \\ &C_V^+(q) \equiv 2q^+, \quad C_V^-(q) \equiv 2\frac{\boldsymbol{q}_\perp^2}{q^+} - 2q^-, \quad \boldsymbol{C}_V^\perp(q) \equiv 2\boldsymbol{q}_\perp . \end{aligned} \tag{3}$$

Using $C_L^\mu = C_U^\mu + C_V^\mu/2$, we can relate the two vector functions above to the Lipatov effective vertex [10–12].

The remaining ingredients in eq. (2) are the Wilson lines U and V, where

$$U(\boldsymbol{x}_\perp) = \mathcal{P}_+ \exp\left[\mathrm{i}g \int \mathrm{d}z^+ A_A^-(z^+, \boldsymbol{x}_\perp)\cdot T\right], \tag{4}$$

$$V(\boldsymbol{x}_\perp) = \mathcal{P}_+ \exp\left[\frac{\mathrm{i}g}{2} \int \mathrm{d}z^+ A_A^-(z^+, \boldsymbol{x}_\perp)\cdot T\right], \tag{5}$$

These objects are color rotation matrices in the adjoint representation, which arise from multiple scatterings of nuclear gluons over the proton gluon field. In the above, the operator $\mathcal{P}_+$ denotes light cone time-ordering in the x^+ direction. The uncommon $1/2$ factor in the V Wilson line is due to V being a gauge artefact. Therefore, terms including V will not be present in the final result for the cross-section. It is interesting to remark that to cancel these contributions, one needs to include a singular gluon field, which accounts for the creation of the pair *inside* the nucleus. A complete discussion of these cancellations is given for $q\bar{q}$ production in ref. [6] and for this process in ref. [9].

Finally, for the computation of process III in fig. 1, one needs to understand how quarks propagate in the presence of the field A_A. The fermion propagator can be expressed as in ref. [13],

$$G_F(p,k) = (2\pi)^4\delta^4(k)G_F^0(p) + G_F^0(q+k)\mathcal{T}(k,p)G_F^0(p) \tag{6}$$

where $\mathcal{T}(k,p)$ is an effective vertex which contains the multiple scatterings of the quark off the nucleus. It is given here explicitly by

$$\mathcal{T}(k,p) = \begin{cases} 2\pi\delta(k^+)\gamma^+ \displaystyle\int_{\boldsymbol{x}_\perp} \mathrm{e}^{\mathrm{i}\boldsymbol{k}_\perp\cdot\boldsymbol{x}_\perp}\left[\tilde{U}(\boldsymbol{x}_\perp) - 1\right] & (p^+ > 0), \\ -2\pi\delta(k^+)\gamma^+ \displaystyle\int_{\boldsymbol{x}_\perp} \mathrm{e}^{\mathrm{i}\boldsymbol{k}_\perp\cdot\boldsymbol{x}_\perp}\left[\tilde{U}^\dagger(\boldsymbol{x}_\perp) - 1\right] & (p^+ < 0), \end{cases} \tag{7}$$

for a momentum exchange k. $\tilde{U}(\boldsymbol{x}_\perp)$ is the Wilson line in the fundamental representation.

3. Calculation of the cross-section

In the Lorentz gauge, there are a total of 14 non-vanishing diagrams that contribute to the amplitude of the process . For the complete set diagrams, and a detailed derivation, see ref. [9]). Summing over all of them, one obtains the amplitude,

$$\begin{aligned} \mathcal{M}^\mu(\boldsymbol{p},\boldsymbol{q},\boldsymbol{k}_\gamma) = -q_f e g^2 \int_{\boldsymbol{k}_\perp \boldsymbol{k}_{1\perp}} \int_{\boldsymbol{x}_\perp \boldsymbol{y}_\perp} \frac{\rho_p^a(\boldsymbol{k}_{1\perp})}{\boldsymbol{k}_{1\perp}^2} \mathrm{e}^{\mathrm{i}\boldsymbol{k}_\perp\cdot\boldsymbol{x}_\perp + \mathrm{i}(\boldsymbol{P}_\perp - \boldsymbol{k}_\perp - \boldsymbol{k}_{1\perp})\cdot\boldsymbol{y}_\perp} \\ \times \bar{u}(\boldsymbol{q})\left\{T_g^\mu(\boldsymbol{k}_{1\perp})U(\boldsymbol{x}_\perp)^{ba}t^b + T_{q\bar{q}}^\mu(\boldsymbol{k}_\perp,\boldsymbol{k}_{1\perp})\tilde{U}(\boldsymbol{x}_\perp)t^a\tilde{U}^\dagger(\boldsymbol{y}_\perp)\right\}v(\boldsymbol{p}), \end{aligned} \tag{8}$$

where we use the notation $\boldsymbol{P}_\perp = \boldsymbol{p}_\perp + \boldsymbol{q}_\perp + \boldsymbol{k}_{\gamma\perp}$. Here T_g^μ stands for the sum of the fermionic structures proportional to $U(\boldsymbol{x}_\perp)^{ba}$, scattering over a gluon line, while $T_{q\bar{q}}^\mu$ for the ones proportional to $U(\boldsymbol{x}_\perp)t^a U(\boldsymbol{y}_\perp)^\dagger$, scattering over both quark and antiquark lines. Therefore, one can easily identify that these two pieces in eq. (8) are the amplitudes of two independent processes: (i) a gluon from the proton goes through the nucleus and multiple scatters before splitting into a $q\bar{q}$ pair, and (ii) the gluon splits into a $q\bar{q}$ dipole, which pierces the nucleus and multiple scatters from it.

It is interesting to remark that the amplitude matches exactly to the present one when calculated in the light-cone gauge [9].

One can then find the cross-section by squaring the amplitude, averaging over the proton and nuclear sources [2] and finally integrating over the impact parameter. The differential cross-section for this process is then explicitly given by

$$\begin{aligned}
\frac{\mathrm{d}\sigma^\gamma}{\mathrm{d}^2\boldsymbol{k}_{\gamma\perp}\mathrm{d}\eta_{k_\gamma}\mathrm{d}^2\boldsymbol{q}_\perp\mathrm{d}\eta_q\mathrm{d}^2\boldsymbol{p}_\perp\mathrm{d}\eta_p} &= \frac{\alpha_e\alpha_S^2\, q_f^2}{256\pi^8 C_F}\int_{\boldsymbol{k}_{1\perp}\boldsymbol{k}_{2\perp}}(2\pi)^2\delta^{(2)}(\boldsymbol{P}_\perp-\boldsymbol{k}_{1\perp}-\boldsymbol{k}_{2\perp})\,\frac{\varphi_p(\boldsymbol{k}_{1\perp})}{\boldsymbol{k}_{1\perp}^2\boldsymbol{k}_{2\perp}^2}\\
&\times\Bigg\{\int_{\boldsymbol{k}_\perp\boldsymbol{k}'_\perp}\mathrm{tr}\left[(\not{q}+m)T_{q\bar{q}}(\boldsymbol{k}_\perp,\boldsymbol{k}_{1\perp})(m-\not{p})\gamma^0 T^\dagger_{q\bar{q}}(\boldsymbol{k}_\perp,\boldsymbol{k}_{1\perp})\gamma^0\right]\,\phi_A^{q\bar{q},q\bar{q}}(\boldsymbol{k}_\perp,\boldsymbol{k}_{2\perp}-\boldsymbol{k}_\perp;\boldsymbol{k}'_\perp,\boldsymbol{k}_{2\perp}-\boldsymbol{k}'_\perp)\\
&+\int_{\boldsymbol{k}_\perp}\mathrm{tr}\left[(\not{q}+m)T_{q\bar{q}}(\boldsymbol{k}_\perp,\boldsymbol{k}_{1\perp})(m-\not{p})\gamma^0 T_g^\dagger(\boldsymbol{k}_{1\perp})\gamma^0\right]\,\phi_A^{q\bar{q},g}(\boldsymbol{k}_\perp,\boldsymbol{k}_{2\perp}-\boldsymbol{k}_\perp;\boldsymbol{k}_{2\perp})+\mathrm{h.c.}\\
&+\mathrm{tr}\left[(\not{q}+m)T_g(\boldsymbol{k}_{1\perp})(m-\not{p})\gamma^0 T_g^\dagger(\boldsymbol{k}_{1\perp})\gamma^0\right]\,\phi_A^{g,g}(\boldsymbol{k}_{2\perp})\Bigg\}.
\end{aligned}\tag{9}$$

The function $\varphi_p(\boldsymbol{k}_{1\perp},Y_p)$ is the unintegrated gluon distribution function of the proton and is here given by $2\pi N_c C_F\, g^2\langle\rho_p^a(\boldsymbol{k}_{1\perp})\rho_p^{\dagger b}(\boldsymbol{k}_{1\perp})\rangle=\delta^{ab}\boldsymbol{k}_{1\perp}^2\,\varphi_p(\boldsymbol{k}_{1\perp},Y_p)$. The functions $\phi_A^{q\bar{q},q\bar{q}}$, $\phi_A^{q\bar{q},g}$ and $\phi_A^{g,g}$ (see refs. [9, 14]) are unintegrated gluon distribution functions of the nucleus. These are the Fourier transforms of four, three and two point Wilson line correlators and correspond to all twist corrections to the gluon distributions. In these unintegrated distributions are encoded the saturation effects which should be observable in the inclusive photon cross-section.

The expression in eq. (9) can be used to calculate the correlation of a quarkonium state, like the J/ψ with a single photon. The information of such a correlation function would give new and valuable information on non trivial aspects of saturation, like angular dependence of final states. However, such a measurement is unfortunately still out of reach, even at LHC energies. Nonetheless, the inclusive prompt photon cross-section has been measured in the RHIC kinematic window, and new measurements are expected at RHIC and LHC. For this, we need to integrate over the $q\bar{q}$ pair phase space to find

$$\begin{aligned}
\frac{\mathrm{d}\sigma^\gamma}{\mathrm{d}^2\boldsymbol{k}_{\gamma\perp}\mathrm{d}\eta_{k_\gamma}} &= \frac{\alpha_e\alpha_S^2 q_f^2}{16\pi^4 C_F}\int_0^\infty\frac{\mathrm{d}q^+}{q^+}\frac{\mathrm{d}p^+}{p^+}\int_{\boldsymbol{k}_{1\perp}\boldsymbol{k}_{2\perp}\boldsymbol{q}_\perp\boldsymbol{p}_\perp}(2\pi)^2\delta^{(2)}(\boldsymbol{P}_\perp-\boldsymbol{k}_{1\perp}-\boldsymbol{k}_{2\perp})\,\frac{\varphi_p(\boldsymbol{k}_{1\perp})}{\boldsymbol{k}_{1\perp}^2\boldsymbol{k}_{2\perp}^2}\\
&\times\Bigg\{\int_{\boldsymbol{k}_\perp\boldsymbol{k}'_\perp}\mathrm{tr}\left[(\not{q}+m)T_{q\bar{q}}(\boldsymbol{k}_\perp,\boldsymbol{k}_{1\perp})(m-\not{p})\gamma^0 T^\dagger_{q\bar{q}}(\boldsymbol{k}_\perp,\boldsymbol{k}_{1\perp})\gamma^0\right]\,\phi_A^{q\bar{q},q\bar{q}}(\boldsymbol{k}_\perp,\boldsymbol{k}_{2\perp}-\boldsymbol{k}_\perp;\boldsymbol{k}'_\perp,\boldsymbol{k}_{2\perp}-\boldsymbol{k}'_\perp)\\
&+\int_{\boldsymbol{k}_\perp}\mathrm{tr}\left[(\not{q}+m)T_{q\bar{q}}(\boldsymbol{k}_\perp,\boldsymbol{k}_{1\perp})(m-\not{p})\gamma^0 T_g^\dagger(\boldsymbol{k}_{1\perp})\gamma^0\right]\,\phi_A^{q\bar{q},g}(\boldsymbol{k}_\perp,\boldsymbol{k}_{2\perp}-\boldsymbol{k}_\perp;\boldsymbol{k}_{2\perp})+\mathrm{h.c.}\\
&+\mathrm{tr}\left[(\not{q}+m)T_g(\boldsymbol{k}_{1\perp})(m-\not{p})\gamma^0 T_g^\dagger(\boldsymbol{k}_{1\perp})\gamma^0\right]\,\phi_A^{g,g}(\boldsymbol{k}_{2\perp})\Bigg\},
\end{aligned}\tag{10}$$

which is the final result for this process.

[2] The average of a source operator $\mathcal{O}$ is given by $\langle\mathcal{O}[\rho_p,\rho_A]\rangle=\int\mathcal{D}\rho_p\,\mathcal{D}\rho_A\,W_p[\rho_p]\,W_A[\rho_A]\,\mathcal{O}[\rho_p,\rho_A]$ .One can see that for this cross-section, the averages factor out in the proton and nuclear ones.

IOP Conf. Series: Journal of Physics: Conf. Series **1024** (2018) 012014 doi:10.1088/1742-6596/1024/1/012014

4. Conclusions

We have presented here our derivation for the cross-section of the process $g \to q\bar{q}\gamma$ in the CGC formalism. With the result in eq. (9) we conclude the calculation of single prompt photon production at NLO in the CGC formalism. The NLO processes studied here are sensitive to saturation, and therefore are an important complement to quarkonium production studies [15, 16], especially because of appearance of the same multi-parton correlators in the cross-sections for both processes. Moreover, this study may contribute to the understanding of the excess signal of soft photons in hadron-hadron collision experiments [17, 18]. Future work will be oriented to the numerical resolution of the inclusive photon cross-section.

Acknowledgements

S. B. was supported by the European Union Seventh Framework Programme (FP7 2007-2013) under grant agreement No. 291823, Marie Curie FP7-PEOPLE-2011-COFUND NEWFELPRO Grant No. 48. K. F. was supported by MEXT-KAKENHI Grant No. 15H03652 and 15K13479. R. V. is supported under DOE Contract No. DE-SC0012704. K. F. and R. V. would like to thank the Institut für Theoretische Physik at Heidelberg University for kind hospitality during the course of this work. K. F. thanks the Extreme Matter Institute (EMMI) for support. This work is part of and supported by the DFG Collaborative Research Centre SFB 1225 (ISOQUANT).

References

[1] McLerran L D and Venugopalan R 1994 *Phys. Rev.* **D49** 2233–2241 (*Preprint* hep-ph/9309289)
[2] McLerran L D and Venugopalan R 1994 *Phys. Rev.* **D49** 3352–3355 (*Preprint* hep-ph/9311205)
[3] McLerran L D and Venugopalan R 1994 *Phys. Rev.* **D50** 2225–2233 (*Preprint* hep-ph/9402335)
[4] Gelis F, Iancu E, Jalilian-Marian J and Venugopalan R 2010 *Ann. Rev. Nucl. Part. Sci.* **60** 463–489 (*Preprint* 1002.0333)
[5] Iancu E and Venugopalan R 2003 The Color glass condensate and high-energy scattering in QCD *In *Hwa, R.C. (ed.) et al.: Quark gluon plasma* 249-3363* (*Preprint* hep-ph/0303204)
[6] Blaizot J P, Gelis F and Venugopalan R 2004 *Nucl. Phys.* **A743** 13–56 (*Preprint* hep-ph/0402256)
[7] Gelis F and Jalilian-Marian J 2002 *Phys. Rev.* **D66** 014021 (*Preprint* hep-ph/0205037)
[8] Benic S and Fukushima K 2017 *Nucl. Phys.* **A958** 1–24 (*Preprint* 1602.01989)
[9] Benic S, Fukushima K, Garcia-Montero O and Venugopalan R 2017 *JHEP* **01** 115 (*Preprint* 1609.09424)
[10] Lipatov L N 1976 *Sov. J. Nucl. Phys.* **23** 338–345 [Yad. Fiz.23,642(1976)]
[11] Kuraev E A, Lipatov L N and Fadin V S 1977 *Sov. Phys. JETP* **45** 199–204 [Zh. Eksp. Teor. Fiz.72,377(1977)]
[12] Balitsky I I and Lipatov L N 1978 *Sov. J. Nucl. Phys.* **28** 822–829 [Yad. Fiz.28,1597(1978)]
[13] McLerran L D and Venugopalan R 1999 *Phys. Rev.* **D59** 094002 (*Preprint* hep-ph/9809427)
[14] Blaizot J P, Gelis F and Venugopalan R 2004 *Nucl. Phys.* **A743** 57–91 (*Preprint* hep-ph/0402257)
[15] Ma Y Q, Venugopalan R and Zhang H F 2015 *Phys. Rev.* **D92** 071901 (*Preprint* 1503.07772)
[16] Ma Y Q and Venugopalan R 2014 *Phys. Rev. Lett.* **113** 192301 (*Preprint* 1408.4075)
[17] Chliapnikov P V, De Wolf E A, Fenyuk A B, Gerdyukov L N, Goldschmidt-Clermont Y, Ronzhin V M and Weigend A (Brussels-CERN-Genoa-Mons-Nijmegen-Serpukhov) 1984 *Phys. Lett.* **141B** 276–280
[18] Belogianni A *et al.* 2002 *Phys. Lett.* **B548** 129–139

IOP Conf. Series: Journal of Physics: Conf. Series **1024** (2018) 012015 doi:10.1088/1742-6596/1024/1/012015

QCD Critical point from a black hole engineered EoS

Israel Portillo

Department of Physics, University of Houston, Houston TX 77204, USA.

Abstract. We construct an holographic model to map the fluctuations of baryon charge in the strongly coupled quark gluon plasma into a gravitational problem involving the charge fluctuations of holographic black holes. This holograph approach successfully reproduces the baryon number fluctuations calculated on the lattice at zero μ_B and makes new quantitatively predictions of those fluctuations at arbitrary μ_B. The model displays a critical end point that is located within the reach of the next generation heavy ion experiments.

1. Introduction

Exploring the phase diagram of QCD at extremely high temperatures (T) and baryonic chemical potential (μ_B) is a prime goal of heavy ion experiments at the Relativistic Heavy Ion Collider at BNL and the Large Hadron Collider at CERN. An important feature of the QCD phases is the line that separates the low-T hadronic matter from the high-T quark-gluon-plasma (QGP). First principles lattice QCD calculations have found that this transition is a rapid crossover at zero μ_B [1]. This crossover is expected to become a first order transition at a critical end point (CEP). The existence and location of the CEP is a fundamental question subject of intense study in present heavy ion experiments and one of the main motivations for the construction of the Facility for Antiproton and Ion Research, at GSI. Unfortunately, lattice QCD calculations cannot be performed at finite μ_B due to the fermion sign problem and, to guide the experimental search for the CEP in heavy ion experiments, effective approaches must be used.

In this contribution we summarize the results obtained in Ref. [2] with a model constructed using the holographic gauge/gravity correspondence [3], a well-known tool developed in string theory that has been successfully used to study properties of the strongly interacting QGP at finite density [4, 5].

2. Holographic Black Hole Model

The action of the holographic model is given by [6, 7]

$$\mathcal{S} = \frac{1}{16\pi G_5}\int \mathrm{d}x^5\sqrt{-g}\left[\mathcal{R} - \frac{1}{2}(\partial_M\phi)^2 - V(\phi) - \frac{1}{4}f(\phi)F_{MN}^2\right]. \tag{1}$$

where $\mathcal{R}$ is the Ricci tensor, ϕ is a scalar field coupled to the metric $g_{\mu\nu}$, and A_N is a Maxwell field, which introduces the baryonic charge effects through the Maxwell tensor $F_{MN} = \partial_M A_N - \partial_N A_M$. The non-conformal behavior of the model is controlled by a scalar potential $V(\phi)$, used to fix the thermodynamics of the model at zero μ_B. Another input parameter is the function $f(\phi)$, which corresponds to the coupling between A_N and ϕ. This function determines the response of the system to a finite μ_B. $V(\phi)$ and $f(\phi)$ are constructed in such a way that the black hole model reproduces two crucial observables obtained from lattice QCD calculations at zero μ_B: the entropy density (s) and the second order baryon density (χ_2) respectively. Therefore, every other observable calculated with this model is a prediction of our model. For instance, Fig. 1 (left) shows the agreement at zero μ_B between the trace anomaly of the black hole model and the one calculated on the lattice. The right panel of this figure also shows χ_2 computed in the black hole model considering three different sets of parameters for $f(\phi)$ (see Ref. [2] for details). The best fit of the model corresponds to the solid line. The other two sets of parameters take into account the uncertainty coming from the lattice calculation of χ_2 in Ref. [8] in order to estimate the sensibility of the CEP to the T-dependence of χ_2.

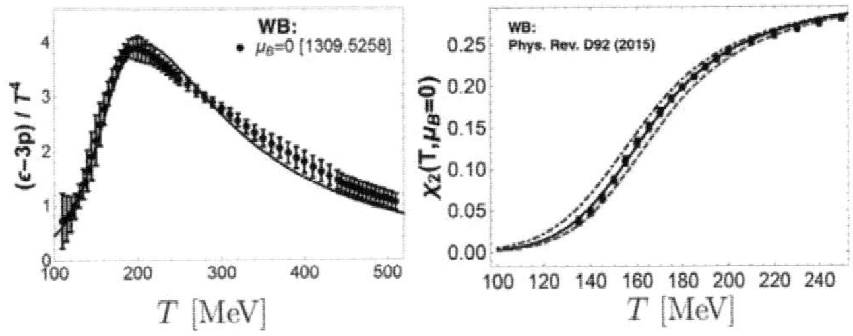

Figure 1. Trace anomaly [9] (left) and second baryonic susceptibility [8] (right) from lattice QCD calculations at $\mu_B = 0$ and compared to our black hole model as a function of temperature.

3. Results

The n-th baryon number susceptibility $\chi_n = \chi_n(T, \mu_B)$ is defined as $\chi_n = \frac{\partial^n}{\partial(\mu_B/T)^n}\left(\frac{P}{T^4}\right)$. Susceptibilities provide essential information about the effective degrees of freedom of a system, and are directly related to the moment of the distribution measured on an event-by-event basis in particle colliders. In the vicinity of a CEP, the susceptibilities χ_n's scale with different powers of the (diverging) correlation length ξ. One can show that the high order susceptibilities diverge with higher powers of ξ [10]. In our black hole model, we numerically calculate up to fourth order baryonic susceptibilities at finite μ_B, and up to eighth order at $\mu_B = 0$.

The black hole susceptibilities are shown in Fig. 2. The error-band on our predictions for χ_6/χ_2 and χ_8 indicates the uncertainty derived by the numerical calculation. One can see that χ_2 and χ_4 begin to develop a peak for large chemical potentials, which will then evolve into a divergence at the CEP. The figure also shows the available lattice results for χ_2 [8], χ_4 [11] and χ_6/χ_2 [12] as a function of T. Our predictions for χ_4 and χ_6/χ_2, at zero μ_B, have a noticeable agreement with lattice QCD results. As for χ_8, our prediction exhibits the expected features from universality [13], which can be compared to future lattice QCD calculations.

Using the higher order susceptibilities calculated at $\mu_B = 0$, one can reconstruct the system's pressure P and baryon density ρ_B as a Taylor series in powers of μ_B/T as follows

$$\frac{P(T,\mu_B) - P(T,\mu_B = 0)}{T^4} = \sum_{n=1}^{\infty} \frac{1}{(2n!)} \chi_{2n}(T) \left(\frac{\mu_B}{T}\right)^{2n}, \tag{2}$$

$$\frac{\rho_B(T,\mu_B)}{T^3} = \sum_{n=1}^{\infty} \frac{1}{(2n-1)!} \chi_{2n}(T) \left(\frac{\mu_B}{T}\right)^{2n-1}. \tag{3}$$

The pressure difference in Eq. (2), and the baryon density in Eq. (3) calculated in the holographic model with no truncations, are compared to the lattice QCD results from Ref. [12] in Fig. 3. The reconstructed holographic pressure truncated at order $\mathcal{O}(\mu_B^6)$ and $\mathcal{O}(\mu_B^8)$ is also shown.

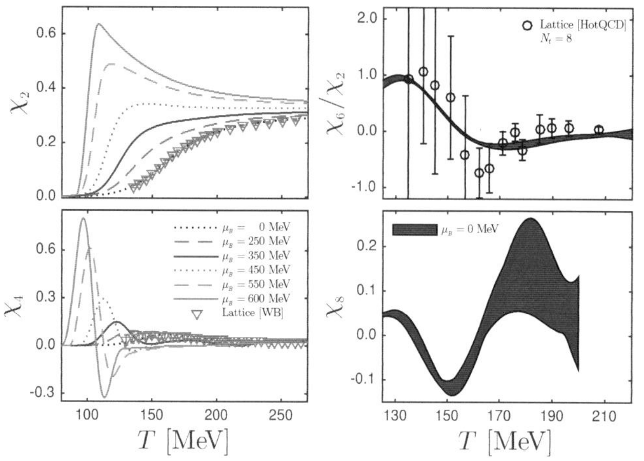

Figure 2. Baryon number susceptibilities (χ_n) as functions of the temperature (T) for different values of the baryon chemical potential (μ_B) computed using holographic model. The lattice data for χ_2 and χ_4 are obtained from Ref. [11], and χ_6/χ_2 from Ref. [13].

Our analysis confirms the applicability of the $\mathcal{O}(\mu_B^6)$ truncation done in [12] for $\mu_B/T \leq 2$, and it also predicts that the inclusion of $\chi_8(T)$ into the expansion extends the domain of applicability of the Taylor series to at least $\mu_B/T \sim 2.5$.

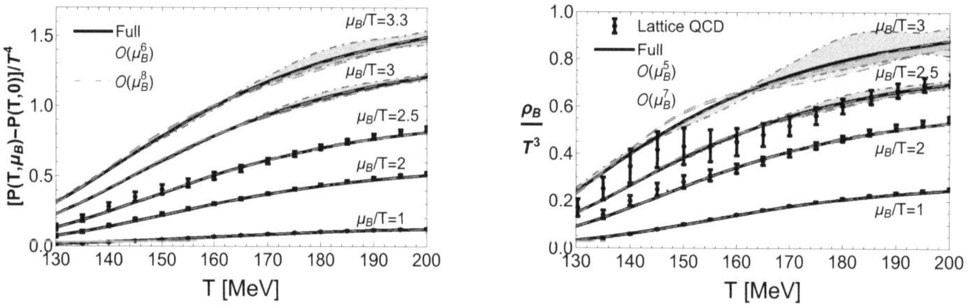

Figure 3. The μ_B-dependent contribution to the pressure (left) and the baryon density (right) as functions of T for different values of μ_B/T. The solid curves correspond to the full holographic result. The lattice points correspond to the reconstructed Taylor series up to $\mathcal{O}(\mu_B^6)$ for the pressure and $\mathcal{O}(\mu_B^5)$ for ρ_B computed in Ref. [12].

The best set of parameters for the holographic mode is used to calculate $\chi_2(T, \mu_B)$, which is plotted in Fig. 4 (left). In the upper plane of the figure, the QCD phase diagram. In order to identify the crossover between hadronic matter and the QGP, we calculate two observables sensitive to this transition: the inflection point of χ_2 and the minimum in the speed of sound squared c_s^2, which is shown in the QCD diagram. The CEP is found as divergence in χ_2 at $T_{CEP} = 89$ MeV and $\mu_B^{CEP} = 724$ MeV. Fig. 4 (right) shows the phase diagram with our critical point including the uncertainties from using the three set of parameters in our black hole model. Our CEP is placed inside the region that has not yet been excluded using different approaches [12, 14]. Regions where $T > 155$ MeV are unlikely to display a CEP due to the known behavior of the curvature of transition line [15].

The chemical freeze-out points, shown in the phase diagram in the upper plane of Fig. 4 (left), were extracted through a comparison of susceptibilities computed in our black hole model and the corresponding net-proton fluctuations from [16] (see Refs. [2, 17, 18] for details). They

IOP Conf. Series: Journal of Physics: Conf. Series **1024** (2018) 012015 doi:10.1088/1742-6596/1024/1/012015

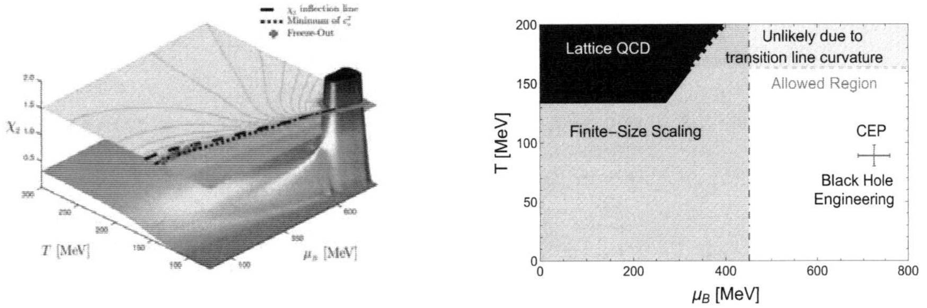

Figure 4. (left) Second baryon susceptibility χ_2 in the (T, μ_B) plane determined from the black hole model. (right) Black hole model CEP in the (T, μ_B) plane taking into account uncertainties. The shaded areas indicate regions in the QCD phase diagram where the presence of a critical point has been excluded by other approaches.

were found to lie along the transition line defined by the minimum of c_s^2 when $\sqrt{s} \geq 27$ GeV.

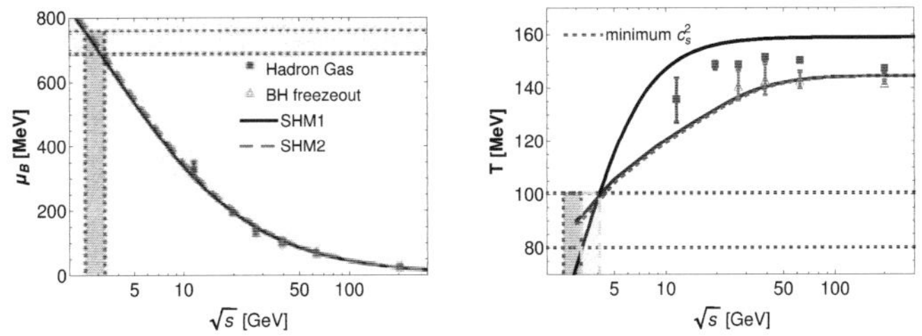

Figure 5. Collision energy dependence of the baryon chemical potential (left) and temperature (right) at the chemical freeze-out.

In the following, we extrapolate the behavior of the calculated freeze-out points towards smaller collision energies in order to provide an estimate for the heavy-ion collision center-of-mass energy that could probe the values of T_{CEP} and μ_B^{CEP} found in our black hole model. We use two parametrizations defined using the statistical hadronization model calculations of [19] (SHM1) and [20] (SHM2). Fig. 5 shows both parametrization as a function of μ_B (left) and as a function of T (right). The freeze-out points extracted in the black hole model via a comparison between the χ_1/χ_2 and χ_3/χ_2, and those obtained by hadron resonance gas comparisons to net-proton and net-electric charge fluctuations from [21] are also shown in this figure.

By consistently extrapolating this behavior towards smaller collision energies, we find that the CEP of the model could be probed using heavy ion experiments with center-of-mass energy in the range $\sqrt{s} = 2.5 - 4.1$ GeV.

4. Conclusions

We have used a holographic approach to construct black hole solutions to study the properties of the phases of strongly interacting matter. Our model predicts the existence of a CEP at $T_{CEP} = 89$ MeVand $\mu_B^{CEP} = 724$ MeV. Since the CEP is located along the line $\mu_B/T \sim 8.1$ in the phase diagram, it is beyond the reach of current lattice QCD calculations where $\mu_B/T \lesssim 2$ [12].

Also, we provide an estimate for the heavy-ion collision center-of-mass energy that could probe the values of T_{CEP} and μ_B^{CEP}. We found that the CEP could be probed using heavy ion experiments with center-of-mass energy in the range $\sqrt{s} = 2.5 - 4.1$ GeV. These collision energies are within the reach of the HADES experiment [22], the planned Fixed Target (FXT) program also at RHIC [23], and the future Compressed Baryonic Matter (CBM) experiment at FAIR [24].

Acknowledgments
I thank my co-authors for their collaboration and input in the present proceedings. This material is based upon work supported by the National Science Foundation under Grants No. PHY-1513864, PHY-1654219 and OAC-1531814 and by the U.S. Department of Energy, Office of Science, Office of Nuclear Physics, within the framework of the Beam Energy Scan Theory (BEST) Topical Collaboration. The use of the Maxwell Cluster and the advanced support from the Center of Advanced Computing and Data Systems at the University of Houston are gratefully acknowledged.

References
[1] Aoki Y, Endrodi G, Fodor Z, Katz S D and Szabo K K 2006 *Nature* **443** 675–678 (*Preprint* hep-lat/0611014)
[2] Critelli R, Noronha J, Noronha-Hostler J, Portillo I, Ratti C and Rougemont R 2017 *Phys. Rev. D* **96**(9) 096026 URL https://link.aps.org/doi/10.1103/PhysRevD.96.096026
[3] Maldacena J M 1999 *Int. J. Theor. Phys.* **38** 1113–1133 [Adv. Theor. Math. Phys.2,231(1998)] (*Preprint* hep-th/9711200)
[4] Rougemont R, Noronha J and Noronha-Hostler J 2015 *Phys. Rev. Lett.* **115** 202301 (*Preprint* 1507.06972)
[5] Rougemont R, Critelli R, Noronha-Hostler J, Noronha J and Ratti C 2017 *Phys. Rev.* **D96** 014032 (*Preprint* 1704.05558)
[6] DeWolfe O, Gubser S S and Rosen C 2011 *Phys. Rev.* **D83** 086005 (*Preprint* 1012.1864)
[7] DeWolfe O, Gubser S S and Rosen C 2011 *Phys. Rev.* **D84** 126014 (*Preprint* 1108.2029)
[8] Borsanyi S, Fodor Z, Katz S D, Krieg S, Ratti C and Szabo K 2012 *JHEP* **01** 138 (*Preprint* 1112.4416)
[9] Borsanyi S, Fodor Z, Hoelbling C, Katz S D, Krieg S and Szabo K K 2014 *Phys. Lett.* **B730** 99–104 (*Preprint* 1309.5258)
[10] Stephanov M A 2009 *Phys. Rev. Lett.* **102** 032301 (*Preprint* 0809.3450)
[11] Bellwied R, Borsanyi S, Fodor Z, Katz S D, Pasztor A, Ratti C and Szabo K K 2015 *Phys. Rev.* **D92** 114505 (*Preprint* 1507.04627)
[12] Bazavov A *et al.* 2017 *Phys. Rev.* **D95** 054504 (*Preprint* 1701.04325)
[13] Karsch F 2009 *Prog. Part. Nucl. Phys.* **62** 503–511
[14] Fraga E S, Palhares L F and Sorensen P 2011 *Phys. Rev.* **C84** 011903 (*Preprint* 1104.3755)
[15] Bellwied R, Borsanyi S, Fodor Z, Gnther J, Katz S D, Ratti C and Szabo K K 2015 *Phys. Lett.* **B751** 559–564 (*Preprint* 1507.07510)
[16] Adamczyk L *et al.* (STAR) 2014 *Phys. Rev. Lett.* **112** 032302 (*Preprint* 1309.5681)
[17] Portillo I 2017 *26th International Conference on Ultrarelativistic Nucleus-Nucleus Collisions (Quark Matter 2017) Chicago,Illinois, USA, February 6-11, 2017* (*Preprint* 1705.01021) URL http://inspirehep.net/record/1597600/files/arXiv:1705.01021.pdf
[18] Portillo I 2017 *J. Phys. Conf. Ser.* **832** 012041 (*Preprint* 1610.09981)
[19] Andronic A, Braun-Munzinger P and Stachel J 2009 *Phys. Lett.* **B673** 142–145 [Erratum: Phys. Lett.B678,516(2009)] (*Preprint* 0812.1186)
[20] Cleymans J, Oeschler H, Redlich K and Wheaton S 2006 *Phys. Rev.* **C73** 034905 (*Preprint* hep-ph/0511094)
[21] Alba P, Alberico W, Bellwied R, Bluhm M, Mantovani Sarti V, Nahrgang M and Ratti C 2014 *Phys. Lett.* **B738** 305–310 (*Preprint* 1403.4903)
[22] Agakishiev G *et al.* (HADES) 2016 *Eur. Phys. J.* **A52** 178 (*Preprint* 1512.07070)
[23] Meehan K (for the STAR) 2017 (*Preprint* 1704.06342)
[24] Ablyazimov T *et al.* (CBM) 2017 *Eur. Phys. J.* **A53** 60 (*Preprint* 1607.01487)

IOP Conf. Series: Journal of Physics: Conf. Series **1024** (2018) 012016 doi:10.1088/1742-6596/1024/1/012016

Determination of electric dipole transitions in heavy quarkonia using potential non-relativistic QCD

Jorge Segovia and Sebastian Steinbeißer
Physik-Department, Technische Universität München,
James-Franck-Str. 1, 85748 Garching, Germany
E-mail: `jorge.segovia@tum.de`, `sebastian.steinbeisser@tum.de`

Abstract. The electric dipole transitions $\chi_{bJ}(1P) \to \gamma\Upsilon(1S)$ with $J = 0, 1, 2$ and $h_b(1P) \to \gamma\eta_b(1S)$ are computed using the weak-coupling version of a low-energy effective field theory named potential non-relativistic QCD (pNRQCD). In order to improve convergence and thus give firm predictions for the studied reactions, the full static potential is incorporated into the leading order Hamiltonian; moreover, we must handle properly renormalon effects and re-summation of large logarithms. The precision we reach is $k_\gamma^3/(mv)^2 \times \mathcal{O}(v^2)$, where k_γ is the photon energy, m is the mass of the heavy quark and v its velocity. Our analysis separates those relativistic contributions that account for the electromagnetic interaction terms in the pNRQCD Lagrangian which are v^2 suppressed and those that account for wave function corrections of relative order v^2. Among the last ones, corrections from $1/m$ and $1/m^2$ potentials are computed, but not those coming from higher Fock states since they demand non-perturbative input and are $\Lambda_{\rm QCD}^2/(mv)^2$ or $\Lambda_{\rm QCD}^3/(m^3v^4)$ suppressed, at least, in the strict weak coupling regime. These proceedings are based on the forthcoming publication [1].

1. Introduction

The electric dipole (E1) and magnetic dipole (M1) transitions between heavy quarkonia have been treated for a long time by means of potential models that use non-relativistic reductions of QCD-based quark–(anti-)quark interactions (see, for instance, Ref. [2] for a recent application to the bottomonium system). However, a new large set of accurate experimental data related with electromagnetic reactions in the heavy quark sector is expected to be reported by B-factories (Belle@KEK), τ-charm facilities (BES@IHEP) and even proton–(anti-)proton colliders (LHCb@CERN and PANDA@GSI); therefore, demanding for a systematic and model-independent analysis.

The aim of this manuscript is to compute the E1 transitions $\chi_{bJ}(1P) \to \gamma\Upsilon(1S)$ with $J = 0, 1, 2$ and $h_b(1P) \to \gamma\eta_b(1S)$ using the effective field theory (EFT) called pNRQCD [3, 4] (see Refs. [5, 6] for reviews). This EFT takes full advantage of the hierarchy of scales that appear in the system:[1] $m \gg p \sim 1/r \sim mv \gg E \sim mv^2$, and makes a systematic connection between the underlying quantum field theory and non-relativistic quantum mechanics.

The specific details on the construction of pNRQCD depend on the relative size of the mv scale with respect to $\Lambda_{\rm QCD}$, for $mv \gg \Lambda_{\rm QCD}$ we have the weak-coupling regime [6], and for $mv \gtrsim \Lambda_{\rm QCD}$ the strong-coupling regime [7].

[1] The heavy quark mass is m, the relative momentum of the heavy quarks is p and the bound state energy is denoted by E. The heavy quark velocity, v, is assumed to be $v \ll 1$ which is reasonably fulfilled in bottomonium ($v^2 \sim 0.1$).

Nowadays, there seems to be a growing consensus that the weak-coupling regime works properly for many physical observables in the bottomonium sector. In order to reach this conclusion, one must (i) include the static potential in the leading order Hamiltonian, which produces a more convenient rearrangement of the perturbative series; (ii) treat adequately the renormalon effects, which improves the convergence of the series; and (iii) calculate the resummation of large logarithms, which significantly diminish the factorization scale dependence of the observable.

The improvements mentioned above have been applied already to the determination of M1 transitions between low-lying heavy quarkonium states in Ref. [8]. Therein, good convergence properties and agreement between theory and experiment was obtained for the bottomonium ground state and for the $n=2$ excitation of the bottomonium system in both S- and P-waves. The motivation of the present paper is to repeat the study of Ref. [8] to the case of E1 transitions between bottomonia. In this case, contrary to M1 transitions, the computation of relativistic corrections is technically complicated: in addition to the effects given by higher order terms in the E1 transition operator, one has to calculate many corrections to the initial and final state wave function due to higher order potentials and higher order Fock states. This fact has hindered numerical computations of the E1 transitions between low-lying heavy quarkonium states within pNRQCD (see Refs. [9, 10] for partial calculations) and this contribution aims to close the gap.

2. Theoretical setup

The formulae of the E1 transitions in the weak-coupling regime of pNRQCD have been presented in detail in Ref. [11]. Up to order k_γ^3/m^2, the expressions we use for the decay rates under study are

$$\begin{aligned}\Gamma(n\,^3P_J \to n'\,^3S_1+\gamma) = \Gamma_{E1}^{(0)}\Bigg\{ & 1 + R^{S=1}(J) - \frac{k_\gamma}{6m} - \frac{k_\gamma^2}{60}\frac{I_5^{(0)}(n1\to n'0)}{I_3^{(0)}(n1\to n'0)} \\ & + \left[\frac{J(J+1)}{2}-2\right]\Bigg[-\left(1+\kappa_Q^{\rm em}\right)\frac{k_\gamma}{2m} \\ & + \frac{1}{m^2}(1+2\kappa_Q^{\rm em})\frac{I_2^{(1)}(n1\to n'0)+2I_1^{(0)}(n1\to n'0)}{I_3^{(0)}(n1\to n'0)}\Bigg]\Bigg\}, \end{aligned} \tag{1}$$

$$\Gamma(n^1P_1 \to n'\,^1S_0+\gamma) = \Gamma_{E1}^{(0)}\left\{1+R^{S=0} - \frac{k_\gamma}{6m} - \frac{k_\gamma^2}{60}\frac{I_5^{(0)}(n1\to n'0)}{I_3^{(0)}(n1\to n'0)}\right\}, \tag{2}$$

where $R^{S=1}(J)$ and $R^{S=0}$ include the initial and final state corrections due to higher order potentials and higher order Fock states (see below). The remaining corrections within the brackets are the result of taking into account additional electromagnetic interaction terms in the Lagrangian suppressed by $\mathcal{O}(v^2)$ [11]. We have displayed in the formulae terms proportional to the anomalous magnetic moment, $\kappa_Q^{\rm em}$. These terms are not considered in the numerical analysis because they are at least suppressed by $\alpha_s(m)v^2$ and thus go beyond our accuracy. The LO decay width, which scales as $\sim k_\gamma^3/(mv)^2$, is

$$\Gamma_{\rm E1}^{(0)} = \frac{4}{9}\alpha_{\rm em}\,e_Q^2\,k_\gamma^3\left[I_3^{(0)}(n1\to n'0)\right]^2, \tag{3}$$

with $\alpha_{\rm em}$ the electromagnetic fine structure constant, e_Q the charge of the heavy quarks in units of the electron charge, k_γ the photon energy, and the function

$$I_N^{(k)}(n\ell\to n'\ell') = \int_0^\infty dr\,r^N R^*_{n'\ell'}(r)\left[\frac{d^k}{dr^k}R_{n\ell}(r)\right] \tag{4}$$

IOP Conf. Series: Journal of Physics: Conf. Series **1024** (2018) 012016 doi:10.1088/1742-6596/1024/1/012016

is a matrix element that involves the radial wave functions of the initial and final states. We assume that these states are solutions of the Schrödinger equation

$$H^{(0)}\psi^{(0)}_{nlm}(\vec{r}) = E^{(0)}_n \psi^{(0)}_{nlm}(\vec{r})\,, \tag{5}$$

with the leading order Hamiltonian given by

$$H^{(0)} = -\frac{\vec{\nabla}^2}{2m_r} + V_s^{(N)}(r)\,, \tag{6}$$

and the static potential approximated by a polynomial of order $N+1$ in powers of α_s

$$V_s^{(N)}(r) = -\frac{C_F\alpha_s(\nu)}{r}\left[1 + \sum_{k=1}^{N}\left(\frac{\alpha_s}{4\pi}\right)^k a_k(\nu, r)\right]. \tag{7}$$

In principle, one would like to take N as large as possible; in practice, we truncate the perturbative series at $N = 3$, *i.e.* at $\mathcal{O}(\alpha_s^4)$, including also the leading ultra-soft corrections.

Due to higher order potentials and the presence of ultra-soft gluons that lead to singlet-to-octet transitions, the state in Eq. (5) is not an eigenstate of the complete Hamiltonian. Therefore, one has to consider corrections to the wave function which contribute to the decay rate at the required order of precision. The first kind of corrections to the wave function are those which arise from relativistic corrections to the potential and to the leading order kinetic operator. They can be organized as an expansion in the inverse of the heavy quark mass, m. At the order we are interested in, such expansion covers all the $1/m$ and $1/m^2$ potentials and, at order $1/m^3$, the first relativistic correction to the kinetic energy [11]. The second kind of corrections to the wave function come from diagrams in which a singlet state is coupled to an octet state due to the emission and re-absorption of an ultra-soft gluon. We do not consider these contributions herein because in the (strict) weak-coupling regime, $E \sim mv^2 \gg \Lambda_{\rm QCD}$, they are negligible [11].

3. Results

The panel (a) of Fig. 1 shows the leading order decay rate, $\Gamma^{(0)}_{E1}$, for the $\chi_{b1}(1P) \to \gamma\Upsilon(1S)$ transition computed when the static potential is included order by order in the Schrödinger equation. If one solves the Schrödinger equation numerically with only the Coulomb-like term of the static potential, one gets a leading order (LO) decay rate (solid blue curve) that depends strongly on the factorization scale (see Ref. [12] for details based on an analytical study). However, the ν-scale dependence becomes mild as next-to-leading order (NLO) (dashed orange curve), NNLO (dot-dashed green curve) and NNNLO (dotted red curve) radiative corrections to the static potential are included in the Schrödinger equation.

The panel (b) of Fig. 1 shows the induced effect on the decay width once the correct logarithmically modulated short distance behavior (RGI) of the static potential is included properly. Now, the improvement in the convergence properties of the perturbative series is seen at $\nu \lesssim 1.7\,\text{GeV}$. This is crucial in order to give final results for the decay rates since a low value of ν, compatible with the typical momentum scale of the system under study, has to be chosen ($\nu = 1.25\,\text{GeV}$). It is fair to mention that, in this final scheme, the dependence on the factorization scale for the NNNLO result seems to increase slightly.

The final decay width for the $\chi_{b1}(1P) \to \gamma\Upsilon(1S)$ process is shown in Fig. 2. The leading order non-relativistic decay rate is the dashed blue curve, the dot-dashed curve is taking into account the relativistic contributions stemming from higher order electromagnetic operators (Eq. (1) with $R^{S=1}(J) = 0$) and the solid black one is including also the relativistic corrections to the wave function of the initial and final states (Eq. (1) with $R^{S=1}(J) \neq 0$).

IOP Conf. Series: Journal of Physics: Conf. Series **1024** (2018) 012016 doi:10.1088/1742-6596/1024/1/012016

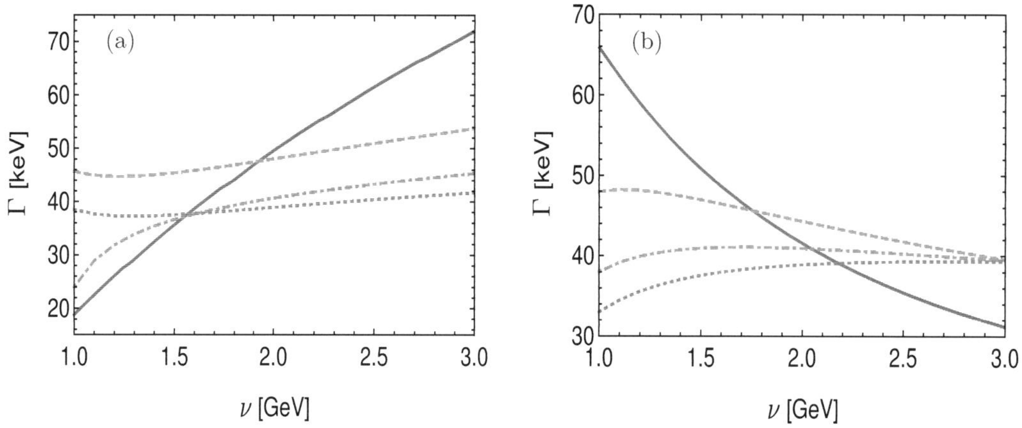

Figure 1. Leading order decay rate, $\Gamma_{E1}^{(0)}$, of the electric dipole transition $\chi_{b1}(1P) \to \gamma\Upsilon(1S)$. The renormalon cancellation is achieved order by order in the static potential which is exactly included when solving the Schrödinger equation. The panel (a) shows the case in which the renormalization group improved (RGI) version of the static potential is not used, whereas the panel (b) shows the results when the RGI potential is considered. The different curves in both panels represent the result coming from taking into account the Coulomb-like (solid blue), NLO (dashed orange), NNLO (dot-dashed green) and NNNLO (dotted red) terms in the static potential.

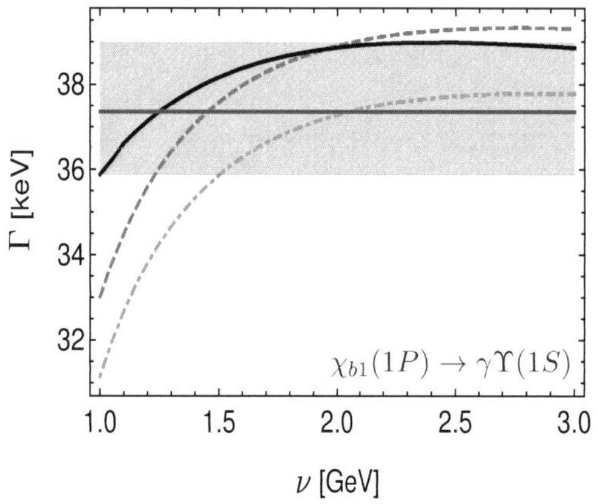

Figure 2. Decay width for the electric dipole transition $\chi_{b1}(1P) \to \gamma\Upsilon(1S)$. The dashed blue curve is the leading order decay rate, the dot-dashed orange curve is including the relativistic contributions stemming from higher order electromagnetic operators (Eq. (1) with $R^{S=1}(J) = 0$) and the solid black curve is the final result including also the relativistic corrections to the wave function of the initial and final states (Eq. (1) with $R^{S=1}(J) \neq 0$). We take our final value at $\nu = 1.25\,\text{GeV}$ and the gray band indicates the associated uncertainty.

One sees in Fig. 2 that the leading order decay width depends weakly on the factorization scale, it varies from $\Gamma \sim 33$ keV at $\nu = 1\,\text{GeV}$ to $\Gamma \sim 39\,\text{keV}$ at $\nu = 3\,\text{GeV}$. This feature is translated to the case in which higher order electromagnetic operators are taken into account and also to the case in which wave function relativistic corrections are included. In fact, somewhat surprisingly, the ν-dependence of our final result, $\sim 3\,\text{keV}$, is weaker than that of the leading order or even the one including higher order electromagnetic operators. A variation of $\sim 3\,\text{keV}$ over a total value of around $\sim 37\,\text{keV}$ represents a relative error of $\sim 8\%$ in our determination of the decay rate, being the biggest source of uncertainty.

It is clearly seen in Fig. 2 that both types of relativistic contributions are under control: they behave smoothly with respect the renormalization scale and produce corrections to the LO decay width which are relatively small. Another interesting feature of Fig. 2 is that the relativistic corrections induced by higher order electromagnetic operators tend to diminish the LO decay rate whereas the effect of the corrected initial and final wave functions is to increase it.

We have performed the same analysis than above to the electric dipole transitions $\chi_{b0}(1P) \to \gamma\Upsilon(1S)$, $\chi_{b2}(1P) \to \gamma\Upsilon(1S)$ and $h_b(1P) \to \gamma\Upsilon(1S)$. Similar conclusions than the ones already mentioned apply to these cases. Our final values for the electric dipole transitions $\chi_{bJ}(1P) \to \gamma\Upsilon(1S)$ with $J = 0, 1, 2$ and $h_b(1P) \to \gamma\eta_b(1S)$ read

$$\Gamma(\chi_{b0}(1P) \to \gamma\Upsilon(1S)) = 28^{+2}_{-1}\ \text{keV}\,, \tag{8}$$
$$\Gamma(\chi_{b1}(1P) \to \gamma\Upsilon(1S)) = 37^{+2}_{-1}\ \text{keV}\,, \tag{9}$$
$$\Gamma(\chi_{b2}(1P) \to \gamma\Upsilon(1S)) = 45^{+1}_{-1}\ \text{keV}\,, \tag{10}$$
$$\Gamma(h_b(1P) \to \gamma\eta_b(1S)) = 63^{+1}_{-1}\ \text{keV}\,. \tag{11}$$

Because of the very mild dependence of our results on the renormalization scale ν, the associated uncertainties are very small.

The Particle Data Group (PDG) [13] only reports the branching fractions of the E1 transitions studied herein. Since the total decay widths of the χ_{bJ} (with $J = 0, 1, 2$) and $h_b(1P)$ mesons are not known, we cannot compare our theoretical results with experimental data. Nevertheless, we can use the branching fractions given by the PDG and our results for the decay rates of the electric dipole transitions to predict their total decay widths. The results are

$$\Gamma(\chi_{b0}(1P)) = 1.6^{+0.3}_{-0.3}\ \text{MeV}\,, \qquad \Gamma(\chi_{b1}(1P)) = 110^{+9}_{-8}\ \text{keV}\,, \tag{12}$$
$$\Gamma(\chi_{b2}(1P)) = 234^{+15}_{-16}\ \text{keV}\,, \qquad \Gamma(h_b(1P)) = 121^{+14}_{-12}\ \text{keV}\,. \tag{13}$$

These numbers could be of special interest for future experimental determinations. For instance, the Belle collaboration has recently reported an upper limit at 90% confidence level on the total decay width of the $\chi_{b0}(1P)$ [14]: $\Gamma(\chi_{b0}(1P)) < 2.4\,\text{MeV}$, which is well within the uncertainty band of our prediction.

4. Summary

We have performed the first numerical analysis of the electric dipole transitions $\chi_{bJ}(1P) \to \gamma\Upsilon(1S)$ with $J = 0, 1, 2$ and $h_b(1P) \to \gamma\eta_b(1S)$ within the weak coupling version of a low-energy effective field theory called potential non-relativistic QCD.

Acknowledgments

This work has been supported by the DFG and the NSFC through funds provided to the Sino-German CRC 110 "Symmetries and the Emergence of Structure in QCD", and by the DFG cluster of excellence "Origin and structure of the universe" (www.universe-cluster.de). J.S. acknowledges the financial support from the Alexander von Humboldt Foundation.

References

[1] Segovia J, Steinbeißer S and Vairo A in preparation: TUM-EFT 86/16
[2] Segovia J, Ortega P G, Entem D R and Fernández F 2016 *Phys. Rev.* **D93** 074027
[3] Pineda A and Soto J 1998 *Nucl. Phys. Proc. Suppl.* **64** 428–432
[4] Brambilla N, Pineda A, Soto J and Vairo A 2000 *Nucl. Phys.* **B566** 275
[5] Brambilla N, Pineda A, Soto J and Vairo A 2005 *Rev. Mod. Phys.* **77** 1423
[6] Pineda A 2012 *Prog. Part. Nucl. Phys.* **67** 735–785
[7] Brambilla N, Pineda A, Soto J and Vairo A 2001 *Phys. Rev.* **D63** 014023
[8] Pineda A and Segovia J 2013 *Phys. Rev.* **D87** 074024
[9] Pietrulewicz P 2013 PoS ConfinementX, 135 (2012)
[10] Martinez H E 2016 *AIP Conf. Proc.* **1701** 050008
[11] Brambilla N, Pietrulewicz P and Vairo A 2012 *Phys. Rev.* **D85** 094005
[12] Steinbeißer S and Segovia J 2017 *EPJ Web Conf.* **137** 06026
[13] Patrignani C *et al.* (Particle Data Group) 2016 *Chin. Phys.* **C40** 100001
[14] Abdesselam A *et al.* (Belle) 2016 (*Preprint* 1606.01276)

IOP Conf. Series: Journal of Physics: Conf. Series **1024** (2018) 012017 doi:10.1088/1742-6596/1024/1/012017

Calculations of $\bar{p}$-nuclear states within the Paris $\bar{N}N$ potential model

J. Hrtánková[1,2], **J. Mareš**[1]

[1] Nuclear Physics Institute, 25068 Řež, Czech Republic

[2] Czech Technical University in Prague, Faculty of Nuclear Sciences and Physical Engineering, Břehová 7, 115 19 Prague 1, Czech Republic

E-mail: hrtankova@ujf.cas.cz

Abstract. We performed calculations of $\bar{p}$-nuclear quasi-bound states using an optical potential derived from the 2009 version of the Paris $\bar{N}N$ potential model. Self-consistent treatment of strong energy dependence of the considered S- and P-wave $\bar{p}N$ scattering amplitudes is essential for proper evaluation of $\bar{p}$ binding energies and widths. The P-wave potential has very small effect on the calculated $\bar{p}$ binding energies, however, it reduces the corresponding widths in large measure. The Paris S-wave potential supplemented by a phenomenological P-wave term yields the $\bar{p}$ binding energies and widths very close to those obtained within the RMF model consistent with $\bar{p}$-atom data.

1. Introduction

Analyses of $\bar{p}$-atom data [1, 2, 3] and $\bar{p}$ scattering off nuclei at low energies [4, 5] revealed that the antiproton interaction with the nuclear medium near threshold and at low densities is dominated by $\bar{p}$ annihilation. The experimental data could be well fitted by a $\bar{p}$-nuclear optical potential, the imaginary part of which considerably exceeds the attractive real part. But deep in the nuclear interior, the phase space for $\bar{p}$ annihilation is significantly reduced. This could result in relatively long lifetime of $\bar{p}$ in the nuclear medium [6].

The knowledge of $\bar{p}$-nucleus interactions at various densities and kinematical conditions will be employed and further improved in forthcoming experiments with $\bar{p}$-beams at FAIR [7]. Simulations of the considered processes in a wide range of $\bar{p}$-beam momenta are being performed within the Giessen Boltzmann-Ueling-Uhlenbeck (GiBUU) transport model [8] in which the $\bar{p}$ optical potential in nuclear matter serves as input. The study of the $\bar{p}$ interaction with the nuclear medium is thus topical.

In this work, we report on our very recent calculations of $\bar{p}$ quasi-bound states in selected nuclei across the periodic table [9], using the $\bar{p}N$ scattering amplitudes stemming from the 2009 version of the Paris potential [10]. This potential was confronted by Friedman et al. [11] with $\bar{p}$-atom data and measured low-energy $\bar{p}$ elastic scattering and annihilation cross sections. Their analysis disclosed the necessity to include the P-wave part of the $\bar{p}N$ potential. The Paris S-wave potential supplemented by a phenomenological P-wave term was found to fit the data on low-density, near-threshold $\bar{p}$-nucleus interaction. We were thus tempted to explore the effect of the P-wave interaction on the calculated binding energies and widths of $\bar{p}$-nuclear states.

In Section 2, we briefly introduce the model applied in our calculations. Few representative results of our study are presented and discussed in Section 3.

IOP Conf. Series: Journal of Physics: Conf. Series **1024** (2018) 012017 doi:10.1088/1742-6596/1024/1/012017

2. Model

The binding energies $B_{\bar{p}}$ and widths $\Gamma_{\bar{p}}$ of $\bar{p}$-nuclear quasi-bound states are determined by solving self-consistently the Dirac equation

$$[-i\vec{\alpha}\cdot\vec{\nabla} + \beta m_{\bar{p}} + V_{\rm opt}(r)]\psi_{\bar{p}} = \epsilon_{\bar{p}}\psi_{\bar{p}} \;, \tag{1}$$

where $m_{\bar{p}}$ is the mass of the antiproton and $\epsilon_{\bar{p}} = -B_{\bar{p}} - {\rm i}\Gamma_{\bar{p}}/2$, $(B_{\bar{p}} > 0)$. The complex optical potential $V_{\rm opt}(r)$ is of the form

$$2E_{\bar{p}}V_{\rm opt}(r) = q(r) + 3\vec{\nabla}\cdot\alpha(r)\vec{\nabla} \;. \tag{2}$$

The S-wave part $q(r)$ is constructed in a '$t\rho$' form as

$$q(r) = -4\pi\frac{\sqrt{s}}{m_N}\left[F_0\frac{1}{2}\rho_p(r) + F_1\left(\frac{1}{2}\rho_p(r) + \rho_n(r)\right)\right] \;, \tag{3}$$

where $E_{\bar{p}} = m_{\bar{p}} - B_{\bar{p}}$ and $\rho_p(r)$ $[\rho_n(r)]$ is the proton (neutron) density calculated within the Relativistic Mean Field (RMF) model NL-SH [12]. The isospin 0 and 1 in-medium amplitudes F_0 and F_1 are constructed from the free-space $\bar{p}N$ amplitudes using the multiple scattering approach of Wass et al. [13] (WRW) which accounts for Pauli correlations in the nuclear medium

$$F_1 = \frac{f^{\rm S}_{\bar{p}n}(\delta\sqrt{s})}{1+\frac{1}{4}\xi_k\frac{\sqrt{s}}{m_N}f^{\rm S}_{\bar{p}n}(\delta\sqrt{s})\rho(r)} \;, \quad F_0 = \frac{[2f^{\rm S}_{\bar{p}p}(\delta\sqrt{s}) - f^{\rm S}_{\bar{p}n}(\delta\sqrt{s})]}{1+\frac{1}{4}\xi_k\frac{\sqrt{s}}{m_N}[2f^{\rm S}_{\bar{p}p}(\delta\sqrt{s}) - f^{\rm S}_{\bar{p}n}(\delta\sqrt{s})]\rho(r)} \;. \tag{4}$$

Here, $f^{\rm S}_{\bar{p}n}$ ($f^{S}_{\bar{p}p}$) denotes the free-space $\bar{p}n$ ($\bar{p}p$) S-wave scattering amplitude as a function of $\delta\sqrt{s} = \sqrt{s} - E_{\rm th}$ with $E_{\rm th} = m_N + m_{\bar{p}}$; $\rho(r) = \rho_p(r) + \rho_n(r)$ is the nuclear density distribution and ξ_k is the Pauli correlation factor (for details see [9]).

The P-wave part is written as

$$\alpha(r) = 4\pi\frac{m_N}{\sqrt{s}}\left(f^P_{\bar{p}p}(\delta\sqrt{s})\rho_p(r) + f^P_{\bar{p}n}(\delta\sqrt{s})\rho_n(r)\right) \;, \tag{5}$$

where $f^P_{\bar{p}p}(\delta\sqrt{s})$ and $f^P_{\bar{p}n}(\delta\sqrt{s})$ denote the P-wave $\bar{p}p$ and $\bar{p}n$ free-space scattering amplitudes, respectively. We do not consider medium modifications of the P-wave amplitudes because the P-wave interaction contributes mainly near the nuclear surface (due to its gradient form) where the nuclear densities are relatively low. Friedman et al. [11] revealed that the potential constructed from the Paris S- and P-wave amplitudes fails to fit the $\bar{p}$-atom data. On the contrary, the potential based on the Paris S-wave and phenomenological P-wave amplitude $f^P_{\bar{p}N} = 2.9 + {\rm i}1.8$ fm^3 fits the data well [11]. We performed calculations using both the Paris and phenomenological P-wave interactions.

In Fig. 1, we demonstrate strong energy dependence of the Paris $\bar{p}N$ scattering amplitudes used in our calculations. In the upper panel, the in-medium $\bar{p}p$ (left) and $\bar{p}n$ (right) S-wave amplitudes evaluated at the saturation density $\rho_0 = 0.17$ fm^{-3} are plotted as a function of the energy shift $\delta\sqrt{s} = E - E_{\rm th} \leq 0$. While the in-medium $\bar{p}p$ amplitude is attractive in the entire energy range, the real part of the in-medium $\bar{p}n$ amplitude is attractive for $\delta\sqrt{s} \leq -70$ MeV with slightly repulsive dip near threshold. The lower panel of Fig. 1 shows the Paris free-space P-wave amplitudes below threshold.

The strong energy dependence of the $\bar{p}N$ amplitudes shown in Fig. 1 calls for proper self-consistent evaluation of the energy shift $\delta\sqrt{s}$. It is expressed using the Mandelstam variable

$$s = (E_N + E_{\bar{p}})^2 - (\vec{p}_N + \vec{p}_{\bar{p}})^2 \;, \tag{6}$$

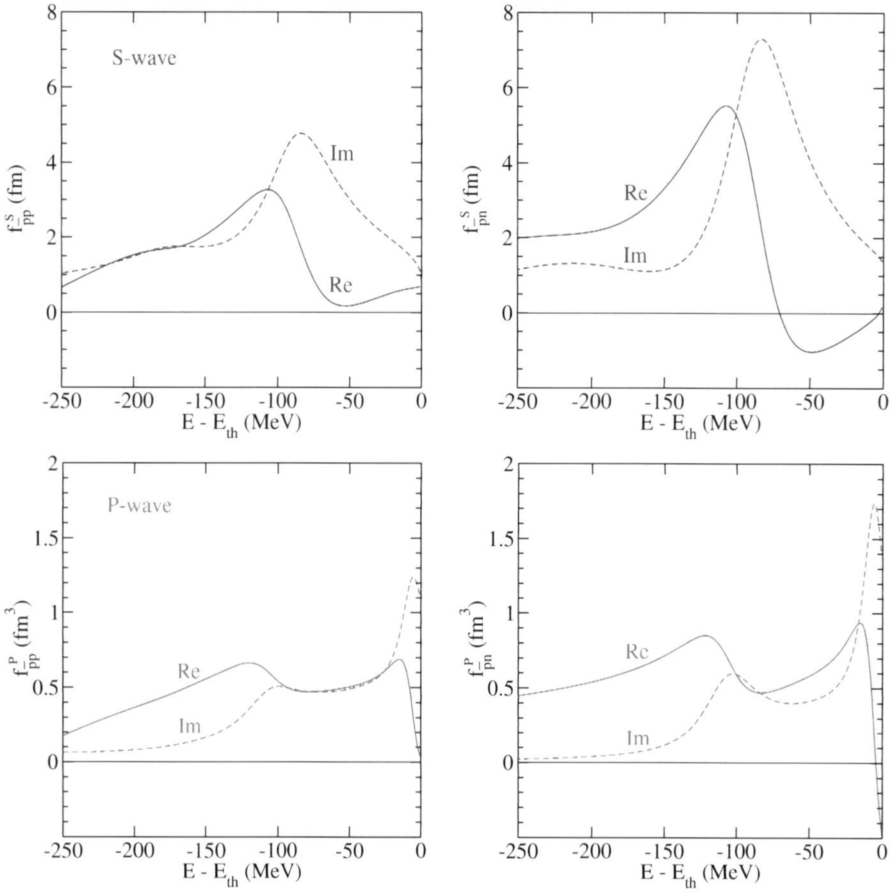

Figure 1. Energy dependence of the Paris 09 $\bar{p}p$ and $\bar{p}n$ two-body cm scattering amplitudes used in the present calculations. Upper panel: in-medium (Pauli blocked) S-wave amplitudes for $\rho_0 = 0.17$ fm^{-3}; lower panel: free-space P-wave amplitudes.

where $E_N = m_N - B_{N\mathrm{av}}$ with $B_{N\mathrm{av}} = 8.5$ MeV being the average binding energy per nucleon. When the interaction of $\bar{p}$ with a nucleon takes place in a nucleus, the momentum dependent term in Eq. (6) does not vanish and gives rise to an additional downward energy shift [14, 15, 16]. The shift $\delta\sqrt{s}$ can then be approximated as

$$\delta\sqrt{s} = E_{\mathrm{th}}\left(1 - \frac{2(B_{\bar{p}} + B_{N\mathrm{av}})}{E_{\mathrm{th}}} + \frac{(B_{\bar{p}} + B_{N\mathrm{av}})^2}{E_{\mathrm{th}}^2} - \frac{T_{\bar{p}}}{E_{\mathrm{th}}} - \frac{T_{N\mathrm{av}}}{E_{\mathrm{th}}}\right)^{1/2} - E_{\mathrm{th}}\ , \quad (7)$$

where the kinetic energy per nucleon $T_{N\mathrm{av}}$ and the $\bar{p}$ kinetic energy $T_{\bar{p}}$ were calculated as the corresponding expectation values of the kinetic energy operator [9].

3. Results

We employed the above introduced in-medium S-wave and free-space P-wave scattering amplitudes derived from the Paris $\bar{N}N$ potential, as well as the phenomenological P-wave potential fitted by Friedman and Gal to $\bar{p}$ atom data [11] and constructed the $S + P$-wave $\bar{p}$-nucleus optical potential [Eq. (2)]. The potential was then used in the self-consistent calculations of $\bar{p}$-nuclear quasi-bound states in selected nuclei across the periodic table.

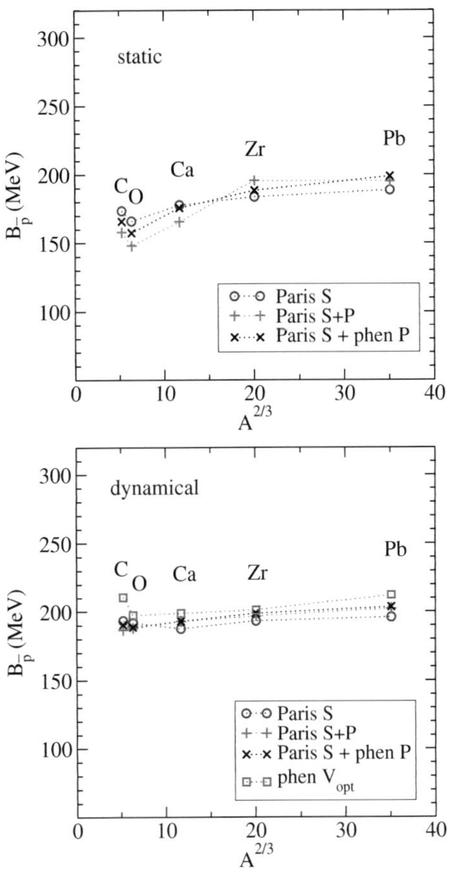

Figure 2. $1s$ $\bar{p}$ binding energies (left) and widths (right) in various nuclei, calculated statically (upper panel) and dynamically (lower panel) using the S-wave Paris potential (blue empty circles), including the Paris P-wave potential (red pluses), and the phenomenological P-wave potential (black crosses). Results for the phenomenological RMF potential from Ref. [16] (phen $V_{\rm opt}$, green empty squares) are shown for comparison.

We performed static, as well as dynamical calculations. While in the static calculations, the nuclear core remains unaffected by the presence of the antiproton, in the dynamical calculations, $\bar{p}$ polarizes the nuclear core, causing changes in the nuclear density distribution. It is to be noted that the response of the nuclear core to $\bar{p}$ is not instant and the characteristic time could be longer than the lifetime of $\bar{p}$ inside a nucleus [17, 18]. As a result, the antiproton probably annihilates before the nuclear core is fully polarized. Present static and dynamical calculations should be considered as two limiting scenarios.

In Fig. 2, we present $1s$ $\bar{p}$ binding energies (left) and widths (right) as a function of mass number A, calculated statically (upper panel) and dynamically (lower panel) with the Paris S-wave (circles), Paris $S+P$-wave (pluses), and Paris S-wave + phen. P-wave (crosses) potentials. The binding energies calculated dynamically, $B_{\bar{p}}^{\rm dyn} \approx 200$ MeV, are somewhat larger than those obtained in static calculations, $B_{\bar{p}}^{\rm stat} \sim 150-200$ MeV; the polarization effects decrease with the mass number A. In dynamical and static calculations alike, both the Paris and phenomenological P-wave interaction terms do not affect much the $\bar{p}$ binding energies which are comparable with the binding energies evaluated using merely the S-wave potential.

On the other hand, the $\bar{p}$ widths calculated dynamically are noticeably larger than the widths calculated statically. It is caused by the increase of the central nuclear density, which exceeds the decrease of the $\bar{p}N$ amplitudes due to the larger energy shift with respect to threshold ($\delta\sqrt{s} \sim -255$ MeV in the dynamical case vs. $\delta\sqrt{s} \sim -200$ MeV in the static case).
The $\bar{p}$ widths exhibit considerably larger dispersion for the different potentials. The Paris S-wave potential yields sizable widths in all nuclei, $\Gamma_{\bar{p}} \geq 250$ MeV. When the P-wave interaction term is included the $\bar{p}$ widths are reduced significantly. The effect is even more pronounced for the Paris P-wave interaction, when the widths $\Gamma_{\bar{p}} \leq 150$ MeV, exhibiting strong A-dependence. On the contrary, the widths calculated with the phenomenological P-wave term [11], as well as only with the S-wave potential vary much less with A.
It is to be noted that the form of the $S + P$-wave potential is a result of delicate interplay between the S- and P-wave parts which are linked together, as well as the balance between the real and imaginary parts of the P-wave amplitudes since they control the range of the potential. It was demonstrated by Friedman and Gal [11] that the Paris S-wave potential supplemented by the contribution of the Paris P-wave amplitudes fails to achieve reasonable fit to $\bar{p}$-atom data. The real and imaginary parts of the Paris P-wave had to be scaled by different factors in order to obtain satisfactory description of the data. We may infer that the real and imaginary parts of the Paris P-wave amplitudes are not well balanced in the energy region relevant to $\bar{p}$-nuclear states calculations.

Finally, Fig. 2 shows that the Paris S-wave + phen. P-wave potential yields very similar $\bar{p}$ binding energies and widths as the phenomenological RMF approach [16]. The widths are in the range $\sim 200 - 230$ MeV, i.e., larger or at least comparable with the corresponding binding energies. The agreement between the phenomenological $\bar{p}$-optical potential [16], parameters of which were tuned to fit the $\bar{p}$-atom data, and the Paris S-wave potential supplemented by the phenomenological P-wave term is quite impressive.

Acknowledgements

We thank E. Friedman and A. Gal for valuable discussions, and B. Loiseau for providing us with the $\bar{N}N$ amplitudes.
This work was supported by the GACR Grant No. P203/15/04301S. J. Hrtánková acknowledges financial support from the CTU-SGS Grant No. SGS16/243/OHK4/3T/14.

References

[1] Batty C J, Friedman E and Gal A 1997 *Phys. Rep.* **287** 385
[2] Friedman E and Gal A 2007 *Phys. Rep.* **452** 89
[3] Friedman E, Gal A and Mareš J 2005 *Nucl. Phys.* A **761** 283
[4] Walker G E, Goodman Ch D and Olmer C (Eds.) 1985 *Antinucleon– and Nucleon–Nucleus Interaction* (New York: Plenum Press)
[5] Klempt E, Batty C, Richard J M 2005 *Phys. Rep.* **413** 197
[6] Bürvenich T J, Greiner W, Mishustin I N, Satarov L M and Stöcker H 2005 *Phys. Rev.* C **71** 035201
[7] *The FAIR facility* [online] URL:<http://www.fair-center.eu/index.php?id=1>
[8] *The GiBUU model* [online] URL:<http://gibuu.physik.uni-giessen.de/GiBUU>
[9] Hrtánková J and Mareš J 2018 *Nucl. Phys.* A **969** 45
[10] El-Bennich B, Lacombe M, Loiseau B and Wycech S 2009 *Phys. Rev.* C **79** 054001
[11] Friedman E, Gal A, Loiseau B and Wycech S 2015 *Nucl. Phys.* A **934** 101
[12] Sharma M M, Nagarajan M A and Ring P 1993 *Phys. Lett.* B **312** 377
[13] Wass T, Rho M and Weise W 1997 *Nucl. Phys.* A **617** 449
[14] Cieplý A, Friedman E, Gal A, Gazda D and Mareš J 2011 *Phys. Lett.* B **702** 402
[15] Friedman E and Gal A 2013 *Nucl. Phys.* A **899** 60
[16] Hrtánková J and Mareš J 2016 *Nucl. Phys.* A **945** 197
[17] Larionov A B, Mishustin I N, Satarov L M and Greiner W 2008 *Phys. Rev.* C **78** 014604
[18] Larionov A B, Mishustin I N, Satarov L M and Greiner W 2010 *Phys. Rev.* C **82** 024602

IOP Conf. Series: Journal of Physics: Conf. Series **1024** (2018) 012018 doi:10.1088/1742-6596/1024/1/012018

Constraints on the hadronic spectrum from Lattice QCD

P Parotto

Department of Physics, University of Houston, Houston, TX 77204, USA

E-mail: pparotto@uh.edu

Abstract. The spectrum of hadronic resonances continually receives updates from the Particle Data Group, which lists every state with a status representing how established the state is. Moreover, the existence of additional states is predicted by relativistic quark models. It has been suggested that further states might need to be included in the hadronic spectrum in order to improve the agreement between the hadron resonance gas model predictions and lattice QCD data. Such an inclusion would also affect the results of many areas of heavy-ion collision physics that make use of hadronic degrees of freedom, such as hydrodynamical simulations afterburners. However, for some selected observables, the inclusion of further states worsens the agreement with the lattice results. We propose new observables, sensitive to the spectrum content divided by quantum numbers, which allow us to gauge the contribution of additional states. The comparison of Lattice QCD results and predictions from the Hadron Resonance Gas model for these observables, helps to clarify the situation and determine how many, and which new states are needed.

1. Introduction

The precision reached by Lattice simulations of QCD thermodynamics has allowed to study with increasing level of detail the properties of strongly interacting matter. Observables like fluctuations of conserved charges [1, 2], sensitive to the chemical composition of a strongly interacting system can now be calculated with sufficient precision, allowing for an in-depth analysis of the relevant degrees of freedom near the QCD transition [3, 4].

For temperatures below the QCD transition, in the hadronic phase, most Lattice results can be reproduced successfully by the Hadron Resonance Gas (HRG) model, whose underlying assumption is that a gas of interacting hadrons in the ground state can be well approximated by a gas on non-interacting hadrons and resonances. The original idea of Rolf Hagedorn, was that a hadronic system responds to an arbitrary increase in energy density, not through and indefinite increase in the temperature, but rather populating the system via the creation of resonances [5]. Thus, the concept of a limiting temperature for strongly interacting matter naturally arises, which now is well understood to be the limiting temperature for *hadronic* matter, hence connected to the transition temperature of QCD [6–8]. For this reason, it is reasonable to expect the effects of resonance proliferation to become more relevant at the occurrence of such transition.

The HRG model containing no interaction, provides the most advantageous setup, being the evaluation of any thermodynamic quantity a simple sum over the hadron and resonance spectrum of the theory. In this sense, the model is effectively parameter free, and depending

IOP Conf. Series: Journal of Physics: Conf. Series **1024** (2018) 012018 doi:10.1088/1742-6596/1024/1/012018

only on the content of the spectrum of hadronic states that is included in the sum. However, the content of the hadronic spectrum, especially in the strange sector, is far from being established. The Particle Data Group (PDG) continuously updates the list of experimentally discovered resonances [9], and the existence of each resonance is assigned a confidence level, where **** indicate the most established, and * indicates those with the least experimental confirmation.

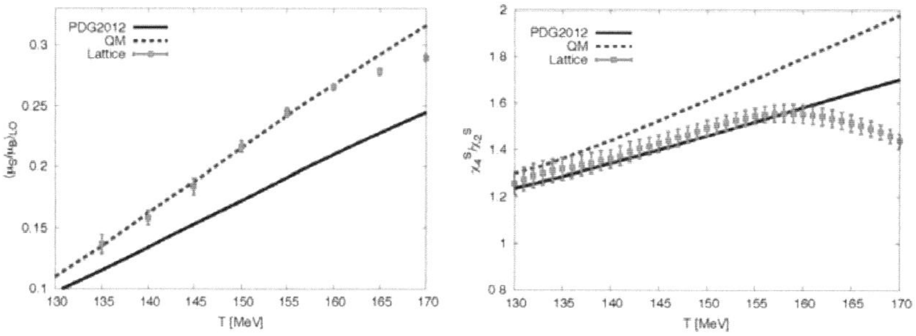

Figure 1. Comparison of Lattice QCD and HRG model calculations based on the PDG2012 (solid black line) and QM (dashed red line) lists, for the ratio μ_S/μ_B at leading order (left) and for the ratio χ_4^S/χ_2^S (right).

With the increasing precision achieved in Lattice QCD calculations of more differential observables, some discrepancy with the HRG results appeared [10]. It was suggested that the inclusion of additional states in the spectrum, not yet experimentally discovered, but predicted by relativistic quark models (QM) [11–13] or Lattice QCD spectroscopy, might be necessary and cure the problem. However, the inclusion of additional states in the spectrum not always improves the agreement between HRG predictions and Lattice results [14], as can be seen in Fig.1.

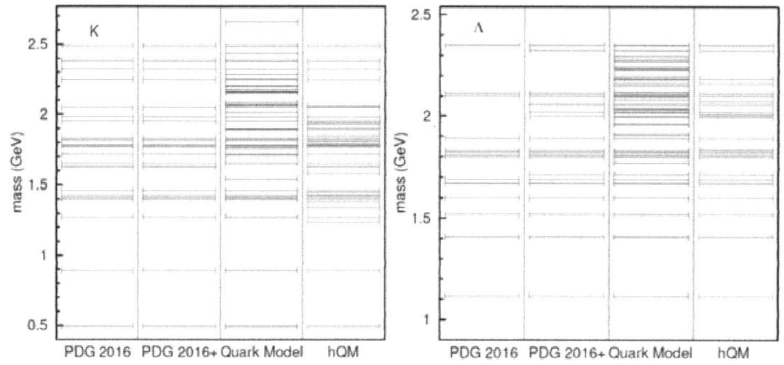

Figure 2. Strange mesonic states and Lambda resonances experimentally established in the PDG2016 (green), PDG2016 including also one star states (red) and predicted by the QM (blue) and the hQM (magenta).

Therefore, a more systematic analysis of the content of the hadronic spectrum is needed, taking into consideration different possible choices. In Fig.2 we compare, for illustrative reasons, the strange meson and Lambda states content of different lists: PDG2016 [9] (all PDG states with **, *** and ****), PDG2016+ [9] (including also * states), the original Quark Model (QM) [11, 12] and a more recent and refined hypercentral Quark Model [13] (hQM), where an interaction is included between the constituent quarks in the baryons. For hadrons containing only u, d and s quarks, including both particles and anti-particles, as well as isospin multiplicity, the number of states for the lists extracted from the PDG are 608 in the case of PDG2016 and 738 in the case of PDG2016+ respectively. A large increase in the number of states comes from including QM resonances: the QM list contains 1517 states, whereas the hQM one contains 985.

In the following, comparison between HRG and Lattice QCD results for several strangeness related observables is performed [14], which will depict a rather clear picture of whether, how many, and which states are most likely still missing in the hadronic spectrum established from experiment.

2. HRG and Lattice: Partial pressures

In the formulation of the HRG with no interactions, thermodynamical quantities can be calculated in the grand-canonical ensemble as a function of temperature and baryonic chemical potential [15]. The expression for the total pressure of the gas is

$$P_{tot}(T,\mu) = \sum_k (-1)^{B_k+1} \frac{d_k T}{(2\pi)^3} \int d^3\vec{p} \ln\left(1 + (-1)^{B_k+1} \exp\left[-\frac{(\sqrt{\vec{p}^2 + m_k^2} - \mu_k)}{T}\right]\right),$$

where the sum runs over all hadrons and resonances included in the spectrum. Every quantity carrying the index k depends on the specific state, hence on the specific term in the sum. The single particle chemical potential is defined with respect to the conserved charges of strong interactions (baryon number B, strangeness S and electric charge Q) as $\mu_k = B_k\mu_B + S_k\mu_S + Q_k\mu_Q$.

In order to systematically analyze the different hadron spectra content, we separate the contribution to the pressures from hadrons grouped according to their quantum numbers. Under the assumption that Boltzmann approximation is valid for QCD thermodynamics in the range of temperatures below the QCD transition, it is possible to write the pressure of the system as

$$\begin{aligned} P(T,\hat{\mu}_B,\hat{\mu}_S) = {} & P_{00}^{BS}(T) + P_{10}^{BS}(T)\cosh(\hat{\mu}_B) + P_{01}^{BS}(T)\cosh(\hat{\mu}_S) + P_{11}^{BS}(T)\cosh(\hat{\mu}_B - \hat{\mu}_S) + \\ & + P_{12}^{BS}(T)\cosh(\hat{\mu}_B - 2\hat{\mu}_S) + P_{13}^{BS}(T)\cosh(\hat{\mu}_B - 3\hat{\mu}_S), \end{aligned}$$

where $\hat{\mu}_i = \mu_i/T$ and in $P_{ij}^{BS}(T)$, i and j refer to the absolute values of baryon number and strangeness respectively. The partial pressures can be calculated on the Lattice using an imaginary chemical potential approach. Indeed, setting $\mu_B = 0$ and $\mu_S = i\mu_I$, the previous expression for the pressure becomes a sum of cosine terms, with the partial pressures being the Fourier coefficients. A combined fit of different observables yields the partial pressures as a function of the temperature. On the other hand, calculating the partial pressures with the HRG model is straightforward: it will be sufficient to sum over particles with the corresponding quantum numbers only.

3. Results and discussion

We show the comparison of Lattice QCD results and HRG model predictions using PDG2016, PDG2016+, QM and hQM lists, for the partial pressures of $|S| = 0$,$|S| = 1$,$|S| = 2$ and $|S| = 3$ baryons, as well as $|S| = 1$ mesons. Note that, the pressure being a positive definite quantity, a prediction from the HRG model that underestimates the Lattice data in a certain sector, for a specific hadron list, would directly imply that the list does not contain enough states in the specific sector considered. Also, if the prediction overestimates the data, then it will follow that too many states are included in such a list.

In Fig.3 we can see the partial pressure for different baryonic sectors. The first thing that catches the eye is that, except for the case on non-strange baryons, the most established states from the PDG are not enough to describe the data. For $|S| = 1$ and $|S| = 2$ baryons, the

IOP Conf. Series: Journal of Physics: Conf. Series **1024** (2018) 012018 doi:10.1088/1742-6596/1024/1/012018

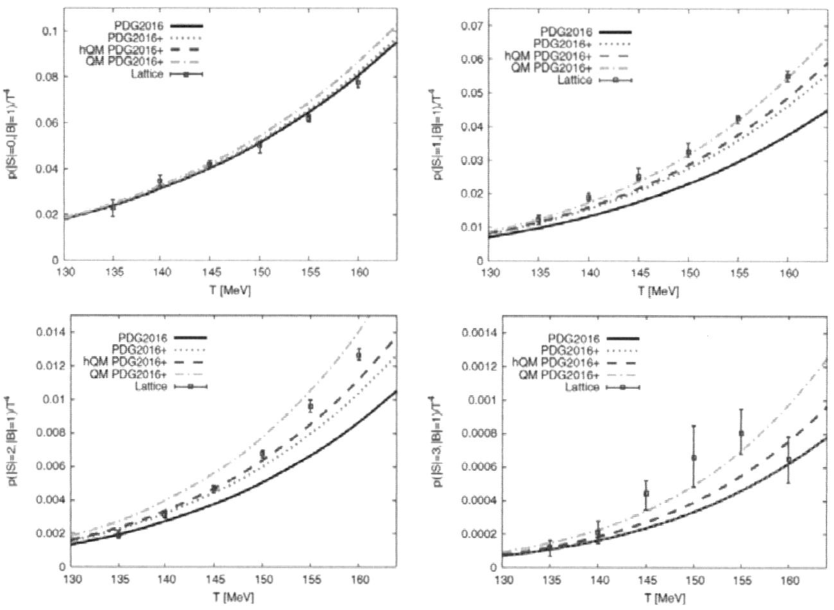

Figure 3. Comparison of Lattice QCD results and HRG model calculations based on the PDG2016 (solid black line), PDG2016+ (dashed red line), hQM (dashed blue line) and QM (dot dashed light blue line) lists, for the partial pressures of $|S| = 0$ (top left), $|S| = 1$ (top right), $|S| = 2$ (bottom left) and $|S| = 3$ (bottom right) baryons. All Lattice results are continuum extrapolated.

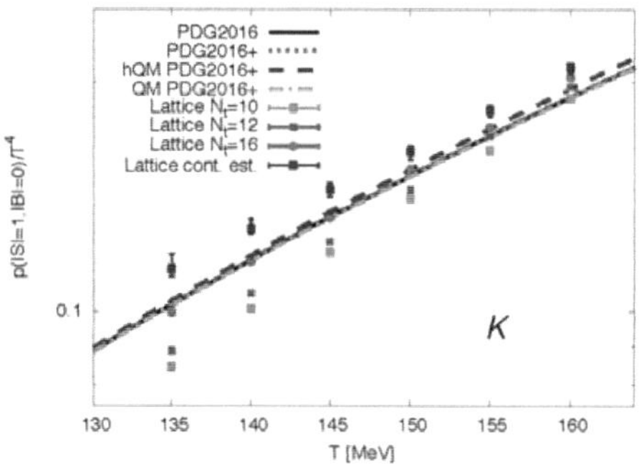

Figure 4. Comparison of Lattice QCD results and HRG model calculations based on the PDG2016 (solid black line), PDG2016+ (dashed red line), hQM (dashed blue line) and QM (dot dashed light blue line) lists, for the partial pressures of $|S| = 1$ mesons. The black dots indicate the continuum estimate.

inclusion of * states sensibly improves the agreement with the data. When futher states from the hQM list are included, an additional slight improvement is noticeable in all the cases. For $|S| = 1$ and $|S| = 3$ baryons, however, it still seems that states predicted by the QM are needed to reproduce the data; on the other hand, these are clearly too many in the case of $|S| = 2$ baryons.

Different considerations need to be made about the strange mesons in Fig.4. First, data are shown for different lattice spacings, and a continuous estimate is shown in place of a continuum extrapolation, which was not achievable being the data not in a clear scaling regime. Secondly, it is remarkable that all the lists seem to underpredict the value of this partial pressure. This may be due to two different reasons: either all the lists do not contain enough states in this sector, or the assumption of the ideal HRG model may not be applicable in the case of strange mesons.

4. Conclusions

In this proceedings we showed the analysis of different hadronic spectra through the comparison of Lattice QCD results and HRG model predictions. In the baryonic sector, it was rather evident that the list of established states listed by the PDG is not sufficient, and additional states are needed for all the families, in a couple cases even up to those from the original quark model. In the mesonic sector, it seemed that additional states are needed in all the considered spectra, if one trusts that the description of an ideal HRG model holds in the case of strange mesons too.

Through the presented analysis, it was possible to discern to what extent, and in what quantum number sectors, additional states need to be included in the hadronic spectra to those listed as established by the Particle Data Group. This can impact any analysis of the QCD transition in heavy-ion collisions that makes use of thermal models, such as the extraction of freeze-out parameters through thermal fits [16, 17], or ratios of fluctuations of conserved charges [1, 2, 15], especially in the study of strangeness freeze-out [3, 18].

Acknowledgements

This material is based upon work supported by the National Science Foundation under Grants No. PHY-1513864, PHY-1654219 and OAC-1531814 and by the U.S. Department of Energy, Office of Science, Office of Nuclear Physics, within the framework of the Beam Energy Scan Theory (BEST) Topical Collaboration. An award of computer time was provided by the INCITE program. This research used resources of the Argonne Leadership Computing Facility, which is a DOE Office of Science User Facility supported under Contract No. DE-AC02-06CH11357. The authors gratefully acknowledge the use of the Maxwell Cluster and the advanced support from the Center of Advanced Computing and Data Systems at the University of Houston.

References

[1] Borsanyi S, Fodor Z, Katz S D, Krieg S, Ratti C and Szabo K 2012 *JHEP* **01** 138 (*Preprint* 1112.4416)
[2] Bazavov A *et al.* (HotQCD) 2012 *Phys. Rev.* **D86** 034509 (*Preprint* 1203.0784)
[3] Bellwied R, Borsanyi S, Fodor Z, Katz S D and Ratti C 2013 *Phys. Rev. Lett.* **111** 202302 (*Preprint* 1305.6297)
[4] Bazavov A *et al.* 2013 *Phys. Rev. Lett.* **111** 082301 (*Preprint* 1304.7220)
[5] Hagedorn R 1965 *Nuovo Cim. Suppl.* **3** 147–186
[6] Cabibbo N and Parisi G 1975 *Phys. Lett.* **59B** 67–69
[7] Aoki Y, Fodor Z, Katz S D and Szabo K K 2006 *Phys. Lett.* **B643** 46–54 (*Preprint* hep-lat/0609068)
[8] Bazavov A *et al.* 2012 *Phys. Rev.* **D85** 054503 (*Preprint* 1111.1710)
[9] Patrignani C *et al.* (Particle Data Group) 2016 *Chin. Phys.* **C40** 100001
[10] Bazavov A *et al.* 2014 *Phys. Rev. Lett.* **113** 072001 (*Preprint* 1404.6511)
[11] Capstick S and Isgur N 1986 *Phys. Rev.* **D34** 2809 [AIP Conf. Proc.132,267(1985)]
[12] Ebert D, Faustov R N and Galkin V O 2009 *Phys. Rev.* **D79** 114029 (*Preprint* 0903.5183)
[13] Ferretti J, Bijker R, Galatà G, García-Tecocoatzi H and Santopinto E 2016 *Phys. Rev.* **D94** 074040 (*Preprint* 1506.07469)
[14] Alba P *et al.* 2017 *Phys. Rev.* **D96** 034517 (*Preprint* 1702.01113)
[15] Bluhm M, Alba P, Alberico W, Bellwied R, Mantovani Sarti V, Nahrgang M and Ratti C 2014 *Nucl. Phys.* **A931** 814–819 (*Preprint* 1408.4734)
[16] Andronic A, Braun-Munzinger P, Redlich K and Stachel J 2011 *J. Phys.* **G38** 124081 (*Preprint* 1106.6321)
[17] Cleymans J, Oeschler H, Redlich K and Wheaton S 2006 *Phys. Rev.* **C73** 034905 (*Preprint* hep-ph/0511094)
[18] Noronha-Hostler J, Bellwied R, Gunther J, Parotto P, Pasztor A, Vazquez I P and Ratti C 2016 (*Preprint* 1607.02527)

IOP Conf. Series: Journal of Physics: Conf. Series **1024** (2018) 012019 doi:10.1088/1742-6596/1024/1/012019

Extraction of Polarization Parameters in the $\bar{\rm p}{\rm p} \rightarrow \bar{\Omega}\Omega$ Reaction

E Perotti[1]

[1] Institutionen för fysik och astronomi, Uppsala Universitet, Box 516, S-75120 Uppsala, SE

E-mail: elisabetta.perotti@physics.uu.se

Abstract. A method to extract the polarization of Ω hyperons produced via the strong interaction is presented. Assuming they are spin 3/2 particles, the corresponding spin density matrix can be written in terms of seven non-zero polarization parameters, all retrievable from the angular distribution of the decay products. Moreover by considering the full decay chain $\Omega \rightarrow \Lambda K \rightarrow p\pi K$ the magnitude of the asymmetry parameters β_Ω and γ_Ω can be obtained. This method, applied here to the specific Ω case, can be generalized to any weakly decaying hyperon and is perfectly suited for the PANDA experiment where hyperon-antihyperon pairs will be copiously produced in proton-antiproton collisions. The aim is to take a step forward towards the understanding of the mechanism that reigns strangeness production in these processes.

1. Introduction

The strangeness creation mechanism remains an open question: how are light quarks replaced by heavier ones? Spin observables allow for much deeper insights than just production cross sections. So far several theoretical models have been proposed in order to describe the production of hyperons, often leading to disagreement among each other and with experimental data (see [1] and references therein). Colliding protons with antiprotons, one has direct access to hyperon-antihyperon pair production. Hyperons do contain at least one strange quark. In the following we will only focus on the $p\bar{p} \rightarrow \Omega\bar{\Omega}$ case, which involves triple-strangeness baryons and the annihilation of all light quarks in the initial system. It will be studied for the first time at the upcoming PANDA experiment. One of the spin observables that could help us understand how strangeness is created starting from strange-free objects is the polarization of the hyperon. Luckily it can be retrieved experimentally without the need of a polarized beam or target. One exploits the fact that hyperons are weakly decaying – parity violating! – particles and therefore their decay products are not isotropically distributed but are emitted along a preferred direction. In brief, from the angular distribution of the decay products one can extract the polarization of the mother hyperon. In weak decays more than one partial wave is populated, i.e. both the one violating parity conservation and the one respecting it. The interference between the two gives rise to the angular dependence in the decay distribution. This is the concrete meaning of the expression (often found in the literature): the weak decays of hyperons are self-analyzing. Exploiting this property we present a method that allows to extract the polarization of the Ω hyperon. The Uppsala group is currently performing simulation studies that will complement the material presented here [2].

2. The Ω spin density matrix

This section is entirely based on the work done by Doncel *et al.* [3, 4]. The reader is referred to those papers for more details. The density matrix of a particle with spin j can be written in terms of the polarization parameters r^L_M as

$$\rho = \frac{1}{2j+1}\mathcal{I} + \sum_{L=1}^{2j} \rho^L \tag{1}$$

with

$$\rho^L = \frac{2j}{2j+1} \sum_{M=-L}^{L} Q^L_M r^L_M \tag{2}$$

where $\mathcal{I}$ is the identity matrix and Q^L_M is a set of hermitian matrices. The sums are over the orbital angular momentum L and its projection M. It follows that the spin density matrix for a particle with $j = 3/2$ depends on fifteen polarization parameters r^L_M. However, for a particle produced via the strong interaction, as it is in the $\bar{p}p \to \bar{\Omega}\Omega$ reaction, parity conservation imposes symmetries on the density matrix [5] so that eight out of the fifteen polarization parameters have to be zero. Therefore the density matrix reduces to

$$\rho(3/2) = \frac{1}{4}\begin{bmatrix} 1+\sqrt{3}r^2_0 & -i\frac{3}{\sqrt{5}}r^1_{-1}+\sqrt{3}r^2_1 -i\sqrt{\frac{6}{5}}r^3_{-1} & \sqrt{3}r^2_2 - i\sqrt{3}r^3_{-2} & -i\sqrt{6}r^3_{-3} \\ i\frac{3}{\sqrt{5}}r^1_{-1}+\sqrt{3}r^2_1 +i\sqrt{\frac{6}{5}}r^3_{-1} & 1-\sqrt{3}r^2_0 & -i2\sqrt{\frac{3}{5}}r^1_{-1}+i3\sqrt{\frac{2}{5}}r^3_{-1} & \sqrt{3}r^2_2 + i\sqrt{3}r^3_{-2} \\ \sqrt{3}r^2_2 + i\sqrt{3}r^3_{-2} & i2\sqrt{\frac{3}{5}}r^1_{-1}-i3\sqrt{\frac{2}{5}}r^3_{-1} & 1-\sqrt{3}r^2_0 & -i\frac{3}{\sqrt{5}}r^1_{-1}+\sqrt{3}r^2_1 -i\sqrt{\frac{6}{5}}r^3_{-1} \\ i\sqrt{6}r^3_{-3} & \sqrt{3}r^2_2 - i\sqrt{3}r^3_{-2} & i\frac{3}{\sqrt{5}}r^1_{-1}+\sqrt{3}r^2_1 +i\sqrt{\frac{6}{5}}r^3_{-1} & 1+\sqrt{3}r^2_0 \end{bmatrix}. \tag{3}$$

In other words the Ω polarization is obtained by measuring the following seven parameters: r^3_{-3}, r^3_{-2}, r^3_{-1}, r^2_2, r^2_1, r^2_0 and r^1_{-1}. These can be retrieved from the angular distributions of the decay products of the Ω and its daughter Λ hyperon, as will be shown in the following. We note in passing that for a spin 1/2 hyperon eq.(3) reduces to a 2×2 matrix depending on a single parameter.

3. Extraction of polarization parameters from decay angular distributions

3.1. Single decay: $\Omega \to \Lambda K$

The normalized decay T matrix introduced below acts on the density matrix according to $\rho_{\text{fin}} = T\rho_{\text{in}}T^{\dagger}$. Conservation of total spin implies that the orbital angular momentum of the final state ΛK can be $L = 1$ or 2, with parity given by $(-1)^L$. Since this is a parity violating weak decay, both the P- and D-wave are allowed and the components of the T matrix can be written in terms of their amplitudes:

$$T(3/2 \to 1/2, 0) = \frac{1}{\sqrt{8\pi}} \begin{bmatrix} \begin{array}{c} -\sqrt{3}(T_d + T_p) \\ e^{i\phi} \sin\theta \cos\frac{\theta}{2} \end{array} & \begin{array}{c} \frac{1}{2}(T_d + T_p) \\ (\cos\frac{\theta}{2} + 3\cos\frac{3\theta}{2}) \end{array} & \begin{array}{c} -\frac{1}{2}(T_d + T_p)e^{-i\phi} \\ (\sin\frac{\theta}{2} - 3\sin\frac{3\theta}{2}) \end{array} & \begin{array}{c} \sqrt{3}(T_d + T_p) \\ e^{-2i\phi} \sin\theta \sin\frac{\theta}{2} \end{array} \\ \begin{array}{c} -\sqrt{3}(T_d - T_p) \\ e^{2i\phi} \sin\theta \sin\frac{\theta}{2} \end{array} & \begin{array}{c} \frac{1}{2}(-T_d + T_p)e^{i\phi} \\ (\sin\frac{\theta}{2} - 3\sin\frac{3\theta}{2}) \end{array} & \begin{array}{c} -\frac{1}{2}(T_d - T_p) \\ (\cos\frac{\theta}{2} + 3\cos\frac{3\theta}{2}) \end{array} & \begin{array}{c} \sqrt{3}(-T_d + T_p) \\ e^{-i\phi} \sin\theta \cos\frac{\theta}{2} \end{array} \end{bmatrix}. \tag{4}$$

The decay angles θ, ϕ of the Λ hyperon are measured in the Ω rest frame, where the xz-plane constitutes the Ω production plane:

$$\hat{x} = \hat{y} \times \hat{z} \qquad \hat{y} = \frac{\vec{p}_\Omega \times \vec{p}_{beam}}{|\vec{p}_\Omega \times \vec{p}_{beam}|} \qquad \hat{z} = \hat{p}_\Omega. \tag{5}$$

Here we adopt the Jacob-Wick decomposition in partial waves [6] which refers to helicities. This choice avoids complications when considering also the subsequent decay of the Λ hyperon.

The angular distribution is then obtained by taking the trace of the final state density matrix [7]:

$$I = \mathrm{Tr}(\rho_{\mathrm{fin}}) = \mathrm{Tr}(T\rho_{\mathrm{in}}T^\dagger) = \mathrm{Tr}(\rho_{\mathrm{in}}T^\dagger T). \tag{6}$$

The normalized P- and D-amplitudes are combined into the following parameters:

$$\begin{aligned} \alpha_\Omega &= 2\mathrm{Re}(T_d^* T_p) \\ \beta_\Omega &= 2\mathrm{Im}(T_d^* T_p) \\ \gamma_\Omega &= |T_d|^2 - |T_p|^2 \end{aligned} \tag{7}$$

which quantify the asymmetries in the decay angular distribution. By construction they fulfill $\alpha^2 + \beta^2 + \gamma^2 = |T_p|^2 + |T_d|^2 = 1$. The angular distribution then reads:

$$\begin{aligned} I(\theta, \phi) &= \mathrm{Tr}(\rho(3/2)T^\dagger T(3/2 \to 1/2, 0)) \\ &= \frac{1}{4\pi}\Bigg[1 + \frac{\sqrt{3}}{2}(1 - 3\cos^2\theta)r_0^2 - \frac{3}{2}\sin^2\theta\cos 2\phi\, r_2^2 - \frac{3}{2}\sin 2\theta\cos\phi\, r_1^2 \\ &+ \frac{1}{40}\alpha_\Omega \sin\theta\Big(8\sqrt{15}r_{-1}^1 \sin\phi - 9\sqrt{10}r_{-1}^3(3 + 5\cos 2\theta)\sin\phi \\ &- 30(3r_{-2}^3 \sin 2\phi \sin 2\theta + \sqrt{6}r_{-3}^3 \sin 3\phi \sin^2\theta)\Big)\Bigg]. \end{aligned} \tag{8}$$

Of the three asymmetry parameters only the α_Ω parameter shows up in the angular distribution of the Ω decay, whereas the β_Ω and γ_Ω parameters will appear when looking also into the subsequent decay of the daughter Λ hyperon. The asymmetry parameter α_Ω has been measured experimentally for the $\Omega \to \Lambda K$ decay ($\alpha_\Omega = 0.0180 \pm 0.0024$ [8]), while β_Ω and γ_Ω are unknown. Note that all seven non-zero polarization parameters show up in eq.(8) but only three of them are not multiplied by the parameter α_Ω. For the moment we focus on these ones only. They can be retrieved if the following expectation values are measured:

$$\begin{aligned} \langle \sin\theta_\Lambda \rangle &= \int I(\theta_\Lambda, \phi_\Lambda) \times \sin\theta_\Lambda \, d\Omega_\Lambda = \frac{\pi}{32}(8 + \sqrt{3}r_0^2) \\ \langle \cos\phi_\Lambda \cos\theta_\Lambda \rangle &= \int I(\theta_\Lambda, \phi_\Lambda) \times \cos\phi_\Lambda \cos\theta_\Lambda \, d\Omega_\Lambda = -\frac{3\pi}{32}r_1^2 \\ \langle \sin^2\phi_\Lambda \rangle &= \int I(\theta_\Lambda, \phi_\Lambda) \times \sin^2\phi_\Lambda \, d\Omega_\Lambda = \frac{1}{4}(2 + r_2^2) \end{aligned} \tag{9}$$

IOP Conf. Series: Journal of Physics: Conf. Series **1024** (2018) 012019 doi:10.1088/1742-6596/1024/1/012019

where $d\Omega_\Lambda = \sin\theta_\Lambda d\theta_\Lambda d\phi_\Lambda$ and $\theta_\Lambda \in [0,\pi]$, $\phi_\Lambda \in [0,2\pi]$. This technique is known as the Method of Moments [9]. We have slightly modified the angle notation ($\theta \to \theta_\Lambda$, $\phi \to \phi_\Lambda$) to clearly indicate that these angles belong to the first decay, i.e. $\Omega \to \Lambda K$. Similar expressions can be obtained also for the remaining four r's but we postpone their derivation to the next section.

3.2. *Full decay chain: $\Omega \to \Lambda K \to p\pi K$*

At this point four non-zero polarization parameters remain to be extracted. The reason why we have not done it so far is that we do not want the small parameter α to appear in their expressions. We expect to gain additional information by studying the angular distribution of the subsequent decay of the daughter Λ hyperon. This angular distribution can be obtained by letting the $\tilde{T}$ matrix for the $\Lambda \to p\pi$ decay act on ρ_{fin}, i.e.

$$I = \text{Tr}(\tilde{T}\rho_{\text{fin}}\tilde{T}^\dagger) = \text{Tr}(\tilde{T}T\rho T^\dagger\tilde{T}^\dagger) = \text{Tr}(\rho T^\dagger\tilde{T}^\dagger\tilde{T}T) \tag{10}$$

with T and $\tilde{T}$ referring respectively to the first and second decay. For the $\Lambda \to p\pi$ weak decay, conservation of total spin implies $L = 0$ or 1 in the final state. Since this is a parity violating decay the $\tilde{T}$ matrix is given in terms of both the S- and P-amplitudes:

$$\tilde{T}(1/2 \to 1/2, 0) = \frac{1}{\sqrt{4\pi}} \begin{bmatrix} (\tilde{T}_s + \tilde{T}_p)(\cos\frac{\theta_p}{2}) & (\tilde{T}_s + \tilde{T}_p)e^{-i\phi_p}(\sin\frac{\theta_p}{2}) \\ (-\tilde{T}_s + \tilde{T}_p)e^{i\phi_p}(\sin\frac{\theta_p}{2}) & (\tilde{T}_s - \tilde{T}_p)(\cos\frac{\theta_p}{2}) \end{bmatrix}. \tag{11}$$

Following our notation, the angles in (11) carry p as subscript, to indicate that they belong to the second decay, i.e. $\Lambda \to p\pi$. They are measured in the rest system of the daughter Λ hyperon, defined by:

$$\hat{x}' = \hat{y}' \times \hat{z}' \qquad \hat{y}' = \frac{\vec{p}_\Lambda \times \hat{y}}{|\vec{p}_\Lambda \times \hat{y}|} \qquad \hat{z}' = \hat{p}_\Lambda \tag{12}$$

with $\hat{y}$ perpendicular to the Ω production plane. The S- and P-amplitudes are combined into the asymmetry parameters according to (7), by simply replacing the D-amplitude with the S-amplitude. Calculating the trace in (10) and integrating over the angles of the first decay one gets the angular distribution of the second decay ($\alpha_\Lambda = 0.642 \pm 0.013$ [8]):

$$I(\theta_p, \phi_p) = \frac{1}{4\pi}\left[1 + \alpha_\Omega\alpha_\Lambda\cos\theta_p - \alpha_\Lambda\left(\sqrt{\frac{3}{5}}r^1_{-1} - \frac{1}{2\sqrt{10}}r^3_{-1}\right)(\beta_\Omega\cos\phi_p + \gamma_\Omega\sin\phi_p)\sin\theta_p\right] \tag{13}$$

where a linear combination of two polarization parameters shows up. Moreover also β_Ω and γ_Ω appear for the first time.

The full expression for the joint angular distribution $I(\phi_\Lambda, \phi_\Lambda, \theta_p, \phi_p)$ obtained from (10) depends on all four angles of the two decays. It contains the asymmetry parameters α_Ω, β_Ω, γ_Ω, α_Λ and all seven polarization parameters. Using the Method of Moments as before one can access the remaining four polarization parameters:

$$\begin{aligned}
\langle \sin\phi_\Lambda \cos\phi_\mathrm{p}\rangle &= \int I(\theta_\Lambda,\phi_\Lambda,\theta_\mathrm{p},\phi_\mathrm{p}) \times \sin\phi_\Lambda\cos\phi_\mathrm{p}\, \mathrm{d}\Omega_\Lambda \mathrm{d}\Omega_\mathrm{p} = -\frac{3\pi^2\alpha_\Lambda\gamma_\Omega r^3_{-2}}{1024} \\
\langle (3\cos\theta_\Lambda - 1)\sin\phi_\mathrm{p}\rangle &= \int I(\theta_\Lambda,\phi_\Lambda,\theta_\mathrm{p},\phi_\mathrm{p}) \times (3\cos\theta_\Lambda - 1)\sin\phi_\mathrm{p}\, \mathrm{d}\Omega_\Lambda \mathrm{d}\Omega_\mathrm{p} = -\frac{\pi\alpha_\Lambda\gamma_\Omega r^3_{-1}}{4\sqrt{10}} \\
\langle \sin\phi_\mathrm{p}\rangle &= \int I(\theta_\Lambda,\phi_\Lambda,\theta_\mathrm{p},\phi_\mathrm{p}) \times \sin\phi_\mathrm{p}\, \mathrm{d}\Omega_\Lambda \mathrm{d}\Omega_\mathrm{p} = \frac{\pi\alpha_\Lambda\gamma_\Omega}{160}\left(-4\sqrt{15}r^1_{-1} + \sqrt{10}r^3_{-1}\right) \\
\langle \sin\phi_\Lambda\cos\phi_\Lambda\cos\phi_\mathrm{p}\rangle &= \int I(\theta_\Lambda,\phi_\Lambda,\theta_\mathrm{p},\phi_\mathrm{p}) \times \sin\phi_\Lambda\cos\phi_\Lambda\cos\phi_\mathrm{p}\, \mathrm{d}\Omega_\Lambda \mathrm{d}\Omega_\mathrm{p} \\
&= \frac{\pi\alpha_\Lambda\gamma_\Omega}{640}\left(5\sqrt{6}r^3_{-3} + 4\sqrt{15}r^1_{-1} - \sqrt{10}r^3_{-1}\right).
\end{aligned} \tag{14}$$

Inverting these expressions, the polarization parameters can be extracted. Moreover the ratio β/γ can be reconstructed for example from

$$\frac{\beta_\Omega}{\gamma_\Omega} = \frac{\langle\cos\phi_\mathrm{p}\rangle}{\langle\sin\phi_\mathrm{p}\rangle} \tag{15}$$

or by directly fitting the angular distribution (13) to the function

$$a + b\sin\theta_\mathrm{p}\cos\phi_\mathrm{p} + c\sin\theta_\mathrm{p}\sin\phi_\mathrm{p} \tag{16}$$

and then looking at the ratio $b/c = \beta_\Omega/\gamma_\Omega$. Recalling that the asymmetry parameters satisfy $\alpha^2 + \beta^2 + \gamma^2 = 1$ and the value of α_Ω is known, one can for the first time determine up to a sign γ_Ω and β_Ω which is used for testing CP violation in hyperon weak decays [10, 11]. It follows that also the last four polarization parameters (14) can only be determined up to a sign since γ_Ω appears in their expressions. It is however enough to know their absolute value to get the total polarization of the Ω hyperon defined by

$$d(\rho) = \sqrt{\sum_{L=1}^{2j}\sum_{M=-L}^{L}(r^L_M)^2}\quad . \tag{17}$$

Acknowledgments

This work builds heavily on E. Thomé's PhD thesis [12]. It has been supported by the Swedish Research Council. I thank K. Schönning and S. Leupold for their guidance through this work. I also acknowledge valuable discussions with A. Pilloni.

References

[1] Ortega P G, Entem D R and Fernandez F 2011 *Phys. Lett.* B **696** 352–358
[2] Ikegami Andersson W, Perotti E, Leupold S, Schönning K and the PANDA collaboration 2017 *In preparation*
[3] Doncel M G, Mery P, Michel L, Minnaert P and Wali K C 1973 *Phys. Rev.* D **7** 815–835
[4] Doncel M G, Michel L and Minnaert P 1972 *Nucl. Phys.* B **38** 477–528
[5] Pilkuhn H 1967 *The Interactions of Hadrons* (Amsterdam: North-Holland)
[6] Jacob M and Wick G C 1959 *Annals Phys.* **7** 404–428
[7] Koch W ed Nikolic M 1968 *Analysis of Scattering and Decay* (New York-London-Paris: Gordon and Breach)
[8] Patrignani C *et al.* (Particle Data Group) 2016 *Chin. Phys.* C **40** 100001
[9] Frodesen A G *et al.* 1979 *Probability and Statistics in Particle Physics* (Universitetsforlaget)
[10] Donoghue J F, Holstein B R and Valencia G 1986 *Phys. Lett.* B **178** 319–323
[11] He X G, Steger H and Valencia G 1991 *Phys. Lett.* B **272** 411–418
[12] Thomé E 2012 *Multi-Strange and Charmed Antihyperon-Hyperon Physics for PANDA* Ph.D. thesis Uppsala University

The NLO Chiral Lagrangian from the meson-baryon interaction in the $S = -1$ sector.

A Feijoo and V K Magas and A Ramos

Departament de Física Quàntica i Astrofísica and Institut de Ciències del Cosmos, Universitat de Barcelona, Martí Franquès 1, E08028 Barcelona, Spain

E-mail: feijoo@fqa.ub.edu

Abstract. The present study consists of establishing the role played by the next-to-leading order (NLO) terms as well as the Born terms of a chiral $SU(3)$ Lagrangian in the $S = -1$ meson-baryon interaction. This necessarily leads to the determination of the low energy constants present in the NLO contribution, which have not been fixed so far. In order to constrain them, we employ scattering data from the K^-p interaction, paying a special attention to $K^-p \to K\Xi$ reactions and to isospin filtering processes such as $K^-p \to \eta\Lambda, \eta\Sigma^0$. In addition, the stability and the accuracy of all the parameters involved in the model have been examined by means of the inclusion of high spin hyperonic resonances.

1. Theoretical background and procedure

In the last years, the interest in studying the meson-baryon interaction in the $S = -1$ sector has been sparked by the availability of more precise data coming from the measurement of the energy shift and width of the 1s state in kaonic hydrogen by the SIDDHARTA collaboration [1]. All these works were carried out as a response to the need to extend the approach to higher orders and energies aiming for greater accuracy in data description. Since this scenario requires a non-perturbative resummation of Chiral Perturbation Theory (ChPT), the authors employed a chiral $SU(3)$ Lagrangian up to next-to-leading order (NLO) and implemented unitarization in coupled channels. Unitarized Chiral Perturbation Theory (UChPT) consists in solving the Bethe-Salpeter equation (BSE), which can be converted into a system of algebraic equations using the on shell factorization [2,3]: $T_{ij} = (1 - V_{il}G_l)^{-1}V_{lj}$, where T_{ij} is a scattering amplitude for a given incoming i-channel and an outgoing j-channel, and G_l is a loop function, which after dimensional regularization introduces the so called subtraction constants (a_l) which are normally fitted to the experimental data. Considering the isospin symmetry, in this sector the subtraction constants become 6. On the other hand, V_{ij}, the interaction kernel, can be derived from the chiral Lagrangian. The leading order of the chiral Lagrangian at low momenta consists of the famous Weimberg-Tomozawa (WT) interaction kernel and the Born vertices.

$$\mathcal{L}_{\phi B}^{(1)} = i\langle \bar{B}\gamma_\mu[D^\mu, B]\rangle - M_0\langle \bar{B}B\rangle - \frac{1}{2}D\langle \bar{B}\gamma_\mu\gamma_5\{u^\mu, B\}\rangle - \frac{1}{2}F\langle \bar{B}\gamma_\mu\gamma_5[u^\mu, B]\rangle \ , \qquad (1)$$

where M_0 is the common baryon octet mass in the chiral limit. The constants D, F denote the axial vector couplings of the baryons to the mesons. B is the baryon octet field and the symbol $\langle \ldots \rangle$ stands for the trace in flavour space, while the pseudoscalar meson octet field ϕ enters in a more complicated way: $u_\mu = iu^\dagger \partial_\mu U u^\dagger$, where $U(\phi) = u^2(\phi) = \exp\left(\sqrt{2} i\phi/f\right)$ (see [4]). At NLO, the contributions of the Lagrangian to meson-baryon scattering in S wave can be written as:

$$\begin{aligned} \mathcal{L}_{\phi B}^{(2)} &= b_D\langle \bar{B}\{\chi_+, B\}\rangle + b_F\langle \bar{B}[\chi_+, B]\rangle + b_0\langle \bar{B}B\rangle\langle\chi_+\rangle + d_1\langle \bar{B}\{u_\mu, [u^\mu, B]\}\rangle \\ &\quad + d_2\langle \bar{B}[u_\mu, [u^\mu, B]]\rangle + d_3\langle \bar{B}u_\mu\rangle\langle u^\mu B\rangle + d_4\langle \bar{B}B\rangle\langle u^\mu u_\mu\rangle \ , \end{aligned} \tag{2}$$

where $\chi_+ = 2B_0(u^\dagger \mathcal{M} u^\dagger + u\mathcal{M}u)$ breaks chiral symmetry explicitly via the quark mass matrix $\mathcal{M} = \mathrm{diag}(m_u, m_d, m_s)$ and $B_0 = -\langle 0|\,\bar{q}q\,|0\rangle/f^2$ relates to the order parameter of spontaneously broken chiral symmetry, with f being the pseudoscalar decay constant in the chiral limit. The coefficients b_D, b_F, b_0 and d_i ($i = 1,\ldots,4$) are the low energy constants. In principle, the first three coefficients, involved in terms proportional to the χ_+ field, should fulfill constraints related to the mass splitting of baryons. However, since our study goes beyond tree level and incorporates higher order terms via coupled channel scattering equations, we will relax those constraints and we will fit these b-type constants, together with the d_i ones, to experimental data.

Apart from a good reproduction of the experimental data, one of the most challenging aspects of this sort of study is the determination of the non constrained low energy constants involved in the NLO terms of the Chiral Lagrangian. For the sake of extracting information about these coefficients, it would be convenient to explore processes in which the higher order terms of the Lagrangian play a significant role beyond that of fine tuning. With this motivation, in previous studies, we focused on the $K^-p \to K^+\Xi^-$, $K^0\Xi^0$ reactions, because they do not proceed via the WT term of the Chiral Lagrangian and, hence, they are expected to be especially sensitive to the higher order terms.

After checking the effects on the low energy constants of the systematic inclusion of terms beyond the WT one, our models demonstrated the sensitivity of the $K^-p \to K\Xi$ reactions to the NLO and Born terms and, therefore, more reliable values of the low energy constants were obtained.

Another interesting finding of [5] is that models that include or do not include the Born terms give very different isospin projections of the $K^-p \to K\Xi$ total cross sections while offering very similar reproduction of the $K^-p \to K\Xi$ scattering data. This fact points towards the idea that processes that filter a single isospin component are an essential requirement to get much more reliable values for the NLO coefficients. This background marks the next natural step: the extra experimental data from the $K^-p \to \eta\Lambda, \eta\Sigma^0$ reactions in the fitting procedure should be included to discriminate possible ambiguities in the isospin distributions since they are pure $I = 0$ and $I = 1$ processes, respectively.

Finally, in accordance with [4], the accommodation to the experimental $K^-p \to K\Xi$ total cross sections was improved by the incorporation of additional contributions to the pure chiral scattering amplitude coming from high-spin and high-mass resonant diagrams. Complementary to this phenomenological reason, they allow us to study the accuracy and stability of the NLO coefficients. Moreover, the inclusion of these resonant contributions implicitly simulates higher-angular-momenta contributions involving low lying meson-baryon states of the coupled channel problem. This is due to the fact that the low energy constants absorb these contributions in order to reproduce the experimental data at the expense of taking realistic values. Thus, the resonant contributions permit such parameters to relax avoiding a possible overestimation of their values. Consequently, guided by [4] and after an exhaustive study of the effects on the $K^-p \to K\Xi, \eta\Lambda$ total cross sections of the incorporation of all the possible resonances [6] ranging in energy from 1800 to 2300 MeV, we conclude that the candidates which reproduce the experimental data better were $\Sigma(2030)$, $\Sigma(2250)$ and $\Lambda(1890)$ with $J^P = 7/2^+, 5/2^-$ and $3/2^+$, respectively. The resonant contributions have been implemented applying the Rarita-Schwinger effective Lagrangians that allow one to couple 1/2 spins with the higher ones [7,8]. Furthermore, an exponential form factor has been included due to the high momentum dependence of the scattering amplitudes as done in [7]. The resonant terms are only taken into account for the amplitudes connecting K^-p states with $K^+\Xi^-, K^0\Xi^0$ and $\eta\Lambda$ ones replacing the scattering amplitudes by:

$$\begin{aligned} T_{K^-p\to K\Xi} &\to T^{Chiral}_{K^-p\to K\Xi} + T^{3/2^+}_{K^-p\to K\Xi} + T^{5/2^-}_{K^-p\to K\Xi} + T^{7/2^+}_{K^-p\to K\Xi} \\ T_{K^-p\to\eta\Lambda} &\to T^{Chiral}_{K^-p\to\eta\Lambda} + T^{3/2^+}_{K^-p\to\eta\Lambda}, \end{aligned}$$

where T^{Chiral}_{ij} is the scattering amplitude obtained by means of (BSE), and where T^J_{ij} stands for the corresponding resonant term with J^P quantum numbers.

2. Results and discussion

With the aim of improving the reliability of the NLO parameters and discussing the implications of including resonant terms, two fits have been performed. The **Model 1** fit corresponds to a unitarized calculation employing the full chiral Lagrangian, eqs. (1) and (2), which involves 16 fitting parameters: the pseudoscalar decay constant, the 2 axial vector couplings, the 6 subtraction constants, and the 7 NLO low energy constants. While the **Model 2** fit improves upon the first one by using the inclusion of resonant terms in the way described before. It must be mentioned that 13 new parameters have been introduced in the last fit that we avoid giving any further information for reasons of simplicity (see [8] for specific details). Both fitting procedures take into account K^-p elastic and inelastic cross section data for all the two body channels of the sector [10–24] and the precise SIDDHARTA value of the energy shift and width of kaonic hydrogen [1], as well as, branching ratios at threshold [25, 26].

Table 1. Values of the parameters for a regularization scale $\mu = 1$ GeV and the corresponding $\chi^2_{\text{d.o.f.}}$, defined in [4], for both described fits. The error of the parameters are given by the MINUIT minimization procedure.

	$a_{\bar{K}N}$ (10^{-3})	$a_{\pi\Lambda}$ (10^{-3})	$a_{\pi\Sigma}$ (10^{-3})	$a_{\eta\Lambda}$ (10^{-3})	$a_{\eta\Sigma}$ (10^{-3})	$a_{K\Xi}$ (10^{-3})	f/f_π	D	F
Model 1	1.27 ± 0.12	-6.1 ± 12.9	0.68 ± 1.43	-0.67 ± 1.06	8.00 ± 3.26	-2.51 ± 0.99	1.20 ± 0.03	0.70 ± 0.16	0.51 ± 0.11
Model 2	1.52 ± 0.21	-2.6 ± 13.9	2.1 ± 1.2	0.76 ± 1.21	10.1 ± 3.7	-2.01 ± 0.74	1.18 ± 0.03	0.70 ± 0.15	0.40 ± 0.11
	b_0 (GeV^{-1})	b_D (GeV^{-1})	b_F (GeV^{-1})	d_1 (GeV^{-1})	d_2 (GeV^{-1})	d_3 (GeV^{-1})	d_4 (GeV^{-1})	$\chi^2_{\text{d.o.f.}}$ (defined in [5])	
Model 1	0.13 ± 0.04	0.12 ± 0.01	0.21 ± 0.02	0.15 ± 0.03	0.13 ± 0.03	0.30 ± 0.02	0.25 ± 0.03	1.14	
Model 2	-0.07 ± 0.01	0.13 ± 0.01	0.27 ± 0.02	0.14 ± 0.03	0.13 ± 0.01	0.40 ± 0.02	0.02 ± 0.02	0.96	

One of main findings of **Model 1**, when comparing the fitting parameters of this model to the ones from the models in [4,5], is the homogeneity and accuracy achieved by the NLO coefficients and the natural sized values for all the subtraction constants obtained. The fitting parameters are displayed in Table 1, from which, one can appreciate that the most remarkable result is the 16% improvement in the goodness of the fit, as is clearly reflected in $\chi^2_{\text{d.o.f.}}$ values. It can be noted that the fitting parameters are quite stable; this stability being more marked for the NLO coefficients. The reason stems from the fact that we have employed more observables sensitive to the NLO term. In addition, as a consequence of this stability, we still obtain natural sized subtraction constants. Regarding the accuracy, we obtain similar errors associated to the fitting parameters for both models which is in contrast to what was found in [4].

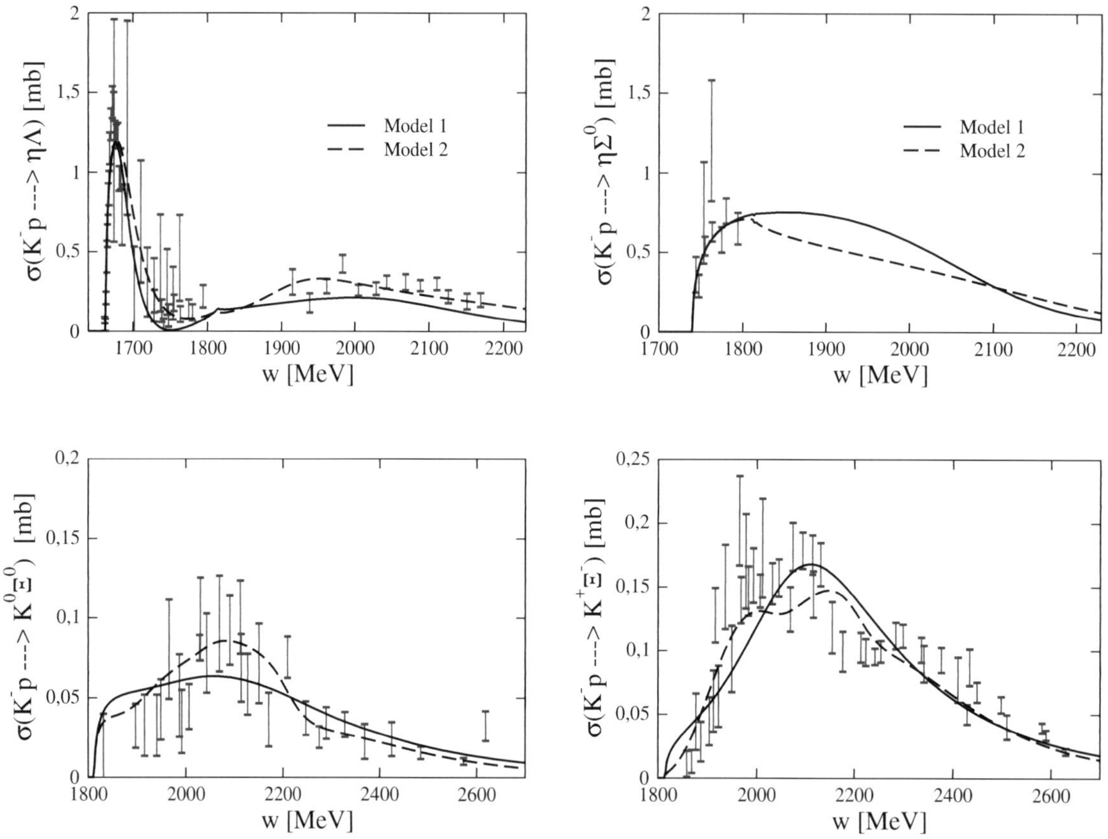

Figure 1. Total cross sections of the $K^-p \to \eta\Lambda, \eta\Sigma^0, K^0\Xi^0, K^+\Xi^-$ reactions obtained from Model 1 (solid line) and Model 2 (dashed line). The experimental data of these total cross sections with their corresponding symmetric error bars are taken from [14–24].

In Figure 1, we only represent those cross sections where the novelty comes, because non significant differences between the models exist for the rest due to the dominance of the WT term. From the solid line (**Model 1**), one sees that the cross sections of the η channels can be properly reproduced, which is not the case with previous models [9], even the resonant structure from $\Lambda(1670)$ seen in $K^-p \to \eta\Lambda$. In the $K\Xi$ channels, one can observe a loss of agreement with the data in order to reproduce the $K^-p \to K^0\Xi^0$ reaction at low energy. Partly, the use of resonant processes was motivated by this fact and to improve the $K^-p \to \eta\Lambda$ cross section around 1950 MeV. From the dashed lines, one appreciates a clear improvement in reproducing the experimental $K^-p \to \eta\Lambda$ cross section in the energies ranging from 1850 to 2200 MeV, and, at the same time, **Model 2** respects the $\Lambda(1670)$ resonant structure. But, the most notable effect is the one observed in the $K^-p \to K^0\Xi^0$ cross section, which reproduces better the experimental data located just above threshold due to the $\Lambda(1890)$ resonance. Moreover, the combined contribution of the $\Sigma(2030)$ and $\Sigma(2250)$ resonances provides a clear bump structure reaching its maximum at around 2100 MeV. A similar behavior for low energies can be noticed in the $K^-p \to K^+\Xi^-$ cross section, while, in the vicinity of the energy where the experimental data shows a maximum, the $\Sigma(2030)$ and $\Sigma(2250)$ resonant contributions produce some structure, together with a slight reduction of strength. With respect to the $K^-p \to \eta\Sigma^0$ cross section, we do not notice any difference in the reproduction of the experimental data, but for energies around 1800 MeV **Model 2** presents a more pronounced slope.

3. Acknowledgements
This work is partly supported by the Spanish Ministerio de Economia y Competitividad (MINECO) under the project MDM-2014-0369 of ICCUB (Unidad de Excelencia 'María de Maeztu'), and, with additional European FEDER funds, under the contract FIS2014-54762-P, by the Generalitat de Catalunya contract 2014SGR-401.

References
[1] Bazzi M *et al.* 2011 *Phys. Lett.* B **704** 113
[2] Oset E and Ramos A 1998 *Nucl. Phys.* A **635** 99
[3] Hyodo T and Jido D 2012 *Progress in Particle and Nuclear Physics* **67** 55
[4] Feijoo A, Magas V and Ramos A 2015 *Phys. Rev.* C **92** 015206
[5] Ramos A, Feijoo A and Magas V 2016 *Nucl. Phys.* A **954** 58
[6] Olive K *et al.* [Particle Data Group Collaboration] 2014 *Chin. Phys.* C **38** 090001
[7] Sharov D, Korotkikh V and Lanskoy D 2011 *Eur. Phys. J.* A **47** 109
[8] Feijoo A 2017 *Meson-Baryon interactions from effective Chiral Lagrangians* (Barcelona: UB PhD Thesis)
[9] Feijoo A, Magas V and Ramos A 2017 *EPJ Web Conf.* **137** 05003
[10] Kim J K 1965 *Phys. Rev. Lett.* **14** 89
[11] Mast T S et al. 1976 *Phys. Rev.* D **14** 13
[12] Bangerter R O et al. 1981 *Phys. Rev.* D **23** 1484
[13] Ciborowski J et al. 1982 *J. Phys.* G **8** 13
[14] Burgun G et al. 1968 *Nucl. Phys.* B **8** 447
[15] Carlson J R et al. 1973 *Phys. Rev.* D **7** 2533
[16] Dauber P M et al. 1969 *Phys. Rev.* **179** 1262
[17] Haque M et al. 1966 *Phys. Rev.* **152** 1148
[18] London G W et al. 1966 *Phys. Rev.* **143** 1034
[19] Trippe T G and Schlein P E 1967 *Phys. Rev.* **158** 1334
[20] Trower W P et al. 1968 *Phys. Rev.* **170** 1207
[21] Starostin A et al. [Crystal Ball Collaboration] 2001 *Phys. Rev.* C **64** 055205
[22] Baxter D F et al. 1973 *Nucl. Phys.* B **67** 125
[23] Jones M et al. 1975 *Nucl. Phys.* B **90** 349
[24] Berthon A et al. 1974 *Nuovo Cim.* A **21** 146
[25] Nowak R J et al. 1978 *Nucl. Phys.* B **139** 61
[26] Tovee D N et al. 1971 *Nucl. Phys.* B **33** 493

IOP Conf. Series: Journal of Physics: Conf. Series **1024** (2018) 012021 doi:10.1088/1742-6596/1024/1/012021

The EXPERT project: part of the Super-FRS Experiment Collaboration

V. Chudoba for the EXPERT project
Institute of Physics, Silesian University in Opava, Czech Republic;
FLNR, Joint Institute for Nuclear Research, Dubna, Russia
E-mail: chudoba@jinr.ru

Abstract. The EXPERT project is subtopic of the Super-FRS Experiment Collaboration, being part of NUSTAR@FAIR. The experiments will be aimed at the nuclear systems in the most outer regions of the nuclear landscape. Special attention will be payed to few-body decays which are expected to be a regular phenomena in the vicinity and beyond the driplines. Studies of rare and not yet observed (multi-)nucleon radioactivity, many-body decays of resonance states of unbound nuclei as well as beta-delayed decays and search for exotic excitation modes will be carried out. The scientific program will be performed using the modular detector system containing the methods for detection of all particle and radiation types.

1. Introduction

The EXPERT project [1] is a part of the scientific program of the Super-FRS Experiment collaboration in the frame of NUSTAR@FAIR [2]. It is aimed at studies of the nuclear landscape beyond the proton and neutron drip-lines and intends to push researches up to limits of nuclear existence. The program will be realized by means of a compact modular detector setup supplementing the Super-FRS facility which will be employed in a separator-spectrometer mode [3]. Employment of the EXPERT instrumentation in different scenarios will allow to investigate phenomena of radioactivity, resonance decays, beta-delayed decays and exotic excitation modes either by means of in-flight or implantation-decay techniques.

2. Physical case

Arbitrary combination of protons and neutrons does not give a stable nucleus. If more and more nucleons of the same type is added to nucleus, a boundary of stability is reached and newly formed nucleus will undergo an immediate emission of the nucleon (proton or neutron). Boundaries of nuclear particle-stability are called driplines. The nuclei beyond the driplines still show individual states and may have half-lives exceeding far the characteristic nuclear time 10^{-21} s. Those unbound states are called resonances, and their lifetimes defined by the centrifugal and Coulomb barriers are strongly affected by nucleon correlations. The main objectives of the EXPERT project are focused on the nuclei in the vicinity of the driplines and may be schematically enumerated as:

(i) Exotic 2p radioactivity studies and search for novel types of radioactive decays: 4p, 2n, 4n.

(ii) Studies of p, 2p, 4p, 6p, 1n, 2n, 4n, 6n resonance decays and spectroscopy of continuum states.

IOP Conf. Series: Journal of Physics: Conf. Series **1024** (2018) 012021 doi:10.1088/1742-6596/1024/1/012021

(iii) Search for systems located far beyond the driplines aimed to answer the basic question: Where is the border-line between a resonant behavior and continuum response of nuclear matter?

(iv) Studies of beta-delayed particle (multi-particle) emission from exotic isotopes.

Radioactivity is a spontaneous decay of elements (it means of long-lived nuclear states) and nuclei with large excess of protons or neutrons become radioactive by their emission. One-proton (1p) and two-proton (2p) radioactivity were predicted already in 1960 [4]. It has first been discovered for ^{45}Fe [5], and further observations of 2p-radioactivity reported for ^{54}Zn [6], ^{19}Mg [7], ^{67}Kr [8], ^{48}Ni [9] have confirmed unexpectedly large half-lives of 2p-decay precursors. The first quantum-mechanical theory of 2p radioactivity which uses a three-body core+p+p model [10] interprets this observation as being due to a considerable influence of Coulomb and centrifugal barriers together with nuclear structure effects, and is able to predict the regular existence of considerably long-living 2p-decay precursors. Number of 2p-radioactivity candidates has been predicted using the theory presented in Ref. [10] for light- and medium-mass isotopes, in particular ^{26}S, ^{34}Ca, ^{38}Ti, ^{41,42}Cr. Such a general feature of proton-unbound nuclei with a three-body structure may be of general interest, e.g. for nuclear astrophysics: the inverse reaction to 2p decay, namely 2p radiative capture, may play an important role in the synthesis of heavy elements in the Universe, possibly bridging some waiting points in the hot rp-process, see e.g. [11, 12]. The measurements of 2p decays are the only way of studies of 2p radiative capture so far.

As for the direct neutron emission, all known nuclear ground states (e.g., ^{5}He, ^{10}Li, ^{13}Be etc.) are either very short-living or exist as virtual states only. The reason of such a difference between proton and neutron decays is in the absence of a Coulomb barrier in the latter case. Thus even small admixture of an s-wave configuration in the neutron precursor (i.e., the system without a centrifugal barrier) causes a dramatic reduction of its life-time [13]. There is a possibility that (multi-) neutron emission may take the form of 2n or 4n radioactivity. The first theoretical estimates for searching of one- two- and four- neutron radioactivity which are expected for exotic extremely neutron-rich nuclei have been proposed in Ref. [13]. The estimated lifetimes for simultaneous emission of two neutrons are much longer compared to the life-times of one-neutron emitters with the same energy due to the higher centrifugal barrier. A similar effect is known for true 2p emission (2p radioactivity) and understood theoretically and the trend towards longer lifetimes continues for true four-nucleon emission.

In light nuclei beyond the proton dripline, resonances dominated by single-particle configurations are usually expected to be very broad due to low Coulomb barriers. However, some states could exist as very narrow resonances due to their more complicated structure. Their structure may be understood as proton orbits built on excited-core configurations whose 1p-decay branches into the excited core are larger than those to the respective ground states. Such a phenomenon may be general for nuclei beyond the proton drip line where 1p, 2p thresholds are very low.

Most beta-delayed decay modes are enhanced at the drip lines since multi-nucleon separation energies are low there. Need in beta-decay data arises in astrophysics for processes where weak interactions play dominating role, either directly as beta-decay rates or indirectly where neutrino interactions are important (see e.g. [14]).

3. Experimental methods

We intend to populate proton and neutron precursors in reactions of one- or two-nucleon knockout on the secondary target located at middle focal plane (FMF2) of the Super-FRS separator. The intended experiments will be generally carried out using either ion-implantation or the in-flight decay method, in dependence on the decay precursors life-time. The implantation

IOP Conf. Series: Journal of Physics: Conf. Series **1024** (2018) 012021 doi:10.1088/1742-6596/1024/1/012021

technique is feasible for the half-live intervals from 10^{-7} to 10^{-2} s. The in-flight decay method by tracking technique may be employed for the half-lifes from 10^{-7} to 10^{-12} s. We are going to take an advantage of employment both of them together with the measurement of gamma coincidences. We want to emphasize the uniqueness of our experiment when employment of these two alternative (and overlapping with regard to boundary time) methods covers a range of 10 orders of magnitude of half-life measurements of the unknown precursors.

When using the ion-implantation method, radioactive nucleus is firstly stopped in a detector and its subsequent activity is measured. With this technique, the 2p-radioactivity of the ^{45}Fe ground state was discovered at GSI and GANIL [5]. We intend to use the similar method with the Optical Time-Projection Chamber (OTPC) which has been developed in the frame of the EXPERT project being part of Super-FRS Experiment Collaboration. Ions at the final focal plane (FHF2), after passing through the standard in-flight identification detectors will be slowed down in a wedge degrader and implanted into the OTPC [15]. The OTPC detector will record charged particles emitted during the decay of the stopped ions. While at the intermediate focal plane the production of nucleus of interest by means of one- or two-nucleon removal proceeds, the unreacted beam ions may be implanted into the OTPC at the final focus plane, where their decays will be recorded. Such an approach was proven efficiently and successfully at the FRS facility [16].

The conventional in-flight decay method aims at detecting fragments of a short-lived precursor in missing-mass or invariant mass measurements. A special in-flight decay method where all fragments are tracked and the decay vertices are recovered from the measured trajectories was first-time performed at GSI [7]. The method gives us possibility to extract rich information on investigated nucleus: life-time in range from 10^{-12} to 10^{-7} s can be ascertained; present few-body decay channels may be identified; and moreover the energy spectrum of precursor can be reconstructed. Such a tracking technique has proven to be a precise and effective tool for 2p-radioactivity studies which led to the discovery of the ^{19}Mg [7, 17] and ^{30}Ar [16] isotopes and their spectroscopy. The more detailed information on the method can be found e.g. in Ref. [18]. We intend to use the already proven technique using new tracking detectors developed within the EXPERT project.

4. Hardware

The nuclei of interest, extremely exotic and mostly unobserved yet, will be best produced in secondary reactions on radioactive ion beams of high energy 1.5 GeV/A impinging a secondary target at the middle focal plane of Super-FRS. The instrumentation of the EXPERT setup (see Fig. 1) will be localized in the middle FMF2 and final FHF1, FRF1 focal planes of the the Super-FRS facility. The decay products of unbound nuclei will be measured by a combination of several compact experimental appliances which will be installed in addition to standard Super-FRS beam diagnostic detectors. Experimental setup of the EXPERT equipment will be adapted to a number of experimental conditions and the main advantage will be taken of the versatile modular structure of particular components.

There are five main detector arrays allowing us to perform different experimental scenarios:

(i) Radiation-hard silicon strip detector (SSD) will provide information on time-of-flight, position and energy loss of ions, and it will be used for tracking of the secondary beam impinging the secondary target. The same type of detector will be used for the Super-FRS beam diagnostics.

(ii) The microstrip silicon (μSi) tracking system is a new generation detection system, similar to the other systems to be used at the several large facilities of FAIR, e.g. by the CBM, EXL, R3B, HISPEC/DESPEC collaborations and at the Super-FRS. Therefore, the unique experience gained by tracking charged particles in a very broad dynamic range (in our case,

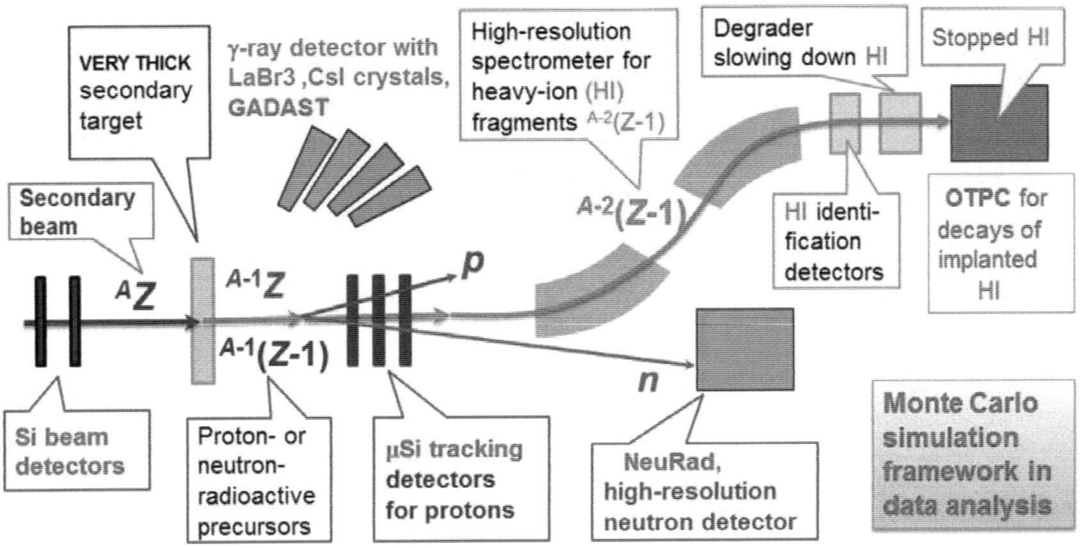

Figure 1. Schematic layout of the proposed experiments for exploratory studies of nuclei beyond the proton and neutron driplines. Two illustrated scenarios suggest that incident ^{A}Z secondary beam is stripped of one neutron or proton on the thick secondary target. These processes populate either particle-unstable proton-rich $^{A-1}$Z precursor or particle-unstable neutron-rich $^{A-1}$(Z-1) precursor. They decay via proton or neutron emission producing an $^{A-2}$(Z-1) heavy-ion fragment. Theoretical and Monte Carlo (MC) simulation framework is mentioned in this scheme as a component of the proposal required in most considered experimental scenarios.

it is the simultaneous detection of several protons and heavy ions) will provide a valuable reference in planning of the mentioned FAIR experiments.

(iii) The NeuRad (Neutron Radioactivity) fine-resolution detector of neutrons. Together with μSi detectors, this small-size $40\times40\times100$ cm^3 neutron detector will be able to provide precise information on angular correlations of decay neutrons with a charged fragment, which is used to derive the decay energy of exotic radioactive decays (for example an unobserved yet phenomenon of neutron radioactivity is suggested to be probed in the decay energy range of 0.1-100 keV).

(iv) The GADAST cluster (GAmma-ray Detectors Around Secondary Target) is the EXPERT component augmenting the tracking system for charged particles. It is a compact array of gamma-ray sensitive modules which will be located in the middle focal plane of the Super-FRS. The principal task of this array in the context of our physics program is to disentangle measurements of (few-)proton radioactivity by tagging the gamma-ray de-excitation of the heavy fragment in the excited state(s). In addition, due to its thickness, it is suitable to be used as the detector of charged particles. It will consist of three arrays of scintillation crystals. Two identical arrays of 64 trapezoidal CsI(Tl) crystals will cover the region characterized by the angles of more than 15 degrees in laboratory system. Around the beam axis, where much better energy resolution and counting rate is required, an annular array of 32 detectors based on $LaBr_3$(Ce) crystals will be positioned.

(v) The OTPC detector main features and principles of operation are described in Ref. [15]. In particular, a new concept of the signal readout using two CCD cameras will be implemented. This new feature is expected to improve the sensitivity and the efficiency of the OTPC to detect both proton and heavy ion at the final focus of the Super-FRS. In comparison with the previous generation of OTPC, the gas mixture will be optimized to reach the best detection of the low-energy protons emitted by beta-delayed decays. This will yield much more precise information on investigated decays and it will facilitate a test of isospin symmetry. The length of the active volume along the beam direction is 35 cm. Primary electrons, resulting from the gas ionization by the stopping ion and by the emitted charged particles, drift with constant velocity toward the amplification section, passing through a gating electrode. At the anode plane electrons stimulate atoms of the gas to emit light which is recorded by a pair of digital cameras (CCD) and a photomultiplier (PMT). Having such an information, we will be able to reconstruct the tracks of all particles in three dimensions.

It is worthy to remark that the construction of all above mentioned components is already in process and their prototypes are tested at ACCULINNA and ACCULINNA-2 facilities [19] at Flerov Laboratory of Nuclear Reactions, JINR Dubna. Completion of all detectors and their test on radioactive ion beams provided at the ACCULINNA-2 facility is planned by end of year 2022. GADAST and OTPC detectors will be employed in experimental studies on proton-rich nuclei (e.g. ^{17}Ne, ^{26,27}S, ^{23}Si) conducted at ACCULINNA-2 within the next few years.

5. Conclusion

The EXPERT project is aimed at research of exotic nuclear systems in the vicinity and beyond the driplines. It will take an advantage of the Super-FRS, the backbone facility for whole NUSTAR Collaboration, which will be supplemented by versatile and modular system of detectors. The proposed scenarios of experimental setup will allow to study properties of both proton- and neutron-rich nuclear systems characterized by few-cluster structure. The conception of measuring techniques, together with their readiness, will allow us to take a part in experimental campaign within FAIR-Phase-0.

6. Acknowledgements

This work was partly supported by the RSF 17-12-01367 Grant and MEYS Projects (Czech Republic) LTT17003 and LM2015049.

References

[1] Geissel H *et al.*, Proc. Int. Symposium on Exotic Nuclei (EXON2014), World Scientific (2015) ISBN 978-981-4699-45-7, 579.
[2] http://www.fair-center.eu/for-users/experiments/nustar.html
[3] Winkler M *et al.*, Nucl. Instr. Meth. B **266** (2008) 4183.
[4] Goldansky V I, Nucl. Phys. **19** (1960) 482.
[5] Pfützner M *et al.*, Eur. Phys. J. A **14** (2002) 279; Giovinazzo J *et al.*, Phys. Rev. Lett. **89** (2002) 102501.
[6] Blank B *et al.*, Phys. Rev. Lett. **94** (2005) 232501; ibid 94 (2005) 249901.
[7] Mukha I *et al.*, Phys. Rev. Lett. **99** (2007) 182501.
[8] Goigoux T *et al.*, Phys. Rev. Lett. **117** (2016) 162501.
[9] Pomorski M *et al.*, Phys. Rev. C **83** (2011) 014306.
[10] Grigorenko L V *et al.*, Phys. Rev. Lett. **85** (2000) 22.
[11] Schatz H *et al.*, Phys. Rep. **294** (1998) 167.
[12] Grigorenko L V *et al.*, Phys. Rev. C **72** (2005) 015803.
[13] Grigorenko L V, Mukha I, Scheidenberg Ch and Zhukov M V, Phys. Rev. C **84** (2011) 021303(R).
[14] Martinez-Pinedo G *et al.*, Nucl. Phys. A **718** (2003) 452c.
[15] Pomorski M *et al.*, Phys. Rev. C **90** (2014) 014311.
[16] Mukha I *et al.*, Phys. Rev. Lett. **115** (2015) 2022501.
[17] Mukha I *et al.*, Phys. Rev. C **77** (2008) 061303(R).
[18] Mukha I *et al.*, Phys. Rev. C **82** (2010) 054315.
[19] Grigorenko L V *et al.*, Phys. Usp. **59** (2016) 321.

IOP Conf. Series: Journal of Physics: Conf. Series **1024** (2018) 012022 doi:10.1088/1742-6596/1024/1/012022

Heavy quark potentials in effective string theory

Sungmin Hwang

Physik-Department, Technische Universität München, James-Franck-Str.1, D-85747 Garching and Max-Planck-Institut für Physik, Föhringer Ring 6, D-80805 München, Germany

E-mail: sungmin.hwang@tum.de

Abstract. We present our recent analytic calculation of the next-to-leading order contribution to the heavy quark-antiquark potentials, by utilizing the effective description of a long string. Discussions on the effective field theory power counting and the method for reducing a number of dimensionful parameters arising from the EST are followed. Furthermore, physical implications on the comparison to the lattice data are briefly discussed.

1. Introduction

Perturbative Quantum Chromodynamics (pQCD) [1] provides an analytic description of the interactions between quarks and gluons at high energy [2, 3, 4]. However, this approach is no longer valid as the coupling α_s becomes a large value at the scale around or below 200 MeV; this feature is called *color confinement* in QCD [5]. A possible solution to this theoretical challenge within a perturbative framework is still to be pursued to this day. Although a complete and consistent theory for the dynamics concerning color-charged particles from the deconfining to confining phase is absent, various kind of low-energy effective descriptions has been developed, and they differ depending on the particle interactions of one's interest.

Chiral perturbation theory (χPT) [6] provides a perturbative framework for the interactions among light mesons (π, σ, ρ, etc), which is valid below the chiral breaking scale ~1 GeV. As for the heavy quarks (charm or bottom), heavy quark effective theory (HQET) [7] gives an effective description of the dynamics of a heavy-light quark (or antiquark) pair, while non-relativistic QCD (NRQCD) [8, 9] and potential NRQCD (pNRQCD) [10] are the effective field theories (EFTs) for a heavy quark-antiquark pair and its bound state, respectively.

On the other hand, lattice gauge theory (or lattice QCD-LQCD) provides numerical values of the mass spectra of quarks and hadrons. While LQCD is useful for extracting physical values at the non-perturbative regime, an analytic method for understanding the bound state between a quark-antiquark pair at long-distance has been developed. This method is called QCD *flux tube model* [11]. In this paper, Nambu proposed that if a quark-antiquark pair is separated to long distance, comparable to that of the confining scale, then the gluodynamics between the pair can be substituted by the dynamics of a fluctuating string. This model is verified when it is compared to the lattice simulation [12]. It shows that the energy density between the pair increases when they are separated, and this increase of density appears in the shape of a *flux tube*.

Lüscher et. al. developed this idea further [13, 14]. Based on the conjecture of the *Wilson loop-string partition function equivalence*, the static potential between the pair at long distance was calculated. Kogut and Parisi, shortly afterwards, derived the analytic expression of the spin-spin interaction part of the heavy quark potential by using the string picture [15].

During the last few decades, significant progress has been made in this line of research. Heavy quark-antiquark potential was shown to be equivalent to the Wilson loop and the gauge field insertions therein, at the large time limit [16, 17]. This was given by the matching calculation. Recently, Brambilla et. al. have calculated all of the heavy quark potentials by using the effective string theory (EST) up to leading order (LO) [18]. As was pointed out here, however, this LO calculation is not fully inclusive, because some of the next-to-leading order (NLO) contribution would contain terms which are of the same order of the expansion parameter. In other words, NLO terms can alter the leading order coefficients of the potentials. Thus, it is necessary for us to understand the proper EFT systematics of the string picture, so that not only the higher order suppression terms are derived, but all of the missing LO terms are also acquired.

This proceeding is organized as the following: In Sec. 2, we briefly introduce NRQCD and pNRQCD for a heavy quark-antiquark pair and its bound state. The equivalence between the Wilson loop expectation value (and gauge field insertions therein) from NRQCD and the potential terms from pNRQCD at the large time limit is shown. In Sec. 3, we introduce the EST, and through the Wilson loop-string partition function equivalence conjecture, the heavy quark potentials are analytically calculated up to NLO. We discuss the implication of this result in Sec. 4. Comparison of our result to the LQCD data as well as the extension of our method to other cases are given at the end.

2. Non-relativistic EFTs of QCD

NRQCD is a non-relativistic EFT for a heavy quark-antiquark pair [8, 9]. As the hierarchy of scales is realized, $M \gg Mv$ (M: heavy quark mass, v: relative velocity), the NRQCD Lagrangian is derived by integrating out M: it is organized by $1/M$ expansion. Then its dynamical degrees of freedom are Pauli spinor fields for the heavy quark and antiquark, SU(3) field strength terms for the soft gluons, and Dirac spinor fields for the light quarks. A series of operators concerning the heavy quark and antiquark fields carry Wilson coefficients. The coefficients contain information at the high-energy regime. They are determined by matching to the underlying theory, pQCD.

An even lower-energy counterpart to NRQCD is pNRQCD [10], which is an effective framework of heavy quark-antiquark bound state. The hierarchy of scales for this system is realized as $M \gg Mv \gg Mv^2$, where Mv^2 is the relative kinetic energy between the quark-antiquark pair. By integrating out the relative momentum Mv, which scales like $\sim 1/r$ (r: relative distance), the pNRQCD Lagrangian is then organized by the multipole expansion in r, in addition to $1/M$ expansion from NRQCD. Then the dynamical degrees of freedom are SU(3) singlet and octet fields for the heavy quarkonium, SU(3) field strength term for ultra-soft gluons, Dirac spinor fields for the light quarks. Note that both the singlet and octet fields depend on time t, relative coordinate $\boldsymbol{r}$, and center-of-mass coordinate $\boldsymbol{R}$, $S(t, \boldsymbol{r}, \boldsymbol{R})$, $O^a(t, \boldsymbol{r}, \boldsymbol{R})$. A schematic form of the pNRQCD Lagrangian is given by (only the bilinear terms of singlet and octet are shown below)

$$\begin{aligned}\mathcal{L}_{\text{pNRQCD}} \ni \int d^3r \Big\{ &\text{Tr}\left[S^\dagger\left(i\partial_0 - V_S(r) + \dots\right)S + O^\dagger\left(iD_0 - V_O(r) + \dots\right)O\right] \\ &gV_A(r)\text{Tr}\left[O^\dagger \boldsymbol{r}\cdot\boldsymbol{E} + S^\dagger \boldsymbol{r}\cdot\boldsymbol{E}O\right] + g\frac{V_B(r)}{2}\text{Tr}\left[O^\dagger \boldsymbol{r}\cdot\boldsymbol{E}O + O^\dagger O\boldsymbol{r}\cdot\boldsymbol{E}\right]\Big\},\end{aligned} \tag{1}$$

in which $V(r)$'s are the heavy quark potentials. They are the Wilson coefficients of pNRQCD, which are to be determined by matching to NRQCD. In this report, we focus on the analytic expression of the singlet potential $V_S(r)$. The potential expanded up to $1/M^2$ is given by (from now on we omit the subscript S for the singlet sector) [10]

$$V(r) = V^{(0)}(r) + \frac{2}{M}V^{(1,0)}(r) + \frac{1}{M^2}\left[V^{(2,0)} + V^{(1,1)}\right] \tag{2}$$

where the superscript (a, b) denotes the order of $1/M_{1,2}$ expansion. Note that the mass of the heavy quark and the antiquark is identical, $M_1 = M_2 = M$. $V^{(2,0)}$ includes terms like $V_{L^2}^{(2,0)}, V_{p^2}^{(2,0)}, V_{LS}^{(2,0)}$, and $V_r^{(2,0)}$, and $V^{(1,1)}$ includes $V_{L^2}^{(1,1)}, V_{L_2S_1}^{(1,1)}, V_{S^2}^{(1,1)}, V_{S_{12}}^{(1,1)}, V_{p^2}^{(1,1)}$, and $V_r^{(1,1)}$.

By comparing the heavy quark-antiquark correlator of NRQCD to the singlet propagator of pNRQCD, the following relation holds at the zeroth order in $1/M$ expansion [16]

$$V^{(0)}(r) = \lim_{T\to\infty} \frac{i}{T} \ln\langle W_\Box\rangle, \quad \text{for} \quad W_\Box = P\exp\left\{-ig \oint_{r\times T} dz^\mu A_\mu(z)\right\}. \tag{3}$$

The angular bracket around the rectangular Wilson loop $W_\Box$ denotes the expectation value over the Yang-Mills action, and P stands for the path-ordering operator in color space. In the similar fashion, we obtain the following relation for the first order relativistic correction [17]:

$$V^{(1,0)}(r) = -\frac{1}{2} \lim_{T\to\infty} \int_0^T dt\, t \langle\langle g\boldsymbol{E}_1(t) \cdot g\boldsymbol{E}_1(0)\rangle\rangle_c\,, \tag{4}$$

where the double angular bracket denotes gauge fields insertions to the normalized Wilson loop expectation value, and the subscript c stands for the connected part, which is defined by $\langle\langle O_1(t_1)O_2(t_2)\rangle\rangle_c = \langle\langle O_1(t_1)O_2(t_2)\rangle\rangle - \langle\langle O_1(t_1)\rangle\rangle\langle\langle O_2(t_2)\rangle\rangle$, for $t_1 \geq t_2$. Also $\boldsymbol{E}_{1,2}(t) = \boldsymbol{E}(t, \pm r/2)$ as the quark and antiquark are located at $(0, 0, \pm r/2)$, respectively. The second order corrections are divided into spin-independent and spin-dependent parts. The spin-independent part is given by,

$$\begin{aligned} V_{p^2}^{(2,0)}(r) &= \frac{i}{2}\hat{\boldsymbol{r}}^i\hat{\boldsymbol{r}}^j \lim_{T\to\infty} \int_0^T dt\, t^2 \langle\langle g\boldsymbol{E}_1^i(t) g\boldsymbol{E}_1^j(0)\rangle\rangle_c\,, \\ V_{p^2}^{(1,1)}(r) &= i\hat{\boldsymbol{r}}^i\hat{\boldsymbol{r}}^j \lim_{T\to\infty} \int_0^T dt\, t^2 \langle\langle g\boldsymbol{E}_1^i(t) g\boldsymbol{E}_2^j(0)\rangle\rangle_c\,, \\ &\cdots \end{aligned} \tag{5}$$

where the ellipsis includes terms like V_{L^2}'s and V_r's. The spin-independent part is given by,

$$\begin{aligned} V_{LS}^{(2,0)} &= -\frac{c_F^{(1)}}{r^2} i\boldsymbol{r} \cdot \lim_{T\to\infty} \int_0^T dt\, t \langle\langle g\boldsymbol{B}_1(t) \times g\boldsymbol{E}_1(0)\rangle\rangle + \frac{c_s^{(1)}}{2r^2}\boldsymbol{r} \cdot \left(\nabla_r V^{(0)}\right), \\ V_{L_2S_1}^{(1,1)}(r) &= \frac{c_F^{(1)}}{r^2} i\boldsymbol{r} \cdot \lim_{T\to\infty} \int_0^T dt\, t \langle\langle g\boldsymbol{B}_1(t) \times g\boldsymbol{E}_2(0)\rangle\rangle_c\,, \\ &\cdots \end{aligned} \tag{6}$$

where the ellipsis includes the spin-spin interaction part of the potentials. Explicit form of these potentials are then derived by calculating the gauge field insertions to the Wilson loop expectation value. In the non-perturbative regime, LQCD provides numerical result [19, 20], which observes Poincaré invariance [21] with high accuracy, but it is limited only to the calculations of two-gauge field insertions. V_r contains three- and four-gauge field insertions, so it is necessary to look for some other ways to address this issue.

3. Effective string theory and heavy quark potentials

Analytic behavior of the potentials at long distance can be investigated by utilizing the effective string theory (EST) [11]. The EST is constructed based on the Wilson loop-string partition function equivalence conjecture [13]:

$$\lim_{T\to\infty} \langle W_\Box\rangle = Z \int \mathcal{D}\xi^1\mathcal{D}\xi^2\, e^{iS_{\text{String}}[\xi^1,\xi^2]}\,, \tag{7}$$

where Z is the normalization constant and $\xi^{1,2}$ are the transversal vibrations of the string. With respect to the following hierarchy of scales: $r\Lambda_{\rm QCD} \gg 1$, and the power counting: $\partial_a \sim 1/r$ and $\xi^l \sim \Lambda_{\rm QCD}^{-1}$ (i.e., $\partial_a \xi^l \sim (r\Lambda_{\rm QCD})^{-1} \ll 1$), as well as the Dirichlet boundary condition: $\xi^l(t, \pm r/2) = 0$, the action of the EST in four-dimensional spacetime is derived from the Nambu-Goto action,

$$S_{\rm String} = -\sigma \int dtdz \sqrt{\det\left(\eta_{ab} + \partial_a \xi^l \partial_b \xi^l\right)} = -\sigma \int dtdz \left(1 - \frac{1}{2}\partial_a \xi^l \partial_b \xi^l + \dots\right). \quad (8)$$

The ellipsis on the last equality contains higher order terms of the derivative expansion, and σ is the string tension ($\sim \Lambda_{\rm QCD}^2$), which is a fundamental parameter of the theory. Truncation has been made up to quadratic order. From this expression, it is clear that the LO potential is of linear order, $V^{(0)} \approx \sigma r$ [13]. Also, the Green's function is exactly computed by solving the equations of motion [22]:

$$G^{lm}(t,t';z,z') = \frac{\delta^{lm}}{4\pi\sigma} \ln\left(\frac{\cosh\left[(t-t')\,\pi/r\right] + \cosh\left[(z+z')\,\pi/r\right]}{\cosh\left[(t-t')\,\pi/r\right] - \cosh\left[(z-z')\,\pi/r\right]}\right). \quad (9)$$

Relativistic corrections to the potential at long distance can analytically be computed when a mapping between QCD and the effective string description is established [18]. In this report, we want to extend this idea, such that the subleading contribution to the potentials are acquired.

As for the NLO calculation of the potential, it turns out that subleading part of the EST action contributes as the next-next-to-leading order. So we only consider the mapping at NLO:

$$\begin{aligned}
\langle\!\langle \dots \boldsymbol{E}^l_{1,2}(t) \dots \rangle\!\rangle &= \langle \dots \Lambda^2 \partial_z \xi^l(t, \pm r/2) + \overline{\Lambda}^2 \partial_z \xi^l \left(\partial\xi\right)^2(t, \pm r/2) \dots \rangle, \\
\langle\!\langle \dots \boldsymbol{E}^3_{1,2}(t) \dots \rangle\!\rangle &= \langle \dots \Lambda''^2 + \overline{\Lambda''}^2 (\partial\xi)^2(t, \pm r/2) \dots \rangle, \\
\langle\!\langle \dots \boldsymbol{B}^l_{1,2}(t) \dots \rangle\!\rangle &= \langle \dots \pm \Lambda' \epsilon^{lm} \partial_t \partial_z \xi^m(t, \pm r/2) \pm \overline{\Lambda'} \epsilon^{lm} \partial_t \partial_z \xi^m (\partial\xi)^2(t, \pm r/2) \dots \rangle, \\
\langle\!\langle \dots \boldsymbol{B}^3_{1,2}(t) \dots \rangle\!\rangle &= \langle \dots \pm \Lambda''' \epsilon^{lm} \partial_t \partial_z \xi^l \partial_z \xi^m(t, \pm r/2) \pm \overline{\Lambda'''} \epsilon^{lm} \partial_t \partial_z \xi^l \partial_z \xi^m (\partial\xi)^2(t, \pm r/2) \dots \rangle.
\end{aligned} \quad (10)$$

Note that the mapping for magnetic field insertion is derived by taking the duality transformation of the electric field. There are two types of dimensionful parameters, Λ's and $\overline{\Lambda}$'s, which are $\sim \Lambda_{\rm QCD}$, and they satisfy $\Lambda \geq \overline{\Lambda}$. Then we perform the similar calculation as in [18], but its difficulty lies on the divergence behavior of the string correlator, such as $\partial_z \partial_{z'} G$, $\partial_t \partial_z \partial_{z'} G$, and $\partial_t \partial_{t'} G$, defined at the same spacetime point. It turns out that this divergence stems from the derivation of the string correlator, which contains an infinite sum over the vibrational modes of the string and integral over the entire Fourier space. The infinite sum and integral are not physically compatible with the effective framework we started with, because the validity of theory does not encompass the entire energy scale. Thus, it is necessary to regulate this divergence and renormalize accordingly. We use *zeta-function regularization* for the infinite sum over the modes: $\sum_{n=0}^{\infty} n = -1/12$ and $\sum_{n=0}^{\infty} n^2 = 0$, and for the Fourier integral, we use *dimensional regularization*: $\int dy [y^2/(y^1+1)] = -2\pi$. In fact, these two regularization schemes are identical to each other: the zeta-function regularization is the discrete version of the dimensional regularization. By applying these schemes, we obtain the analytic result.

However, there are a number of free (dimensionful) parameters remaining in the potential. They are either from Eq. (10) or originate from the time integral for the potential. We exploit Poincaré invariance in QCD, which gives non-trivial relations between the potential terms.

Applying these constraints gives a significant result on the momentum-dependent potentials:

$$\begin{aligned} V_{p^2}^{(2,0)} &= \left(\frac{1}{12\pi}+\frac{\pi}{36}\right)\frac{1}{r}, \quad V_{p^2}^{(1,1)} = \left(\frac{1}{6\pi}-\frac{\pi}{36}\right)\frac{1}{r}, \\ V_{L^2}^{(2,0)} &= -\frac{\sigma r}{6}+\left(\frac{11}{36\pi}+\frac{2\pi}{27}\right)\frac{1}{r}, \quad V_{L^2}^{(1,1)} = \frac{\sigma r}{6}+\left(\frac{1}{9\pi}+\frac{5\pi}{216}\right)\frac{1}{r}. \end{aligned} \tag{11}$$

V_{p^2}'s were absent at LO in the EST [18], and here we observe non-trivial contributions arise at NLO. As for V_{L^2}'s, the linear part is from the LO calculation, which is now added by suppression terms. Note also that these NLO terms do not carry any free parameters due to constraint equations between the potentials (Poincaré invariance in QCD). The rest of the potentials are calculated in a similar fashion, and the full result will appear in our upcoming paper.

4. Outlook

Comparing the EST prediction to the LQCD data gives an interesting result. In [23], the comparison has been made, where the analytic result of V_b, V_c, V_d, and V_e (linear combinations of the momentum-dependent potentials) are given up to LO of the EST. It shows a significant discrepancy between lattice data and the analytic prediction, which implies that higher order terms are required for the improvement. Inclusion of NLO terms as in Sec. 3 shows decrease in the discrepancy, and this will be shown in our upcoming paper.

Method of the EST can be applied to the case of heavy hybrids and baryonic spectrum at long-distance regime. Eq. (7) might be different in this case due to the symmetry reason, but the procedure for deriving the analytic expression is similar. Currently, we are investigating these applications.

Acknowledgments

This work has been supported by the DFG grant BR 4058/2-1 and by the DFG cluster 1023 of excellence"Origin and structure of the univers" (www.universe-cluster.de). I thank Nora Brambilla, Hector Martinez, and Antonio Vairo for useful discussions.

[1] Collins J 2013 *Foundations of perturbative QCD* (Cambridge University Press)
[2] Gross D J and Wilczek F 1973 *Phys. Rev. Lett.* **30** 1343–1346
[3] Gross D J and Wilczek F 1973 *Phys. Rev.* **D8** 3633–3652
[4] Politzer H D 1973 *Phys. Rev. Lett.* **30** 1346–1349
[5] Wilson K G 1974 *Phys. Rev.* **D10** 2445–2459
[6] Leutwyler H 1994 *Annals Phys.* **235** 165–203 (*Preprint* 9311274)
[7] Isgur N and Wise M B 1989 *Phys. Lett.* **B232** 113–117
[8] Caswell W and Lepage G 1986 *Phys. Lett.* **B167** 437–442
[9] Bodwin G T, Braaten E and Lepage G P 1995 *Phys. Rev.* **D51** 1125–1171 ISSN 0556-2821 (*Preprint* 9407339)
[10] Brambilla N, Pineda A, Soto J and Vairo A 1999 *Nucl. Phys.* **B566** 275–310 (*Preprint* 9907240)
[11] Nambu Y 1979 *Phys. Lett.* **B80** 372–376
[12] Bali G S, Schlichter C and Schilling K 1995 *Phys. Rev.* **D51** 5165–5198 (*Preprint* 9409005)
[13] Lüscher M, Symanzik K and Weisz P 1980 *Nucl. Phys.* **B173** 365–396
[14] Lüscher M 1981 *Nucl. Phys.* **B180** 317–329
[15] Kogut J and Parisi G 1981 *Phys. Rev. Lett.* **47** 1089–1092 ISSN 0031-9007
[16] Brambilla N, Pineda A, Soto J and Vairo A 2001 *Phys. Rev.* **D63** 014023
[17] Pineda A and Vairo A 2001 *Phys. Rev.* **D63** 054007 (*Preprint* 0009145)
[18] Brambilla N, Groher M, Martinez H E and Vairo A 2014 *Phys. Rev.* **D90** 114032 (*Preprint* 1407.7761)
[19] Koma Y, Koma M and Wittig H 2007 *PoS* **LAT2007** 111 (*Preprint* 0711.2322)
[20] Koma Y and Koma M 2009 *PoS* **LAT2009** 122 (*Preprint* 0911.3204)
[21] Gromes D 1984 *Z. Phys.* **C26** 401–406
[22] Perez-Nadal G and Soto J 2009 *Phys. Rev.* **D79** 114002 (*Preprint* 0811.2762)
[23] Peset C, Pineda A and Stahlhofen M 2017 *Eur. Phys. J.* **C77** 681 (*Preprint* 1706.03971)

IOP Conf. Series: Journal of Physics: Conf. Series **1024** (2018) 012023 doi:10.1088/1742-6596/1024/1/012023

Infrastructures for Distributed Computing: the case of BESIII

J Pellegrino

Università degli Studi di Torino, Dipartimento di Fisica, Via Pietro Giuria 1, Torino, 10125, IT

E-mail: jacopo.pellegrino@to.infn.it

Abstract. The BESIII is an electron-positron collision experiment hosted at BEPCII in Beijing and aimed to investigate Tau-Charm physics. Now BESIII has been running for several years and gathered more than 1PB raw data. In order to analyze these data and perform massive Monte Carlo simulations, a large amount of computing and storage resources is needed. The distributed computing system is based up on DIRAC and it is in production since 2012. It integrates computing and storage resources from different institutes and a variety of resource types such as cluster, grid, cloud or volunteer computing. About 15 sites from BESIII Collaboration from all over the world joined this distributed computing infrastructure, giving a significant contribution to the IHEP computing facility. Nowadays cloud computing is playing a key role in the HEP computing field, due to its scalability and elasticity. Cloud infrastructures take advantages of several tools, such as VMDirac, to manage virtual machines through cloud managers according to the job requirements. With the virtually unlimited resources from commercial clouds, the computing capacity could scale accordingly in order to deal with any burst demands. General computing models have been discussed in the talk and are addressed herewith, with particular focus on the BESIII infrastructure. Moreover new computing tools and upcoming infrastructures will be addressed.

1. Computing for HEP

1.1. Brief Considerations

High Energy Physics (HEP) community is involved in more and more experiments to investigate various subjects. The continuous increase of HEP experiments, together with the progress of measurement tools, leads to a significant growth of the collected data volume and therefore of the amount of computational power required to analyze them within computing sites distributed all over the world.

This trend has led as well to a significant progress in the computing infrastructures required to process, analyze and store large amounts of data. In this work, the focus is on *distributed computing*, an approach adopted by multiple sites (research labs, universities, ect) which allows for the reduction of the computational pressure on a single site by means of an opportunistic usage of resources made available by other sites. Such process is possible thanks to heterogeneous ways of providing and accessing computational resources between different sites. The term *distributed computing* requires multiple aspects that a site has to accomplish in order to share its computational resources. These tasks span from the management of the physical devices (hardware) to the definition of public and private networks, storage, and replication services in order to provide a reliable infrastructure to users.

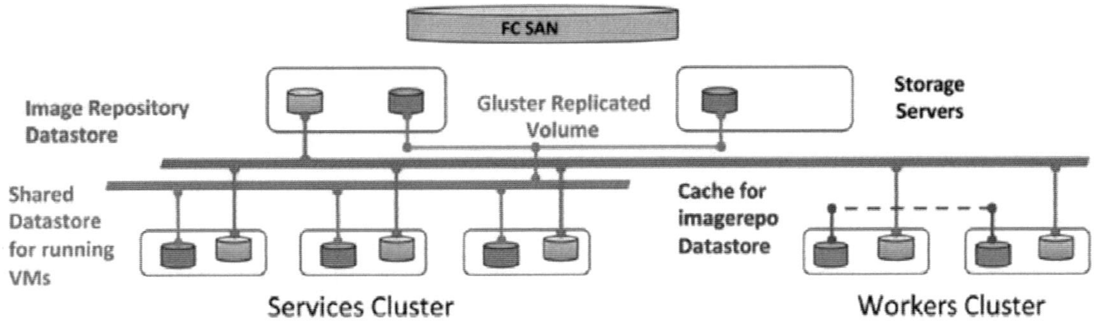

Figure 1. Example of services provided by a computing center.[1]

Computing centers provide and maintain multiple services, the most common ones are depicted in Figure 1.

In order to maintain the sites up to date it is important, on one hand to spread new computing technologies to multiple sites developing tool to optimize the usage of resources, on the other hand it is necessary to think about and build new and more efficient computing centers. Examples of efforts in this direction will be provided respectively in Section 3.1 and 3.2.

1.2. The case of BESIII

BESIII is a spectrometer hosted on the BEPCII e^+e^- collider, situated at the Institute of High Energy Physics (IHEP) of Beijing, and dedicated to the investigation of the Tau-Charm Physics. BESIII has collected around 3PB of data in the last five years. In order to face the computational effort Distributed Computing (DC) was first introduced at IHEP in 2012 and entered in production in 2014, adding external sites (80% batch computing farm, 20% grid). Later in 2015 cloud technology has been adopted as well and gained a significant role involving nowadays around 25% of the resources (the remaining 75% being composed of 10% grid and 65% batch)[3]. Moreover, IHEP computing resources are also provided to other experiments such as JUNO[4], LHAASO[5], and CEPC[6].

The distributed computing of BESIII faced an increasing amount of jobs since its adoption:

- 340k in 2014;
- 665K in 2015;
- 728K in 2016 as shown in Figure 2,

with a maximum number of running jobs of about 2k in 2015. The amount of job data exchange between the various sites is about 300TB/year. These numbers indicate how efficient the DC of BESIII is and they offer an idea of the effort needed to set up and maintain such an complex and distributed infrastructure.

2. Cloud Computing in BESIII

As discussed in Section 1.2, cloud computing is becoming the reference model for HEP computing. The rationale behind this trend can be understood by pointing out a couple of features of this approach: first of all, cloud computing is particularly suitable for different collaborations since, as it will be further discussed in Section 2.1, it allows for provisioning heterogeneous resources in a transparent way from the users' point of view. Moreover it allows for scalability and elasticity according to the actual needs of a site providing access to a shared

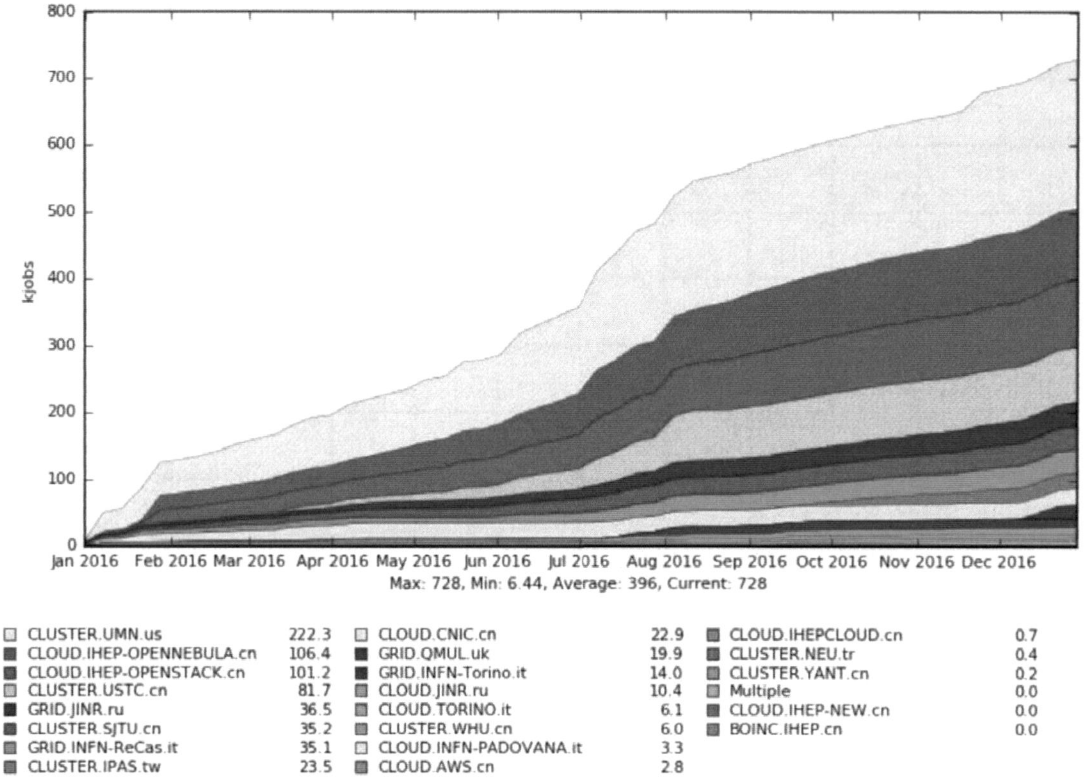

Figure 2. Cumulative jobs by site in 2016[3].

pool of computational and storage resources. Please note, setting a cloud infrastructure up may be not trivial by the providers point of view. A possible solution will be discussed in Section 3.1.

The adoption of cloud computing in BESIII not only involved IHEP but also six other sites such as the JINR in Dubna (Russia) or Torino (Italy) where an OpenNebula-based[7] cloud infrastructure has been prepared. The integration of the cloud approach with preexisting infrastructures has been implemented in an elastic way by means of VMDirac[8], a tool to integrate federated clouds in a transparent manner to the user. Thanks to this system jobs are sent to a VMDirac instance that checks if there are enough resources to create a new VM. If so, a *pilot* job is sent to switch the VM on, which will execute the job and, after that, can be eventually powered off if no further jobs are submitted. The output is anyway transferred the user before the VM is powered off.

Overall, more than 700K jobs have been processed in the past two years by the cloud sites of BESIII, thus indicating the relevance of cloud computing in the BESIII DC.

At the IHEP site three cloud managers are currently supported: OpenStack[9], OpenNebula, and AWS (Amazon Web Service)[10]. Among the others *libcloud*[11] and *rOCCI*[12] are the interfaces used to bind the different cloud infrastructures to the submitting infrastructure in a transparent way. Nevertheless finding a purpose middle-ware to meet all the different cloud-specific requirements remains a non trivial task. While OpenStack and OpenNebula are set up and maintained in site, AWS has been adopted as an attempt of elastic extension to commercial clouds, with the support of Amazon AWS China. The BESIII images were created and uploaded to AWS and VMDirac has been used in combination with AWS API for elastic scheduling.

Several tests revealed that, at the moment, self-maintenance is more convenient compared to the adoption of AWS[3].

2.1. Data Flow and Job Submission

As anticipated in Section 2, several sites provide a cloud infrastructure to BESIII. Among them the IHEP acts as central site providing tools for raw data processing, bulk reconstruction and data analysis. Remote sites are mainly dedicated to MC (Monte Carlo) production (simulation and reconstruction) and, if possible, analysis of data retrieved from IHEP Central Storage. Once the MC or the analysis has been performed, the remote sites send back to IHEP data for backup purposes[3].

The exchange of data also involves non-cloud sites for a total of fifteen sites: eight of them from China Universities and the remaining ones from USA, Italy, Russia, Turkey and Taiwan. Such a data traffic is hosted on the through 10Gb/s network linking either USA, Europe and Taiwan to China.

As far as job submission is concerned, either IHEP or remote cloud sites provide their resources. This tool is the DIRAC extension designed in order to integrate cloud infrastructures to preexisting infrastructures, and allows to perform remote operations on VMs. It acts as a separation layer between heterogeneous resources and users. Among the other tools it is worth to mention GANGA and JSUB for massive job submission, CVMFS to deploy experiment software on remote sites and StoRM for management of the Storage Elements (SE).

3. Developments in Computing

Apart from the BESIII DC, it is important to mention a couple of examples that indicate how relevant distributed and cloud computing is in the HEP context. A tool for cloud infrastructures named *Cloud Toy*[15] and a new computing facility named *Green Cube*[16] will be discussed in the following paragraphs.

3.1. Tools: Cloud Toy

The main idea behind this work is to address two common issues of cloud infrastructures: usability by simplifying the setup and installation process of an OpenNebula hypervisor minimizing the user interaction, and efficiency by making the usage of resources more dynamic, flexible and efficient allowing for an optimal and dynamic use of resources according to actual needs. The aim is to improve existing cloud infrastructures and give sites with limited manpower or cloud specific knowledge an easier access to cloud technologies[15].

The automatic installation takes place through a set of steps as follows: first of all the so-called kickstart file is created where configuration parameters, software packages, disks partitioning information are located. Then, a customized ISO file is prepared starting from a standard netinstall file in order to make it look for the kickstart at boot time. The ISO file is then burnt into a usb drive and installed on the server, which is ready in about one hour.

In this way a ready to use cloud infrastructure could be installed with a very limited user interaction. Such tool is currently under test by BelleII[13] groups and soon will be extended to BESIII[14] Distributed Computing as well.

3.2. Facilities: Green Cube

The Green IT Cube, hosted at the GSI Helmholtz Centre for Heavy Ion Research, is the result of a large investment to build new high-performance computing center. It is a 27 x 30 x 22 meter cube-shaped building on six floors side by side which hosts 768 computer racks with a water-cooled system for an overall of around 300000 CPUs and 100 PB of storage, capable to face a data I/O of over one TB per second[16]. Moreover GPUs will be widely adopted to improve parallel processing, required especially when treating heavy ions collisions (i.e. data from CBM experiment at FAIR). The cooling system is the flagship of the infrastructure: due to its features the energy requirement for cooling is between 5 and 7% of the power required for computing[17].

4. Conclusion

This work aimed to give a brief report of the computing activities of BESIII at IHEP and in remote sites. The overall infrastructure has been presented with a particular focus on the cloud technology and its integration within the preexisting infrastructure. An overview about the data flow process and the job submission system has been given as well.

Among the others, two relevant computing activities have been presented: the Cloud Toy, a tool for spreading cloud technologies, and the Green IT Cube, a futuristic infrastructure, now under construction at FAIR.

5. Acknowledgement

The research activity of this article is funded by the EU and INFN by means of the project BESIIICGEM RISE 645664 within the call H2020-MSCA-RISE-2014, and is supported in part by the NSFC project 11375221.

References

[1] Bagnasco S, Berzano D, Brunetti R, Lusso S, Vallero S, Managing a tier-2 computer centre with a private cloud infrastructure 2014 *J. Phys: Conf. Series* 523 012012
[2] BesIII Experiment, webpage http://bes3.ihep.ac.cn
[3] Zhang XM, Distributed Computing in IHEP 2017 *Spring Hepix* (Budapest, Hungary)
[4] JUNO Jiangmen Underground Neutrino Observatory, webpage http://juno.ihep.cas.cn
[5] ZHEN C, A future project at tibet: the large high altitude air shower observatory (LHAASO) 2010 *Chinese physics C* 34.2: 249
[6] Circular Electron Positron Collider, webpage http://cepc.ihep.ac.cn
[7] Montero S, Moreno-Vozmediano R and Llorente I M 2012 IaaS Cloud Architecture: From Virtualized Datacenters to Federated Cloud Infrastructures *Computer*, vol. 45, no. 12, pp. 65-72 doi:10.1109/MC.2012.76
[8] VMDirac, webpage https://github.com/DIRACGrid/VMDIRAC/wiki
[9] Sefraoui O, Aissaoui M, Eleuldj M OpenStack: toward an open-source solution for cloud computing 2012 *Int. J. of Computer Applications* 55.3
[10] AMAZON, E. C. Amazon web services, website: http://aws.amazon.com
[11] Apache Libcloud, website https://libcloud.apache.org
[12] Edmonds A, Metsch T and Papaspyrou A Open Cloud Computing Interface in Data Management-related Setups 2011 *Springer Grid and Cloud Database Management* pp. 127
[13] Grzymkowski R, Hara T and computing group B 2015 *J. Phys: Conf. Series* 664 022021
[14] Zhang XM, Yan T, Zhao X H, Ma Z T, Yan X F, Lin T, Deng Z Y, Li W D, Belov S, Pelevanyuk I, Zhemchugov A and Cai H 2015 *J. Phys: Conf. Series* 664 032036
[15] Astorino F, Amoroso A, Bagnasco S, Balashov N A, Bianchi F, Destefanis M, Maggiora M, Pellegrino J, Yan L, Yan T, Zhang X and Zhao X, A modular (almost) automatic set-up for elastic multi-tenants cloud (micro)infrastructures, *Proc. Int. Conf. on Computing in High Energy and Nuclear Physics,* (San Francisco, USA)
[16] Green IT Cube, webpage, http://www.fair-center.eu/news-events/news-view/article/green-it-cube.html
[17] Kollegger T, Green Cube GSI FAIR Tier 0 2015 *AIME Big Data* (Budapest, Hungary)

IOP Conf. Series: Journal of Physics: Conf. Series **1024** (2018) 012024 doi:10.1088/1742-6596/1024/1/012024

Higher Order Moments of Proton Number Fluctuations in Au+Au Collisions at 1.23 AGeV studied with HADES

Melanie Szala for the HADES collaboration

Goethe Universität, Frankfurt am Main, Germany

E-mail: m.szala@gsi.de

Abstract.
The HADES (**H**igh **A**cceptance **D**i**E**lectron **S**pectrometer) experiment at the SIS18 accelerator of the GSI Helmholtzzentrum für Schwerionenforschung investigates heavy-ion collisions at kinetic beam energies of a few GeV per nucleon. We have carried out a detailed moment analysis of (net-)proton distributions of Au+Au collisions with 1.23A GeV ($\sqrt{s_{NN}}$=2.41 GeV). We present here the correction techniques used in this investigation.

1. Introduction

Heavy-ion collisions at kinetic beam energies of a few GeV per nucleon provide access to the thermodynamics of QCD in the low T and high μ_B region of the phase diagram. In particular, higher order moments of conserved quantities (e.g. baryon number, charge, strangeness) are predicted to be sensitive to a first order phase transition and especially to the critical point of the QCD phase diagram [1]. Indeed, fluctuations characteristic of the latter features would modify these moments. On the other hand, all kind of detector effects might also modify the measured distributions, hence careful investigations of the efficiency corrections are mandatory. Measuring at SIS18, the HADES detector allows to extend the data taken in the RHIC beam-energy scan [2] to bombarding energies in the few-GeV regime. In this contribution we discuss our investigations of efficiency corrections to the observed fluctuation signal.
The first four moments of a particle number distribution are its mean (M), its variance (σ), its skewness (S) and its kurtosis (κ). To cancel the volume effect of the collision system to first order the scaled skewness $S \cdot \sigma$ and kurtosis $\kappa \cdot \sigma^2$ are often used instead. It is assumed that these experimental observables are related to thermodynamic susceptibility ratios which can be directly calculated in Lattice QCD or in models, e.g. the Hadron Resonance Gas (HRG) model.

2. Experimental Setup

The HADES spectrometer is a fixed target experiment located at the heavy-ion synchrotron SIS18 of the Helmholtzzentrum für Schwerionenforschung GSI in Darmstadt, Germany [3]. Since 2002, the HADES experiment operates with the aim of investigating dense matter in the energy regime of a few GeV per nucleon.
The spectrometer was constructed to have a large geometrical acceptance covering polar angles

from 18 to 85 degrees and almost the full azimuthal angle (see Fig. 1). Further characteristics of the spectrometer are its good momentum resolution and high read-out rate.
In 2012 HADES has recorded about seven billion Au+Au collisions at a bombarding energy of 1.23 GeV/u ($\sqrt{s_{NN}}$ = 2.41 GeV). With these large statistics the higher order moments of proton number distributions can be investigated in great detail.

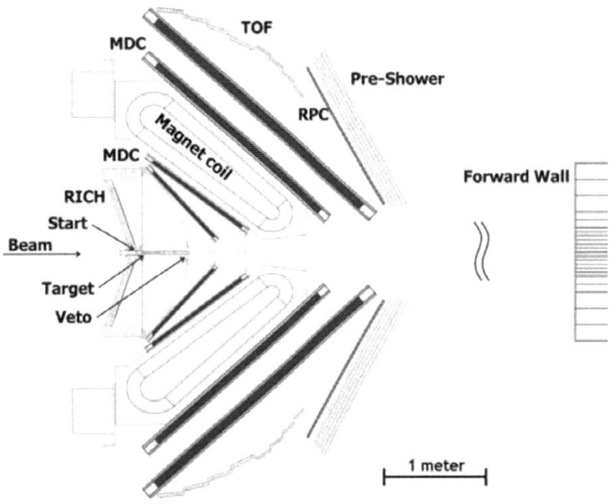

Figure 1. Cross section of the HADES spectrometer. The spectrometer is divided into 6 identical sectors that are placed symmetrically around the beam axis.

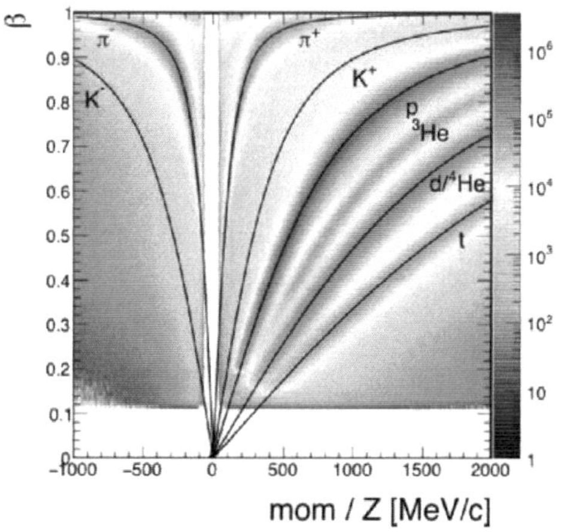

Figure 2. Correlation between the momentum and β of all selected tracks. Black lines correspond to the expected values for different particle species.

The particle identification is based on a velocity vs. momentum x polarity correlation (Fig. 2), where the velocity is determined from the time-of-flight measurement in the TOF and RPC detectors with respect to the time-zero information of the START detector and the tracked flight path. If needed, additional particle discrimination power is gained from the energy loss (dE/dx) information in the MDC and TOF.
Particle multiplicity is usually used in the centrality determination. However, for the moment analysis, a proton-independent centrality determination is necessary to avoid autocorrelations between protons involved in our moment analysis and the centrality definition. Therefore, the projectile spectators of the collision are detected by a forward wall installed at the end of the HADES spectrometer (Fig. 1). It consists of plastic scintillators and covers the small polar angle range (θ = 0.5 - 7.5°). In the scintillators energies deposited by the spectators and the centrality of the collision can be derived therefrom.
Our analysis investigates fluctuations of conserved quantities. When considering the entire reaction system, there will be no fluctuations in the conserved quantities. In the investigation of a sufficiently small subsystem, fluctuations occur because the subsystem can exchange conserved quantities with the rest of the reaction volume. The entire system corresponds to the total rapidity distribution of all particles. The subsystem used for our further analysis is defined by the particles within a smaller subsystem of the total rapidity distribution. In Figure 3 the measured rapidity distribution for protons in Au + Au collisions at $\sqrt{s_{NN}}$= 2.41 GeV is shown. For the further analysis a phase space bin of $y_{accept} = y_{lab} \pm 0.2$ was selected and additionally a transversal momentum of $p_T = 400 - 1600$ MeV/c in order to investigate a reasonably large subsystem avoiding spectator contributions and edge effects of the detector.

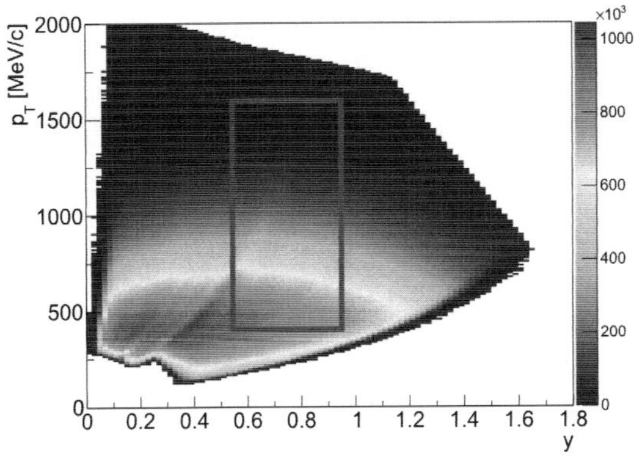

Figure 3. Experimental phase space distribution of protons within HADES and our phase space selection (green rectangle) for Au+Au collisions at $\sqrt{s_{NN}} = 2.41$ GeV.

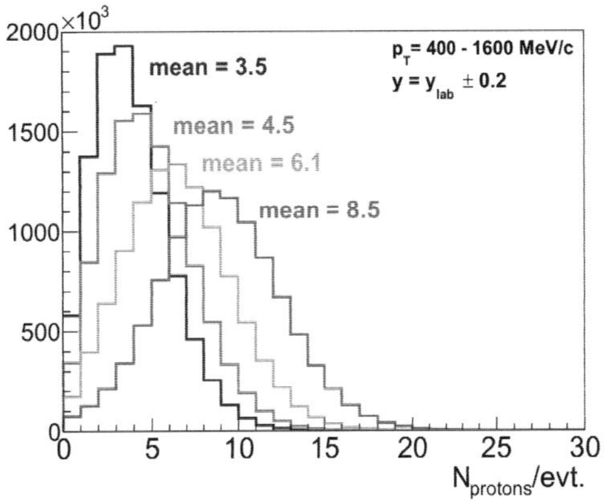

Figure 4. Uncorrected N_p multiplicity distributions within the selection for 0-10% (red), 10-20 % (yellow), 20-30 % (green) and 30-40 % (blue) central collisions.

3. Efficiency Correction

The efficiency correction is an important step in obtaining the true higher order moments of the proton multiplicity distribution. It has been shown in [4] that the moments of the proton multiplicity distributions are drastically influenced by acceptance and efficiency. In order to make valid statements about the properties of the created system the true distributions have to be extracted. For this purpose, detailed investigations of different efficiency correction methods were performed with Au + Au UrQMD events and full scale GEANT detector simulations. Two different approaches were tested: correction of the moments (as proposed in [4, 5]) and unfolding of the distributions [6].

The first method is based on the calculated cumulants of the measured particle distribution (see Fig. 4). To correlate the true cumulants with the experimentally measured cumulants, a correction was derived assuming a binomial probability parameter [4].

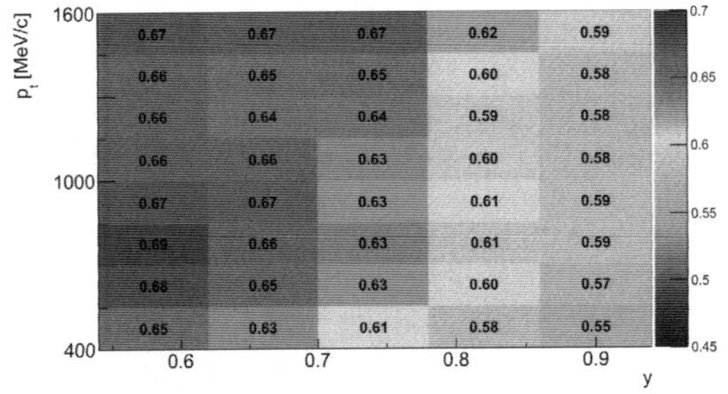

Figure 5. Combined efficiency and acceptance correction factor for protons in the analysed phase space cells in 30-40% central collisions.

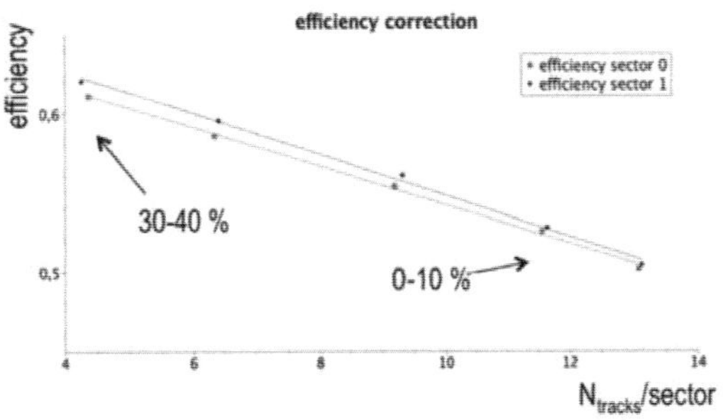

Figure 6. Centrality dependent efficiency shown for two different sectors of the HADES spectrometer. With this correlation, a sector and centrality dependent event-by-event efficiency correction is performed.

The correction was subsequently extended so that the efficiencies depend on the particle species as well as the phase space, such as transverse momentum and rapidity [5]. The efficiencies for the different centrality classes inside the selected phase space window are obtained from UrQMD events and a GEANT detector simulation (see Fig. 5 and 6). A special feature of the HADES spectrometer is its subdivision into 6 identical sectors which are placed symmetrically around the beam axis and are read out separately. Because of anisotropic particle emission in non-central collisions, the efficiencies differ from sector to sector within an event due to the orientation of the event plane. The resulting approach is a dynamic efficiency correction, which is carried out sector-by-sector and event-by-event (E-by-e). For each sector, the average track density is determined for different centrality selections and then linearly interpolated. The centrality dependance is shown in Figure 6 as an example for two different sectors. A dynamic efficiency can now be determined and used for each event in dependence on the centrality, the sector, the track density within the sector and the phase space.
Unfolding is a mathematical method that is used to deconvolute two folded functions. While a convolution can be easily calculated, its inversion is not always possible.
For this purpose, the TUnfold package implemented in ROOT, which is essentially based on a regularized matrix inversion, is used in this analysis. The unfolding is based on the least squares method, including Tikhonov regularisation and an optional area conservation [6]. To determine the strength of the regularization parameter, the L-curve method and the scans of global correlation coefficients are implemented.

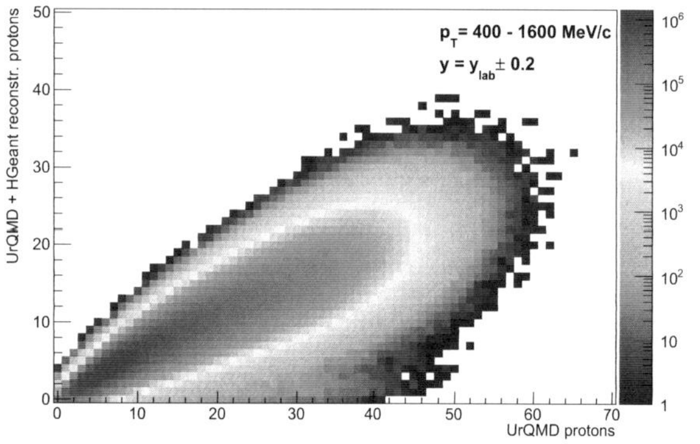

Figure 7. Simulated response matrix for the unfolding correction method.

In the experiment, the proton distributions are measured only within the acceptance of the detector and under the influence of detector and trace reconstruction efficiency. For this purpose, a detector response matrix A is required for the unfolding. In order to determine this response matrix, a realistic detector simulation is used.
The response matrix A for the HADES detector is shown in Figure 7 and is filled using an UrQMD and Geant simulation. On the x-axis, the number of all UrQMD protons per event is shown; it corresponds to the true number of protons. The number of protons per event after particle detection and reconstruction is shown on the y-axis. The response matrix is only filled with particles within the previously selected phase space window of $y_{accept} = y_{lab} \pm 0.2$ and $p_T = 400 - 1600$ MeV/c.
To compare the two methods of efficiency correction, the unfolding method was applied in the same centrality classes and the same phase space bin. Within error bars, the unfolding method delivers a reliable result.
After their successful assessment in the simulation, both methods were applied to the HADES experimental data and the higher moments of the proton distributions were extracted. We find

IOP Conf. Series: Journal of Physics: Conf. Series **1024** (2018) 012024 doi:10.1088/1742-6596/1024/1/012024

that the two methods provide plausible and consistent results in the Au + Au data (see Fig. 8 for preliminary results).

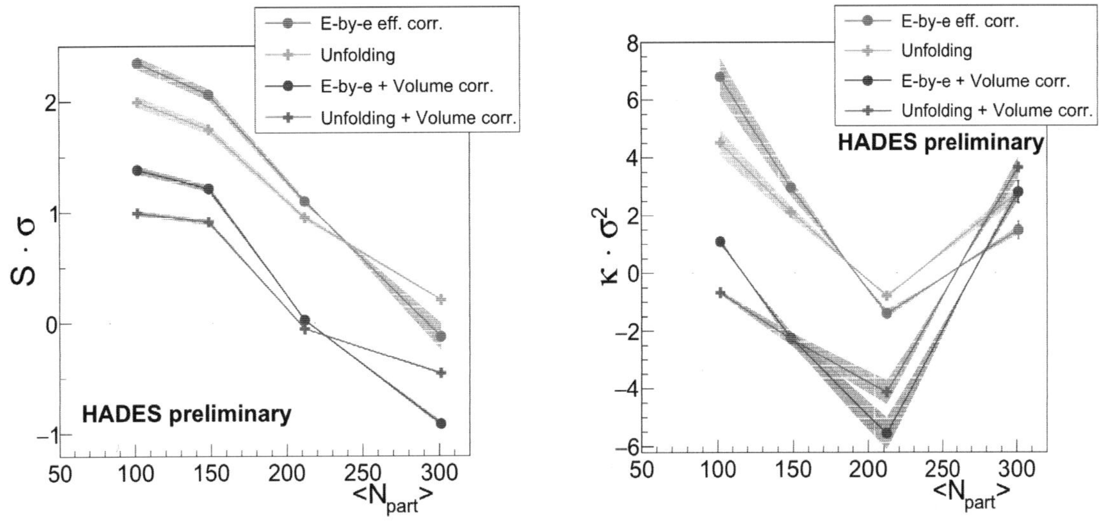

Figure 8. Measured S $\cdot$ σ and $\kappa \cdot \sigma^2$ of protons as function of mean number of participants. Two different methods of efficiency correction were applied, correcting the moments (green) and unfolding (yellow). The results are modified further by volume fluctuation corrections based on a Glauber model (see text).

4. Volume Fluctuations

Experimentally measured dynamical event-by-event fluctuation signals such as cumulants of (net-)particle distributions can also be modified by the fluctuations of the number of participants in a given centrality selection [7, 8]. To correct the data for this effect, we have applied the procedure of Skokov et al. [7] on the higher order moments (see Fig. 8).
For the volume correction we assume that the volume scales like the number of participants and the corresponding N_{part} distributions were modeled with the Glauber and IQMD models. The volume corrections are thus model-depend and investigations of systematic effects are on-going.

5. Summary and Outlook

The results shown here are only preliminary results and the analysis of the higher moments is on-going. Investigations of the influence of the centrality selection and the model-dependance of volume fluctuations on the higher moments are needed to obtain systematic uncertainties. The necessary corrections need to be carefully studied, before any comparison with other experiment and theory becomes meaningful.

References

[1] M.A. Stepanov, Phys. Rev. Lett. **102** (2009) 03230
[2] L. Adamczyk *et al.* [STAR Collaboration], Phys. Rev. Lett. **112** (2014) 032302
[3] G. Agakishiev *et al.* [HADES Collaboration], Eur. Phys. J. A **41** (2009) 243-277
[4] A. Bzdak and V. Koch, Phys. Rev. C **86** (2012) 044904
[5] A. Bzdak and V. Koch, Phys. Rev. C **91** (2015) no.2, 027901
[6] S. Schmitt, JINST **7** (2012) T10003
[7] V. Skokov, B. Friman, K. Redlich, Phys. Rev. C **88** (2013) 034911
[8] P. Braun-Munzinger, A. Rustamov, J. Stachel, Nucl. Phys. A **960** (2017) 114-130

Spinodal instability of baryon-rich quark matter

Feng Li[1] **and Che Ming Ko**[2]

[1] Frankfurt Institute for Advanced Studies, Ruth-Moufang Str. 1, Frankfurt 60438, Germany
[2] Cyclotron Institute and Department of Physics, Texas A&M University, College Station, Texas 77843-3366, USA

E-mail: fengli@fias.uni-frankfurt.de

Abstract. The spinodal instabilities of both confined and expanding baryon-rich quark matter are studied in a transport model derived from the Nambu-Jona-Lasino model. Appreciable higher-order density moments are seen as a result of the first-order phase transition in both cases. The skewness of the quark number event-by-event distribution in a small sub-volume of the system becomes appreciable for the confined quark matter. For the expanding quark matter, the density fluctuations lead to enhanced the dilepton yield.

1. Introduction

The beam energy scan (BES) experiments [1, 2, 3] at the Relativistic Heavy Ion Collider (RHIC) are designed for the search of the critical end point (CEP) in the baryon-rich region of the QCD phase diagram. At CEP, the first-order phase transition between the hadronic matter and the quark-gluon-plasma (QGP) changes to a second-order one. To understand the properties of this phase transition, both schematic models and hydrodynamic approach [4, 5, 6] have been used to study how the spinodal instability of a QGP leads to the self-amplified deviation from its equilibrium state during the phase transition. Using the test particle method, the spinodal instabilities of both the confined and expanding quark matter have been studied in a transport model that is derived from the Nambu-Jona-Lasino (NJL) model [7]. By varying the coupling strength of the vector interaction G_V in the NJL model, different equations of state are obtained for a baryon-rich quark matter. A first-order phase transition occurs if $G_V = 0$ with the CEP at about $T_c \simeq 70$ MeV and $n_c \simeq 0.9$ fm^{-3} [8], and it is absent if $G_V = G_S$. The density fluctuations induced by the spinodal instability can be quantified by the scaled density moments $\langle\rho^N\rangle/\langle\rho\rangle^N$ [6], where $\langle\rho^N\rangle \equiv \int d^3\mathbf{r}\rho(\mathbf{r})^{N+1}/\int d^3\mathbf{r}\rho(\mathbf{r})$. They are all equal to one for a uniform density distribution but become larger if the density is nonuniform. Also, density fluctuations can affect the dilepton yield in heavy ion collisions. Results on these observables for the two cases of with and without a first-order phase transition are compared in the following for both the confined and the expanding quark matter.

2. Quark matter in a box

Shown in the left window of Fig. 1 by the dotted, dashed, and solid lines are the time evolution of scaled density moments for $N = 2$, 4 and 6, respectively, during the first-order phase transition of a quark matter that is confined in a box with an initially uniform net quark density $n_q = 0.5$ fm^{-3} and temperature $T = 20$ MeV. They are obtained by averaging over 1,000 events. The scaled moments are seen to increase during the phase separation and reach their saturated values at

IOP Conf. Series: Journal of Physics: Conf. Series **1024** (2018) 012025 doi:10.1088/1742-6596/1024/1/012025

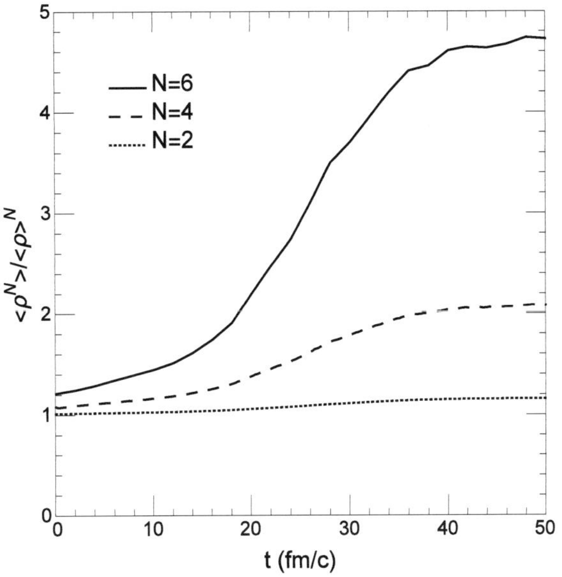

Figure 1. Time evolution of scaled density moments during a first-order phase transition for a quark matter confined in a box with an initially uniform net quark density $n_q = 0.5$ fm^{-3} and temperature $T = 20$ MeV.

about $t = 40$ fm$/c$, when the phase separation almost ends. Also, moments with larger N increase faster and saturate at larger values.

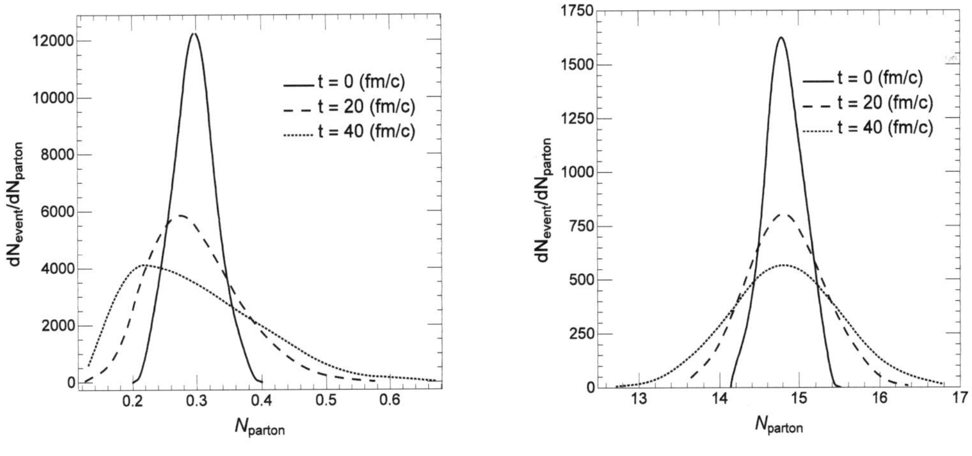

Figure 2. Time evolution of the event-by-event distribution of the number of quarks in a sub-volume of size 0.6 fm^3 (left window) and 30 fm^3 (right window) for a quark matter of temperature $T = 20$ MeV and average net quark density $n_q = 0.5$ fm^{-3} inside the spinodal region. The total number of events is 1000.

The number of quarks inside a sub-volume also varies drastically from event to event during a first-order phase transition as shown in Fig. 2 for the central cell from 1000 events at $t = 0$, 20, and 40 fm$/c$ by the solid, dashed and dotted lines, respectively, for the two cases of sub-volume of size 0.6 fm^3 (left window) and 30 fm^3 (right window). The left panel of Fig. 2 clearly shows

IOP Conf. Series: Journal of Physics: Conf. Series **1024** (2018) 012025 doi:10.1088/1742-6596/1024/1/012025

that the distribution for the small sub-volume becomes asymmetric as time increases, starting with an initial skewness of 0.11 and increasing to 0.60 at 20 fm/c and 0.75 at 40 fm/c, agreeing with the prediction in Ref. [9]. This feature is absent in the right panel of Fig. 2 for the larger sub-volume, where the distribution remains essentially symmetric with increasing time.

3. Expanding quark matter

For an expanding quark matter as in relativistic heavy ion collisions, the initial partons are obtained from a multiphase transport (AMPT) model with vanishing parton scattering cross sections in Zhang's parton cascade (ZPC)[10], with the hadronic afterburner based on a relativistic transport (ART) [11, 12] turned off and with string melting [13] that uses the heavy ion jet interaction generator (HIJING) [14, 15, 16] as the input. In the AMPT, these partons are assumed to be produced on the light cone, i.e., on a thin disk perpendicular to the longitudinal axis or beam direction, which is reasonable in collisions at the top energy of RHIC. For collisions at lower energies such as in the BES experiments, the initial parton distribution is expected to be more extended in the longitudinal direction, and we take this into account by increasing the thickness of the initial disk to $2dm_N/\sqrt{s_{NN}}$ for head-on collisions, where d is the diameter of the colliding nuclei, and assuming that the initial partons are uniformly distributed in this disk. Using 1000 test partons for each physical partons produced from Au+Au collisions at zero impact parameter and a center-of-mass energy $\sqrt{s_{NN}} = 2.5$ GeV and smearing the distribution of initial partons from the AMPT with a longitudinal length of $d/3.5$ to mimic the geometry of the initial quark configuration for collisions at $\sqrt{s_{NN}} = 7$ GeV, we have found that some parts of the system go through the spinodal region when the SU(3) NJL model with $G_V = 0$.

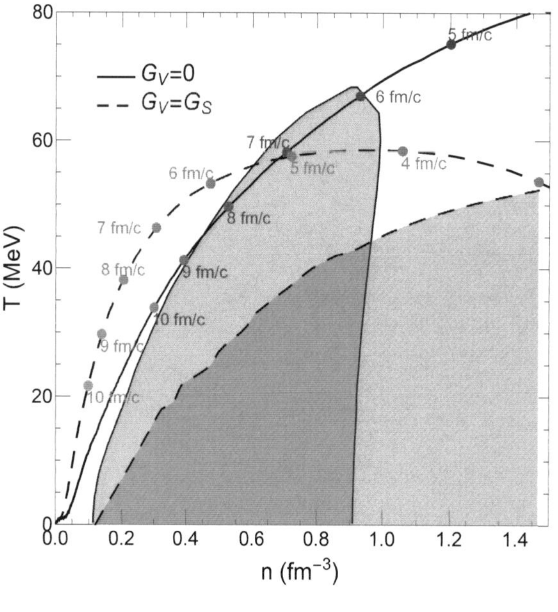

Figure 3. The phase trajectories of the central part of an expanding quark matter obtained by averaging over 100 events for the cases with (solid line) and without (dashed line) a first-order phase transition using the initial parton distribution from the AMPT model that is smeared in the longitudinal direction as described in the text. The spinodal region in the case of $G_V = 0$ is shown by the gray color, and it disappears for $G_V = G_S$. The region where quark matter can be bound in the case of $G_V = 0$ is shown by red color.

As shown by the solid line in Fig. 3, the phase trajectory of the central part of the system,

IOP Conf. Series: Journal of Physics: Conf. Series **1024** (2018) 012025 doi:10.1088/1742-6596/1024/1/012025

obtained by averaging over 100 events for the case with a first-order phase transition, goes into the spinodal instability region, which is colored in gray, at about 6 fm/c after expansion, and moves out of this region at about 8.5 fm/c. For the case without a first-order phase transition, the trajectory is shown by the dashed line in Fig. 3. The temperatures are obtained by solving the equations of state using both the energy density and the baryon number density as the inputs. It's shown in Fig. 3 that the expansion is slowed down by the first-order phase transition. The simulation also shows that in the case with a first-order phase transition, i.e. G_V is set to zero, a disk of high density appears due to the lower pressure as a result of the spinodal instability, which later transforms into several disjointed clusters on a ring at $z = 0$. These cold and dense clusters stay in the spinodal instability region even at $t = 14$ fm/c.

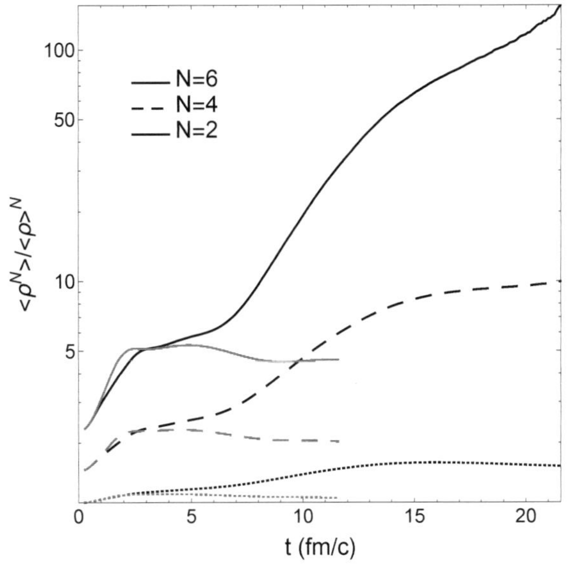

Figure 4. Scaled density moments averaged over 100 events as functions of time for the cases with (black lines) and without (red lines) a first-order phase transition.

These clusters can again be quantified by the event-averaged scaled density moments as shown in Fig. 4 for the 2nd (dotted curve), 4th (dashed curve), and 6th (solid curve) order scaled moments with the black and red colors denoting results obtained with and without a first-order phase transition, respectively. It is seen that the density moments keep on increasing during the spinodal decomposition. Such a feature is more prominent for the higher-order scaled density moments.

Because the rate of the dilepton production through quark-antiquark annihilation is proportional to the product of their densities, it will also be enhanced as the density fluctuations become large. The dilepton invariant mass spectra obtained by averaging over 100 events are plotted in Fig. 5 by the solid and dashed lines for the cases with and without a first-order phase transition, respectively. As expected, more dileptions are produced from a quark matter with a first-order phase transition.

4. Summary

The spinodal instability of a baryon-rich quark matter has been studied in the transport model that is based on the NJL model. Dense clumps are observed in both confined and expanding quark matter, and they lead to the enhancement of higher-order scaled density moments. The skewness of the quark number event-by-event distribution in a small sub-volume also increases

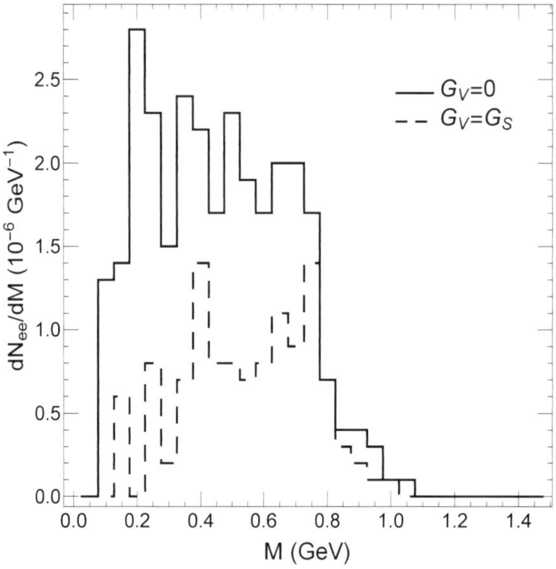

Figure 5. Dilepton yield as a function of the invariant mass $M_{e^+e^-}$ for the cases with (solid line) and without (dashed line) a first-order phase transition in an expanding quark matter with the blast wave initial conditions.

during the first-order phase transition in the confined quark matter, but this feature disappears if the sub-volume is large. The density fluctuations further enhance the dilepton production through quark-antiquark annihilation in the expanding quark matter.

Acknowledgments

This work was supported in part by the US Department of Energy under Contract No. DE-SC0015266 and the Welch Foundation under Grant No. A-1358.

References

[1] Adamczyk L *et al.* (STAR) 2014 *Phys. Rev. Lett.* **112** 032302
[2] Adamczyk L *et al.* (STAR) 2014 *Phys. Rev. Lett.* **113** 092301
[3] McDonald D (STAR) 2015 *EPJ Web Conf.* **95** 01009
[4] Steinheimer J and Randrup J 2012 *Phys. Rev. Lett.* **109**(21) 212301
[5] Herold C, Nahrgang M, Mishustin I and Bleicher M 2014 *Nucl. Phys. A* **925** 14 – 24 ISSN 0375-9474
[6] Steinheimer J and Randrup J 2013 *Phys. Rev. C* **87**(5) 054903
[7] Klevansky S P, Ogura A and Hufner J 1997 *Ann. Phys.* **261** 37–73
[8] Li F and Ko C M 2016 *Phys. Rev. C* **93**(3) 035205
[9] Stephanov M A 2009 *Phys. Rev. Lett.* **102**(3) 032301
[10] Zhang B 1998 *Comput. Phys. Commun.* **109** 193 – 206 ISSN 0010-4655
[11] Li B A and Ko C M 1995 *Phys. Rev. C* **52**(4) 2037–2063
[12] LI B A, Sustich A T, Zhang B and Ko C M 2001 *Int. J. Mod. Phys. E* **10** 267–352
[13] Lin Z W, Ko C M, Li B A, Zhang B and Pal S 2005 *Phys. Rev. C* **72** 064901
[14] Wang X N 1991 *Phys. Rev. D* **43**(1) 104–112
[15] Wang X N and Gyulassy M 1991 *Phys. Rev. D* **44**(11) 3501–3516
[16] Gyulassy M and Wang X N 1994 *Comput. Phys. Commun.* **83** 307 – 331 ISSN 0010-4655

Recent results on small μ_B QCD from the lattice

A. Pásztor

Department of Physics, Wuppertal University, Gaussstr. 20, D-42119 41 Wuppertal, Germany

E-mail: apasztor@bodri.elte.hu

Abstract. I will briefly review our current understanding of the phase diagram of QCD, from the point of view of recent first principle lattice calculations. Whenever possible, I will also make contact with the experimental efforts of the RHIC Beam Energy Scan Program.

1. Introduction

The equilibrium phase diagram of QCD remains vastly unexplored. At $\mu = 0$ there are some well established results from lattice simulations. E.g. the QCD transition at $\mu = 0$ is a smooth crossover [1]. The equation of state (EoS) near and above the cross-over is also known [2, 3] in 2+1 flavour[1] QCD. Recently, the EoS of 2+1+1 flavour[2] QCD, was also calculated [4]. In this conference contribution I will focus on reviewing some of the recent progress on small finite μ_B QCD. The large μ area of the phase diagram still remains unreachable by lattice techniques, due to the complex action problem at finite density.

Probably the most important unresolved issue is about the QCD critical point. It is often conjectured that the crossover line eventually becomes a first order transition line in the $\mu - T$ plane, the two being separated by a critical point. The existence of this critical point is not established however, and if it exists, we do not know where it is. The discovery of the critical point and its location is one of the main challenges in the physics of strong interactions, and one of the main focuses of the RHIC Beam Energy Scan II program, scheduled for 2019 and 2020.

2. Taylor expansion vs. analytical continuation

So far there are only two methods that were used for full QCD (meaning lattice QCD simulations with physical quark masses and also preforming continuum extrapolation). The Taylor expansion method and the analytical continuation method [6, 7]. Both of these start by evaluating fluctuations of conserved charges, i.e. derivatives of the pressure with respect to the chemical potentials:

$$\chi^{u,d,s}_{i,j,k} = \left. \frac{\partial^{i+j+k} \left(p/T^4\right)}{(\partial\hat{\mu}_u)^i(\partial\hat{\mu}_d)^j(\partial\hat{\mu}_s)^k} \right|_{\mu=0} . \tag{1}$$

One can also consider derivatives with respect to μ_B/T, μ_I/T, μ_Q/T instead. The Taylor expansion method calculates the higher derivatives at $\mu = 0$ directly while the analytical

[1] 2+1 here means 2 light quarks with a degenerate mass and a strange quark
[2] 2+1+1 here means 2 light quark with a degenerate mass, a strange quark and a charm quark

continuation method calculates lower derivatives at different values of $\mathrm{Im}\,\mu$, and uses fits in $\mathrm{Im}\,\mu$ to estimate the higher derivatives.

These derivatives are interesting for several reasons. They are sensitive to criticality, They provide stringent tests of the hadron resonance gas (HRG) model at low and resummed perturbation theory at high temperatures[3]. They can also be used for extrapolation of observables to small finite μ.

A major difficulty is that remnants of the sign problem are encountered when evaluating the $\chi^{u,d,s}_{i,j,k}$. More concretely, there are some huge cancellations, as say χ^B_{2k} is finite as $V \to \infty$, but is the sum of terms scaling up to $\mathcal{O}(V^{k-1})$. The Taylor expansion method therefore penalizes calculating higher derivatives at larger volumes. This already leads us to the main differences between the direct Taylor expansion method and the analytical continuation method.

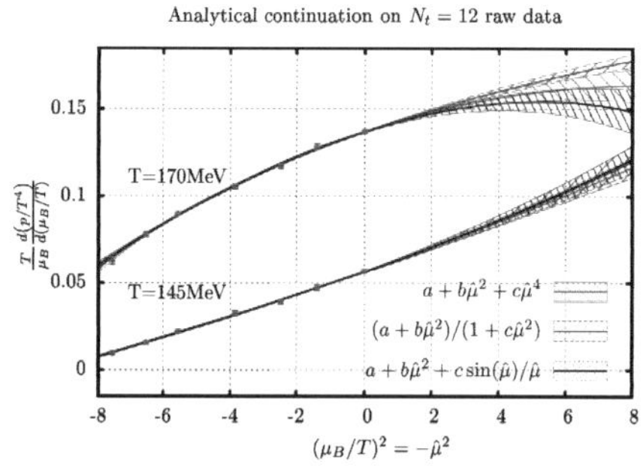

Figure 1. Illustrating analytical continuation for the observable $\frac{T}{\mu_B}\frac{d(p/T^4)}{d(\mu_B/T)}$ for $N_t = 12$ lattice data. The Taylor expansion method only uses, the black $\mu = 0$ point, but with much higher statistics.

The obvious disadvantage of the analytical continuation method is the presence of systematic error, due to the extrapolation to real μ. This is illustrated in Fig. 1. The main advantage is that the analytical continuation method penalizes large volumes much less than the Taylor expansion. This is a consequence of the fact that a lower number of derivatives has to be taken directly.

3. Some recent results

3.1. Taylor coefficients and EoS at small finite μ

The QCD pressure can be written as a Taylor expansion around $\mu_B/T = 0$ [8]:

$$\frac{p(T,\mu_B)}{T^4} = \sum_{n=0}^{\infty} c_{2n}(T)\left(\frac{\mu_B}{T}\right)^{2n}. \tag{2}$$

The coefficients $c_i(T)$ can be obtained from lattice simulations, either directly at $\mu_B = 0$ or by using analytical continuation techniques from imaginary μ_B. The first results for c_2, c_4 and c_6 were obtained at finite lattice spacing and for heavier-than-physical quark masses [8]. The first continuum extrapolated results for c_2 appeared in Ref. [9]; c_4 was obtained for the first time in Ref. [10] at finite lattice spacing. The continuum-extrapolated results for c_6 were calculated for the first time in [11], and later in [12]. Ref. [13] includes a first determination of c_8, at two values of the temperature and for $N_t = 8$. The results from Ref. [11] can be seen in Fig. 2. In our simulations at imaginary baryonic chemical potential, the choice has to be made for

[3] See e.g. [5] for many comparisons.

the strangeness μ_S and electric charge μ_Q chemical potential values to be such that we have vanishing net-strangeness and net-electric charge being 0.4 times the net-baryon number, like it is for the lead or gold nuclei used for heavy ion collision experiments.

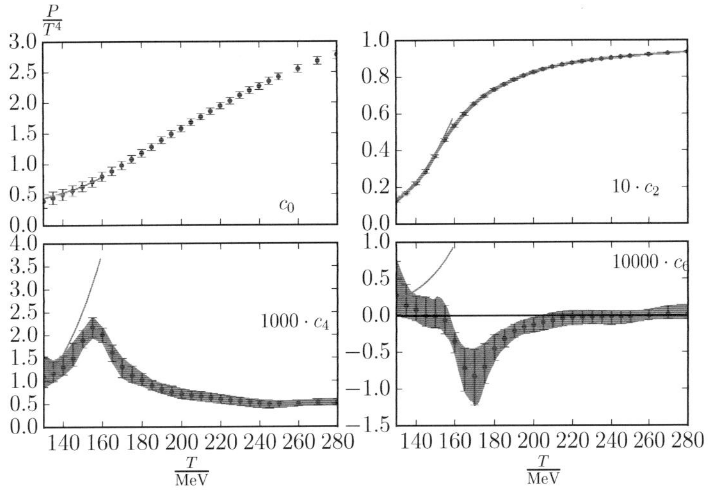

Figure 2. From Ref. [11]. Taylor expansion coefficients $c_0, ...c_6$ for the pressure in the case of strangeness neutrality, as functions of the temperature.

We note that the Taylor coefficients in principle allow us to give constraints of the location of the critical point, due to the low order of the expansion currently available, no reliable estimates can be given at this point.

3.2. Curvature of the crossover temperature

Just like with the equation of state, the transition temperature can also be expanded as a Taylor series around $\mu_B = 0$ [16]:

$$\frac{T_c(\mu_B)}{T_c(0)} = 1 - \kappa \left(\frac{\mu_B}{T_c(0)}\right)^2 + \lambda \left(\frac{\mu_B}{T_c(0)}\right)^4 + ... \quad . \tag{3}$$

The Taylor coefficients κ and λ here can be extracted by either direct Taylor expansion at $\mu = 0$ or analytical continuation from imaginary μ. Here, κ is the curvature of the phase diagram at $\mu_B = 0$. The results available in the literature correspond to different choices for μ_S. A value of $\kappa = 0.020(4)$ [15] has been found when $\mu_s = \mu_{u,d}$, and $m_s/m_{u,d} = 20$. Ref. [14] got $\kappa = 0.0135(20)$ at $\mu_S = 0$. In Ref. [16], a value of $\kappa = 0.0149(21)$ was found in the case of strangeness neutrality ($\langle n_s \rangle = 0$). There are no results on the second coefficient λ as of now.

3.3. Fugacity expansion and hadron chemistry

For simplicity we set $\mu_Q = 0 = \mu_S$. Instead of the Taylor expansion, one can consider the fugacity expansion $p(T, \mu_B)/T^4 = \sum_{k=0}^{\infty} a_k(T) \cosh(k\,\mu_B/T)$. At $\mu_B = 0$, the pressure is a sum of the a_k coefficients, so these can be interpreted as the partial pressures from the sectors of the Hilbert space with a value for the conserved baryon number. If one considers χ_1^B at a imaginary $\mu_B = i\tilde{\mu}_B$, one gets:

$$\chi_1^B(T, i\tilde{\mu}_B) = i \sum_{k=1}^{\infty} b_k(T) \sin(k\,\tilde{\mu}_B/T), \tag{4}$$

where $b_k = ka_k$. If the $\tilde{\mu}_B$-dependence of χ_1^B is known, then the b_k can be calculated by a simple Fourier transform. The real μ_B fugacity expansion coefficients correspond to Fourier coefficients at imaginary μ_B. This makes the fugacity expansion more natural to consider, when working with imaginary μ lattice data, as compared to the Taylor expansion in μ_B. If we consider now a nonzero μ_S as well, we get a double expansion in both $e^{\mu_S/T}$ and $e^{\mu_B/T}$ and we can separate the Hilbert space even further, to get the partial pressure from say strange hadron as opposed to non-strange hadrons. This possibility of decomposing the thermodynamics to partial pressures is especially useful when testing hadronic model near and below T_c, as the contributions from different sectors do not mix. This was pointed out in [17], where the different strangeness sectors were separated, and used to constrain the hadronic spectrum in the HRG model. The strangeness partial pressures obtained in [17] can be seen in Fig. 3. The HRG does not undershoot the full pressure and the strangeless baryonic partial pressure, it does undershoot however every partial pressure with nonzero strangeness, including mesons and baryons.

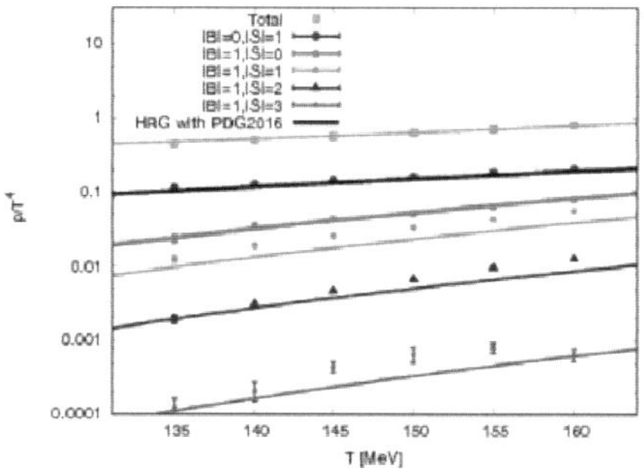

Figure 3. From Ref. [17]. The partial pressures of the different strangeness sectors together with the standard PDG hadron spectrum HRG prediction.

An other application was in [18], where only the baryon number sectors were separated, and used to test repulsive baryonic interactions, introduced as an excluded volume correction. It was found that including repulsive interactions between baryons make hadronic description much better near and slightly above T_c as opposed to the ideal HRG. A summary of the results can be seen in Fig. 4.

These were just some examples of recent progress in small μ physics, a very popular topic in lattice QCD at the moment. Much remains to be done in small μ physics still, but to tackle the real challenge, extending out knowledge to higher μ, fresh ideas are needed.

Acknowledgments

This project was funded by the DFG grant SFB/TR55. This material is based upon work supported by the National Science Foundation through grant number NSF PHY-1513864 and by the U.S. Department of Energy, Office of Science, Office of Nuclear Physics, within the framework of the Beam Energy Scan Theory (BEST) Topical Collaboration. An award of computer time was provided by the INCITE program. This research used resources of the Argonne Leadership Computing Facility, which is a DOE Office of Science User Facility supported under Contract DE-AC02-06CH11357. The authors gratefully acknowledge the Gauss Centre for Supercomputing (GCS) for providing computing time for a GCS Large-Scale Project on the GCS share of the supercomputer JUQUEEN [20] at Jülich Supercomputing Centre (JSC), and at HazelHen supercomputer at HLRS, Stuttgart.

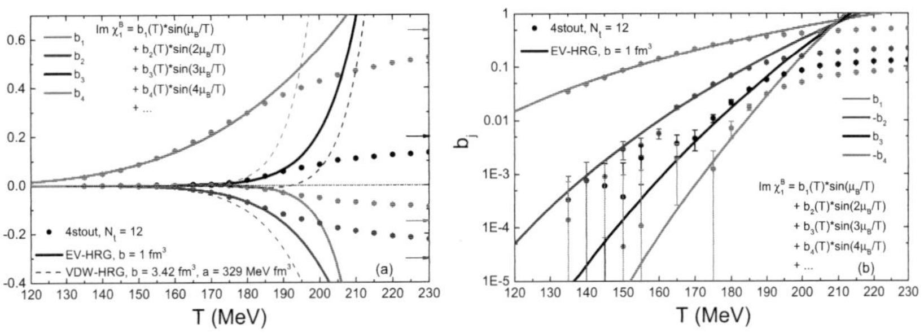

Figure 4. The temperature dependence of the first four Fourier coefficients on $N_t = 12$ lattices, and from the excluded volume HRG model with excluded volume $b = 1$ fm^3 (solid lines) on a linear (left) and logarithmic (right) scale. The dashed lines show the calculations within the Van der Waals HRG model, with van der Waals parameters a and b fixed by the properties of the nuclear ground state [19]. The arrows are the Stefan-Boltzmann limit. Ideal HRG predicts $b_2 = b_3 = b_4 = \cdots = 0$, while EV-HRG captures the sign structure seen in the lattice data.

References

[1] Aoki Y, Endrodi G, Fodor Z, Katz S D and Szabo K K 2006 The Order of the quantum chromodynamics transition predicted by the standard model of particle physics *Nature* **443**, 675

[2] Borsanyi Sz, Fodor Z, Hoelbling C, Katz S D, Krieg S and Szabo K K 2014 Full result for the QCD equation of state with 2+1 flavors *Phys. Lett. B* **730**, 99

[3] Bazavov A *et al.* 2014 Equation of state in (2+1)-flavor QCD *Phys. Rev. D* **90**, 094503

[4] Borsanyi Sz *et al.* 2016 Calculation of the axion mass based on high-temperature lattice quantum chromodynamics *Nature* **539**, no. 7627, 69

[5] Bellwied R, Borsanyi Sz, Fodor Z, Katz S D, Pasztor A, Ratti C and Szabo K K 2015 Fluctuations and correlations in high temperature QCD *Phys. Rev. D* **92**, no. 11, 114505

[6] de Forcrand P and Philipsen O 2002 The QCD phase diagram for small densities from imaginary chemical potential *Nucl. Phys. B* **642**, 290

[7] D'Elia M and Lombardo M P 2003 Finite density QCD via imaginary chemical potential *Phys. Rev. D* **67**, 014505

[8] Allton C R, Doring M, Ejiri S, Hands S J, Kaczmarek O, Karsch D, Laermann E and Redlich K 2005 Thermodynamics of two flavor QCD to sixth order in quark chemical potential *Phys. Rev. D* **71** 054508

[9] Borsanyi Sz, Endrodi G, Fodor Z, Katz S D, Krieg S, Ratti C and Szabo K K 2012 QCD equation of state at nonzero chemical potential: continuum results with physical quark masses at order mu^2 *JHEP* **1208** 053

[10] Hegde P *et at* 2014 The QCD equation of state to $\mathcal{O}(\mu_B^4)$ from lattice QCD *Nucl. Phys. A* **931** 851

[11] Gunther J, Bellwied R, Borsanyi Sz, Fodor Z, Katz S D, Pasztor A and Ratti C 2017 The QCD equation of state at finite density from analytical continuation *EPJ Web Conf.* **137** 07008

[12] Bazavov A *et al.* 2017 The QCD Equation of State to $\mathcal{O}(\mu_B^6)$ from Lattice QCD *Phys. Rev. D* **95** no.5, 054504

[13] D'Elia M, Gagliardi G and Sanfilippo F 2017 Higher order quark number fluctuations via imaginary chemical potentials in $N_f = 2 + 1$ QCD *Phys. Rev. D* **95** no.9, 094503

[14] Bonati C, D'Elia M, Mariti M, Mesiti M, Negro F and Sanfilippo F 2015 Curvature of the chiral pseudocritical line in QCD: Continuum extrapolated results *Phys. Rev. D* **92** no.5, 054503

[15] Cea P, Cosmai L and Papa A 2016 Critical line of 2+1 flavor QCD: Toward the continuum limit *Phys. Rev. D* **93** no.1, 014507

[16] Bellwied R, Borsanyi Sz, Fodor Z, Gnther J, Katz S D, Ratti C and Szabo K K 2015 The QCD phase diagram from analytic continuation *Phys. Lett. B* **751** 559

[17] Alba P *et al.* 2017 Constraining the hadronic spectrum through QCD thermodynamics on the lattice *Phys. Rev. D* **96** no.3, 034517

[18] Vovchenko V, Pasztor A, Fodor Z, Katz S D and Stoecker H 2017 Repulsive baryonic interactions and lattice QCD observables at imaginary chemical potential *Phys. Lett. B* **775** 71

[19] Vovchenko V, Gorenstein M I and Stoecker H 2017 van der Waals Interactions in Hadron Resonance Gas: From Nuclear Matter to Lattice QCD *Phys. Rev. Lett.* **118**, no. 18, 182301

[20] JUQUEEN: IBM Blue Gene/Q Supercomputer System at the Jülich Supercomputing Centre, Tech. Rep. 1 A1

Design studies for the NeuLAND VETO detector

C A Douma[1], K Boretzky[2], I Gasparic[3], N Kalantar-Nayestanaki[1], D Kresan[2], J Mayer[4] and C Rigollet[1], for the R^3B collaboration

[1] KVI-CART, University of Groningen, Groningen, The Netherlands

[2] GSI Helmholtzzentrum für Schwerionenforschung, Darmstadt, Germany

[3] Ruđer Bošković Institute, Zagreb, Croatia

[4] Institut für Kernphysik, Universität zu Köln, Köln, Germany
Supported by the BMBF (05P2015PKFNA). J.M. is supported by the Bonn-Cologne Graduate School of Physics and Astronomy.

E-mail: c.a.douma@rug.nl

Abstract. NeuLAND (New Large-Area Neutron Detector) is the neutron detector for the R^3B-experiment (Reactions with Relativistic Radioactive Beams) at FAIR (Facility for Anti-proton and Ion Research). NeuLAND is a fully active detector composed of plastic scintillator bars. Neutrons are detected by the production of charged particles in the scintillators through hadronic scattering. These charged particles are then detected by their scintillation light. Due to the highly granular design of NeuLAND, the primary neutron interaction points can be accurately reconstructed. These reconstructed points contribute to a kinematically complete reconstruction of reactions with relativistic heavy-ion beams in the target, the goal of the R^3B-experiment. However, charged particles produced by scattering on other parts of the R^3B-setup may provide a significant background. To distinguish the target neutrons from the background, a VETO detector could be placed in front of NeuLAND. This VETO detector is a single plane of thin plastic scintillator bars. It, therefore, provides a high detection efficiency for background particles, but a low detection efficiency for neutrons. For every signal in the VETO, NeuLAND signals can be analyzed with respect to their correlation to the VETO signal and can be eliminated from the further neutron analysis. In this paper, the design of this VETO detector is discussed.

1. Introduction

The R^3B-experiment will be located at the high-energy branch of the Super FRagment Separator (Super-FRS) at the FAIR-facility [1]. The general purpose of this experiment is to obtain a kinematically complete reconstruction of nuclear reactions with radioactive beams. These reactions are an important tool for exploring nuclear structure properties far from stability [2]. During an R^3B-experiment, a radioactive beam from the Super-FRS impinges on a fixed target. The unreacted beam and the charged reaction products are then deflected by the GSI Large Acceptance Dipole Magnet (GLAD). Different detector systems are located around the dipole magnet to provide a spectrometric analysis of the reaction products. Among these detector systems is the gamma spectrometer CALIFA (CALorimeter for the In Flight detection of γ-rays and light charged pArticles), the ion tracking system (including a time-of-flight wall) and the

NeuLAND neutron detector [2, 3, 4] (see Figure 1). A large vacuum chamber is located downstream of the dipole magnet to reduce the background [3].

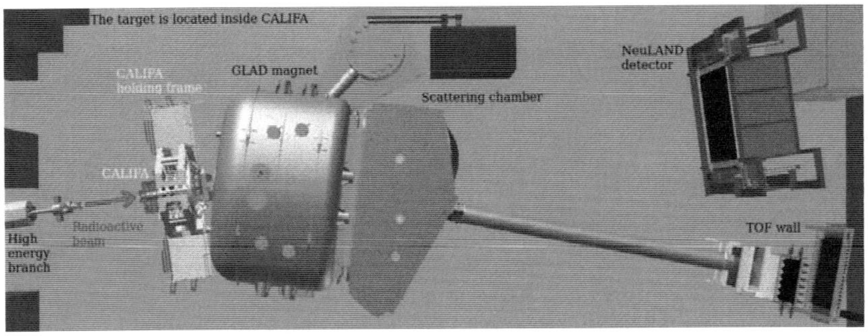

Figure 1: Overview of the phase 0 R^3B experimental setup in Cave C at GSI. Based on the technical drawings of D. Koerper and used with permission.

The complete NeuLAND detector will be composed of 3000 plastic scintillator bars wrapped with reflective aluminum foil and plastic tape. Each scintillator bar has a dimension of 5 cm × 5 cm × 250 cm and is read out by two photomultipliers at the far ends. The scintillator bars will be stacked in 60 planes of 50 bars each with alternating horizontal and vertical orientations, providing a fully active neutron detector with an active area of 250 cm × 250 cm and a depth of 300 cm [4]. A neutron emitted from the projectile after a reaction in the target can then be detected through hadronic interactions in the scintillator material. The time and space coordinates (called a reconstructed first hit) as well as the kinematic properties of the neutron can be obtained in an offline analysis.

The problem we wish to address concerns the background discrimination in NeuLAND. Charged particles produced in the R^3B-setup through secondary scattering may be detected by NeuLAND. To distinguish the neutrons produced in the target from this background, a so-called VETO detector could be placed in front of NeuLAND. This VETO detector should have a large probability to detect charged particles and a small probability to detect neutrons. For every signal in the VETO, NeuLAND signals can be analyzed with respect to their correlation to the VETO signal and can eventually be eliminated from the further neutron analysis. In this paper, the design of this VETO detector is discussed.

2. Simulation procedure

We used R^3BRoot simulations to optimize various parameters for the design of a VETO detector system. R^3BRoot is an integrated simulation framework that contains the geometries of all R^3B detectors, specific analysis software for each R^3B detector, and a ROOT control interface [5, 6]. The Monte Carlo simulations in this paper were performed with the Geant3 module of R^3BRoot. The physics list has been used successfully for neutron simulations in [4, 7]. After performing Monte Carlo simulations, the detector signals were calculated with the NeuLAND digitizer of R^3BRoot. These detector signals were first analyzed with a cluster finding algorithm and then with the neutron tracker software in R^3BRoot to obtain the reconstructed first hits (see [4, 6] for more details). See [8] for the source code of the simulations.

3. Detector design

The VETO detector has been designed as a single wall of non-overlapping thin scintillators with a total surface area of 250 cm × 250 cm, to cover the full frontal active area of NeuLAND. This VETO detector will be placed in front of NeuLAND as shown in Figure 1. The detector design contains three parameters that have to be optimized: the number of scintillators in the wall, the thickness of the scintillators and the distance between the VETO wall and NeuLAND. As a starting point for our optimizations, the geometry of the VETO detector for the NeuLAND demonstrator [9] at the SAMURAI setup [10] was used. The SAMURAI setup is a large-acceptance spectrometer similar to the R^3B-setup (see Figure 1) and is located at RIBF, RIKEN in Tokyo. This VETO contains eight scintillator paddles with a thickness of 1 cm and is placed 32.5 cm in front of the NeuLAND demonstrator.

IOP Conf. Series: Journal of Physics: Conf. Series **1024** (2018) 012027 doi:10.1088/1742-6596/1024/1/012027

During the optimization of the distance between the VETO and NeuLAND, the thickness was kept constant at 1 cm and the number of bars was kept constant at eight. The results from this optimization process are shown in Figures 2 and 3 for a neutron, proton or electron beam with 1000 MeV energy and a particle per event. 25000 events were simulated per run. The black curve in Figure 2 shows the percentage of the events where the VETO gave a signal. The red curve shows the part of the events with a VETO signal (black curve) that is only due to backscattering of charged particles from a hadronic interaction in NeuLAND. The gray curve is computed as the difference between the black curve and the red curve. To obtain the blue curve, the timing information of the hits is used: only the events where the VETO hit occurred before the NeuLAND hit were counted. In Figure 3 conditions are as in Figure 2 but for a proton beam or electron beam instead of a neutron beam.

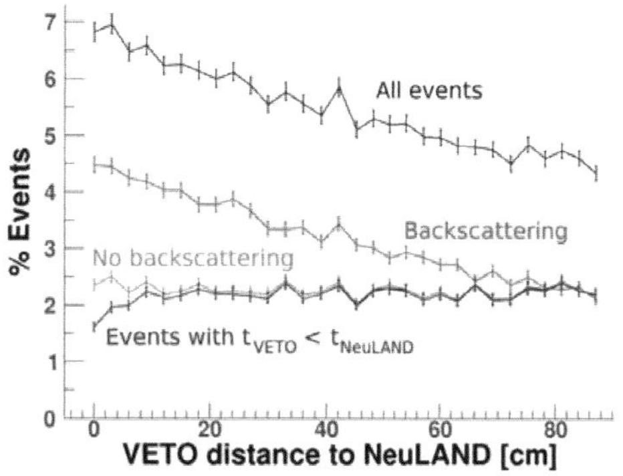

Figure 2: NeuLAND response for 1000 MeV neutrons for different VETO distances.

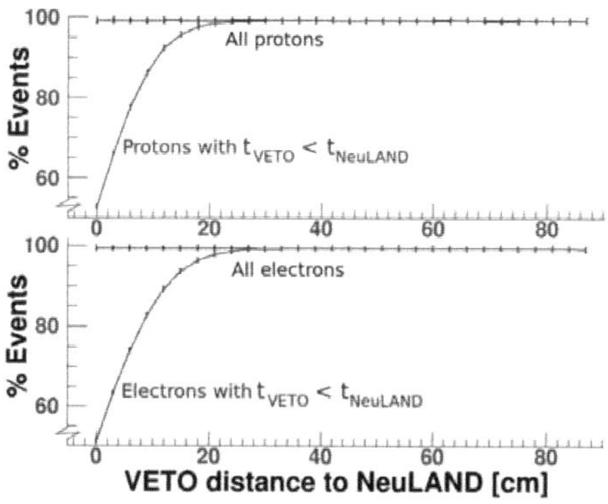

Figure 3: NeuLAND response for 1000 MeV protons (top) and electrons (bottom) for different VETO distances.

Figure 2 shows that the VETO produces a signal for $5\% - 7\%$ of all neutrons. We would like to reduce this as much as possible, since one of the design goals of NeuLAND is the efficient detection and reconstruction of multi neutron events up to a multiplicity of four and more [4]. However, as shown in Figure 2, the backscattering signal, which is more than 50% of the total VETO signals, can be efficiently removed from the VETO elimination by using the timing information of the hits (blue curve). For NeuLAND, a time resolution of $\sigma = 150$ ps was used [4]. A typical time resolution (for scintillators) of $\sigma = 300$ ps was assumed for the VETO. On the other hand, Figure 3 shows that this condition allows a significant fraction of the protons and electrons to pass, unless the distance between NeuLAND and the VETO is at least 30 cm. The simulations show no improvement for distances larger than 30 cm for other charged particles. For particle energies less than 1000 MeV, a distinction based on the timing information is always more accurate than for particles with an energy of 1000 MeV. Therefore, simulation of particles with an energy of 1000 MeV suffices in the optimization of the distance. Hence, based on the simulations, the optimal distance between NeuLAND and the VETO is 30 cm.

To optimize the scintillator thickness, we performed simulations with 1000 MeV protons. The number of bars was again kept constant at eight, but the distance was now kept constant at the optimized value of 30 cm. The results are shown in Figure 4 (a threshold of 1 MeV energy deposition was used). The smallest thickness of the VETO scintillators with which all protons can still be detected is 1.1 cm. This provides the VETO with a 100% detection efficiency for all charged particles while the neutron detection efficiency is around 2% (as illustrated in Figure 2). A detection efficiency for charged particles of near to 100% is mandatory in the design of the VETO, because a typical event of an R^3B experiment contains multiple charged background particles, which all have to be detected (and eliminated) by the VETO.

Figure 2 also shows the justification of our choice of a single wall of non-overlapping scintillators, since it illustrates that to avoid most of the backscattering, a good time resolution is required for the VETO detector (we used $\sigma = 300$ ps in our simulations). Such time resolutions can only be obtained with scintillators, or with Multi-Resistive Plate Chambers (MRPCs) [11]. However, our timing resolution and

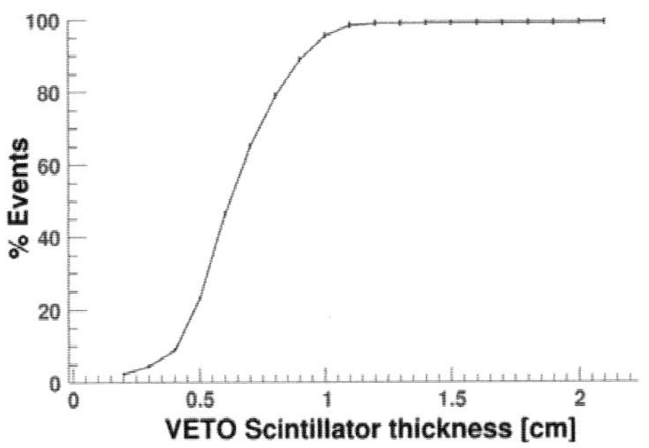

Figure 4: VETO detection efficiency for 1000 MeV protons and for different VETO scintillator thicknesses.

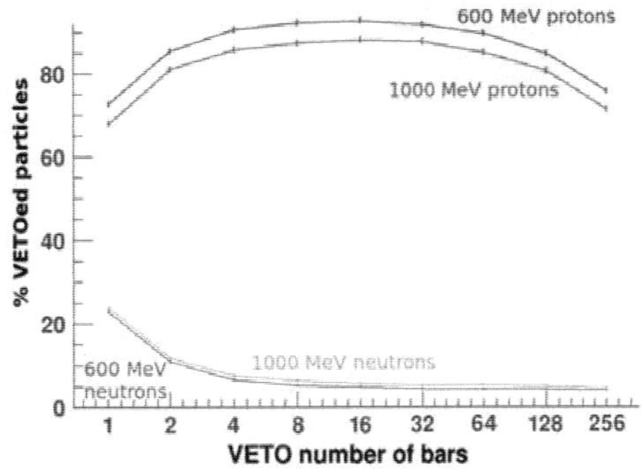

Figure 5: Percentage of VETOed protons and neutrons impinging on NeuLAND in coincidence for different numbers of bars in the VETO.

the requirement that the VETO should detect charged particles with an efficiency of near to 100% would result in a VETO MRPC wall with an areal density that is much bigger than the 1.31 g/cm^2 obtained with a VETO wall consisting of 1.1 cm thick organic BC408 scintillators [4, 12]. This higher areal density results in a larger amount of unfairly eliminated neutrons. The requirement to keep the detector areal density as low as possible is also the reason why the scintillators in the VETO wall should not overlap.

To optimize the number of bars in the VETO, simulations with multiple particles per event and at various beam energies were performed. The distance and thickness parameters were kept constant at their optimized values of 30 cm and 1.1 cm. One reconstructed first hit should then correspond to one incoming particle [4]. This correspondence can be established by minimizing the space and time distances between the reconstructed first hit and the first Monte Carlo interaction point of the incoming track. The proton can then be eliminated by computing straight lines between the reconstructed first hits and the target location. The line that passes closest to the location of the VETO hit is then assumed to correspond to the proton. Figure 5 shows the results of this procedure for a proton and a neutron in coincidence both at 600 MeV or both at 1000 MeV. The simulations show similar behaviour for other energies and for other neutron multiplicities. Therefore, we conclude that the optimal number of scintillator bars in the VETO is 16, although the distribution is rather flat around this number (red and green curves in Figure 5).

4. Efficiency of the VETO wall

To determine the efficiency of the VETO wall under realistic conditions, the entire R^{3}B-setup has to be included in the simulation to model a realistic background of charged particles. In addition, these simulations require the Geant4 module of R^{3}BRoot to model the nuclear reactions in the target [6]. The Geant4 module contains a physics list which is composed of specific physics builders. These builders were chosen to have a best description of hadronic processes at energies relevant for R^{3}B-experiments. This physics list was used in our simulations and was benchmarked against another experiment of the R^{3}B collaboration, S438 [9], to ensure a reasonable agreement with data analysis [13]. Three distinct physics cases were simulated with 10^6 events each: a ^{48}Ca 600 MeV/u-beam on a 1.0 g/cm^2 carbon target, the same beam on a 2.2 g/cm^2 lead target, and a ^{208}Pb 1000 MeV/u-beam on a 500 mg/cm^2 lead target.

The simulations revealed that the neutron background was at least 50% larger than the charged-particle background for all three physics cases. Since the VETO wall is not designed to eliminate neutron signals, an additional method for background reduction was required. Since the neutrons from the target will almost always have an energy close to the beam energy [4], time cuts could be applied to eliminate the neutron background. With proper values for these time cuts, the background of charged particles was reduced by a factor of 5.7 and the neutron background was reduced by a factor of 3.7 for all three physics cases. Only 4% of the neutrons from the target were lost by applying these time cuts [13].

To compare the significance of 4% loss of target neutrons with the significance of the background reduction, the number of successful events was determined. An event was labeled 'successful' if reconstructed first hits were obtained for all neutrons from the target, the overall neutron multiplicity was determined correctly, and all background particles were successfully eliminated. Applying the time cuts increased the number of successful events for all three physics cases by 5% – 25% depending on neutron multiplicity [13].

The efficiency of the VETO wall can be quantified in the same way. The VETO wall allowed 46% – 73% of the events contaminated by background particles to be considered as successful events (depending on neutron multiplicity). In addition, around 1% of the neutrons produced in the target were lost due to the use of the VETO wall (on top of the losses due to the time cuts). These numbers were obtained for the physics case of the carbon target. The other two physics cases produced much less neutrons in the target, so that statistically accurate numbers were very hard to obtain. However, the obtained results were consistent with the carbon target case. The 1% loss of target neutrons is almost consistent with Figure 2 if the difference in neutron energy is taken into account. However, due to the reduction of the charged-particle background by the time cuts, the background of charged particles was very low. Hence, the overall effect of using the VETO wall and the time cuts together is a small decrease in the number of successful events with respect to the usage of the time cuts alone [13].

For the simulations mentioned above, it was assumed that the vacuum chamber downstream of the magnet and the beam lines (see Figure 1) were evacuated, a typical situation in the planned experiments. A simulation was also carried out for the case where the scattering chamber and the downstream beampipe were kept under room conditions (20° C ad 1 atm of air pressure). In this case, the applied time cuts successfully eliminated the low-energetic conversion neutrons originating from the chamber walls. The signal-to-noise ratio also increased dramatically by the VETO, making the VETO essential to have if one uses the VETO under these conditions.

5. Conclusion

In conclusion, this work shows that the optimal design of the NeuLAND VETO detector is a single wall of 16 non-overlapping scintillator bars. The total surface of the wall should be 250 cm × 250 cm. The wall should have an active thickness of 1.1 cm, a time resolution of $\sigma \leq 300$ ps, and it should be placed at a distance of 30 cm in front of NeuLAND. The simulation results in this work also lead to the conclusion that the VETO wall does not provide a significant increase in the overall performance, and also cannot improve the signal-to-noise ratio as long as the scattering chamber and the beam pipes are evacuated. However, the VETO detector will provide a much better signal-to-noise ratio for experiments that require atmosphere in the R^3B scattering chamber.

References

[1] Geissel H *et al* 2009 *Technical Report on the Super-FRS* Tech. rep. GSI and Collaborators
[2] The R^3B collaboration 2005 *Technical Proposal for the Design, Construction, Commissioning and Operation of R^3B.* Tech. rep. GSI Helmholtzzentrum fur Schwerionenforschung
[3] The R^3B collaboration 2014 *Technical Report for the Design, Construction and Commissioning of the Tracking Detectors for R^3B* Tech. rep. GSI Helmholtzzentrum fur Schwerionenforschung
[4] The R^3B collaboration 2011 *Technical Report for the Design, Construction and Commissioning of NeuLAND: The High-Resolution Neutron Time-of-Flight Spectrometer for R^3B* Tech. rep. GSI and Collaborators
[5] Bertini D 2011 *J. Phys.: Conf. Series* **331** 032036
[6] Kresan D R3Broot `http://r3broot.gsi.de/` and `https://github.com/R3BRootGroup/R3BRoot.git`
[7] Blaich T *et al* 1992 *Nucl. Instr. Meth. Phys. Res. A* **314** 136–154
[8] Douma C A *Simulation source code* `https://github.com/ChristiaanAlwin/R3BRoot.git`
[9] Boretzky K *et al* 2015 *NeuLAND - from double-planes to the demonstrator* (*GSI Report* vol 2015-1) (GSI)
[10] Otsu H *et al* 2016 *Nucl. Instr. Meth. Phys. Res. B* **376** 175–179
[11] Yakorev D *et al* (R^3B Collaboration) 2011 *Nucl. Instr. Meth. Phys. Res. A* **654** 79–87
[12] Devi M N, Mondal N K, Satyanarayana B and Shinde R R 2016 *ArXiv:1509.08586v2 [physics.ins-det]*
[13] Douma C A *et al* 2016 *Simulation studies of a NeuLAND VETO wall* (*GSI Report* vol 2016-1) (GSI)

IOP Conf. Series: Journal of Physics: Conf. Series **1024** (2018) 012028 doi:10.1088/1742-6596/1024/1/012028

Systematic errors in transport calculations of shear viscosity using the Green-Kubo formalism

J B Rose[1], J M Torres-Rincon[1], D Oliinychenko[1], A Schäfer[1] and H Petersen[1,2,3]

[1]Frankfurt Institute for Advanced Studies, Ruth-Moufang-Strasse 1, 60438 Frankfurt am Main, Germany

[2]Institute for Theoretical Physics, Goethe University, Max-von-Laue-Strasse 1, 60438 Frankfurt am Main, Germany

[3]GSI Helmholtzzentrum für Schwerionenforschung, Planckstr. 1, 64291 Darmstadt, Germany

E-mail: rose@fias.uni-frankfurt.de

Abstract. The purpose of this study is to provide a reproducible framework in the use of the Green-Kubo formalism to extract transport coefficients. More specifically, in the case of shear viscosity, we investigate the limitations and technical details of fitting the auto-correlation function to a decaying exponential. This fitting procedure is found to be applicable for systems interacting both through constant and energy-dependent cross-sections, although this is only true for sufficiently dilute systems in the latter case. We find that the optimal fit technique consists in simultaneously fixing the intercept of the correlation function and use a fitting interval constrained by the relative error on the correlation function. The formalism is then applied to the full hadron gas, for which we obtain the shear viscosity to entropy ratio.

1. Introduction

There is currently great interest in the extraction of transport coefficients of hot and dense matter in the field of heavy ion physics. At temperatures lower than 170 MeV, in the hadron gas phase, previous studies of shear viscosity have however proven to be inconsistent with each other [1, 2, 3, 4]. In [5], more extensive results are discussed and the discrepancy is resolved by identifying the effect of resonance lifetimes on relaxation dynamics. Within these proceedings, a study of the systematic errors occurring in the application of the Green-Kubo formalism [6, 7] is shown, with the specific aim of providing a reproducible framework for the extraction of shear viscosity. To this end, we will use a newly developed transport model, Simulating Many Accelerated Strongly-interacting Hadrons, or SMASH [8], in which infinite matter simulations for different chemical compositions are carried out.

2. Green-Kubo formalism

The Green-Kubo formalism is a method to extract transport coefficients from fluctuations of given currents around a state of equilibrium in a given system. In the case of shear viscosity,

$$\eta = \frac{V}{T}\int_0^\infty dt\ C^{xy}(t), \tag{1}$$

where V is the volume, T the temperature and t the time. The auto-correlation function $C^{xy}(t)$ is given by

$$C^{xy}(t) = \langle T^{xy}(0)T^{xy}(t)\rangle_{eq} = \lim_{N\to\infty} \frac{1}{N} \sum_{s=0}^{N} T^{xy}(s)\, T^{xy}(s+t), \qquad (2)$$

where T^{xy} is an off-diagonal component of the space-averaged energy-momentum tensor and N the number of timesteps used in taking the average. In all our calculations, this is set to $N = 5000$ for $C^{xy}(0)$, with a timestep length of $\delta t = 0.05$ fm/c, so the total interval used to compute the correlation function is 250 fm/c; a volume of 20^3 fm was used for all simulations. The energy-momentum tensor is computed as

$$T^{\mu\nu} = \frac{1}{V} \sum_{i=1}^{N_{part}} \frac{p_i^{\mu} p_i^{\nu}}{p_i^0}\,, \qquad (3)$$

where N_{part} is the total number of particles and p^{μ} a component of the momentum 4-vector of a given particle.

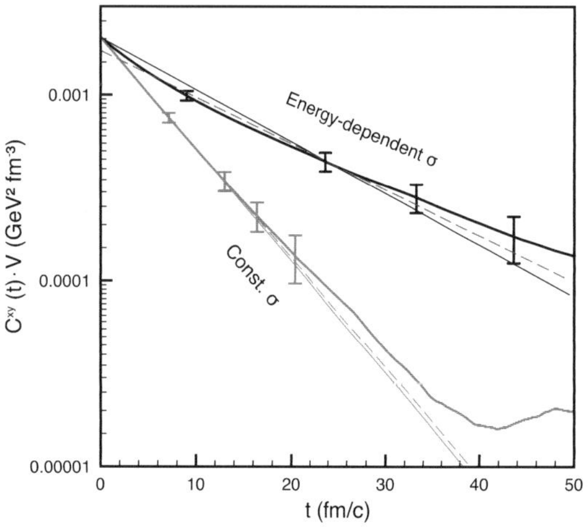

Figure 1. Average volume-independent correlation function as a function of time for a system of particles with constant cross-section (red, bottom set of curves) and energy-dependent pion cross-section (black, top set of curves). Thinner lines are exponential fits, respectively with (solid) and without (dashed) a fixed intercept.

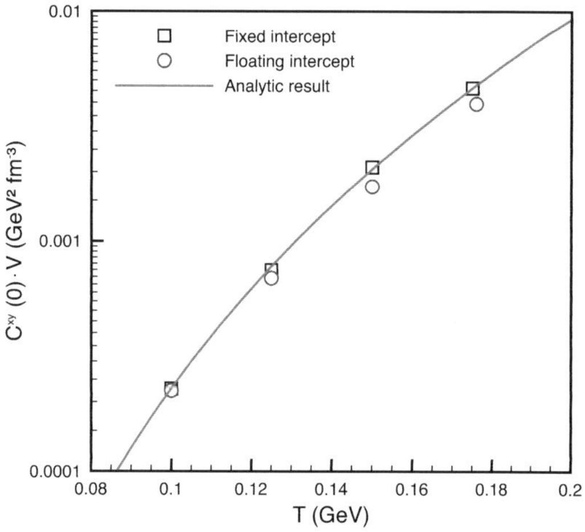

Figure 2. Volume-independent initial value of the correlation function as a function of temperature for a system of particles interacting via the energy-dependent pion cross-section. Statistical error bars are smaller than the symbol size.

3. Correlation functions and shear viscosity of simple systems

As one can see from Eq. (1), the calculation of the viscosity requires integrating C^{xy} from zero to infinity. Numerically, it is challenging to take the limit of $N \to \infty$ in Eq. (2). Consequently, the relative error of any numerical computation of the correlation function necessarily increases rather quickly with t and eventually reaches a state of pure noise, as one can already see in the thick red curve of Fig. 1. To circumvent this limitation, some assumption is made about the analytical shape of the correlation function.

It is generally assumed [2, 9, 10] that for dilute systems, it takes the form of a decaying exponential,

$$C^{xy}(t) = C^{xy}(0)\ e^{-\frac{t}{\tau}}\ , \tag{4}$$

where τ is the relaxation time of the system. For the shear viscosity, it follows that

$$\eta = \frac{C^{xy}(0)V\tau}{T}\ . \tag{5}$$

The thick lines of Fig. 1 correspond to the average correlation functions of two different test systems at $T = 150$ MeV. In the first (red), massive particles ($m = 138$ MeV) interact through a constant isotropic cross-section of $\sigma = 20$ mb, whereas in the second (black), the same particles interact through an isotropic energy-dependent cross-section corresponding to the ρ resonance. In all cases here and further on, the correlation function is obtained by averaging 1000 simulations of the same system. The behavior of the correlation function of these two systems is definitely not the same. While in the case of the constant cross-section interaction, the behavior is quite close to the previously mentioned decaying exponential, we observe a slight deviation from it in the energy-dependent case. This phenomenon is more apparent at high temperatures and densities; the deviation of the shown curve is thus more pronounced than most others used in the rest of this paper, for the purpose of illustration.

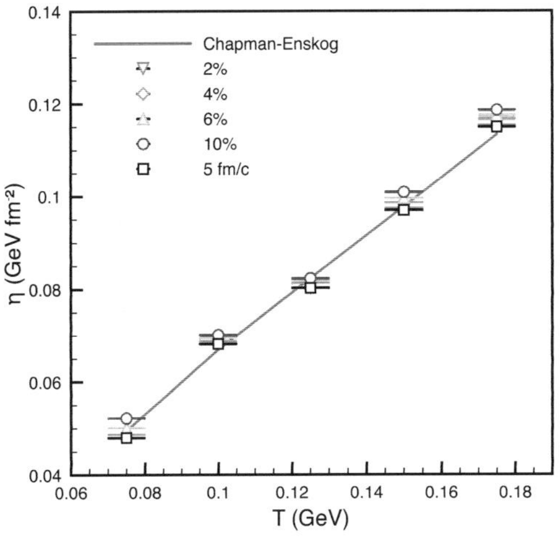

Figure 3. Shear viscosity of a system of particles interacting via constant cross-section for various fitting schemes, compared to a semianalytical Chapman-Enskog calculation. ($\sigma = 20$ mb, $m = 138$ MeV)

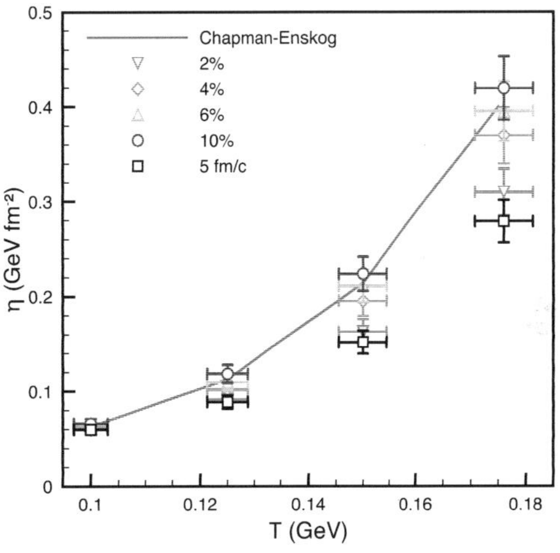

Figure 4. Shear viscosity of a system of particles ($m = 138$ MeV) interacting via the energy-dependent pion cross-section for various fitting schemes, compared to a semianalytical Chapman-Enskog calculation.

For N_{spec} stable particles, the initial value of the correlation function is given by taking the continuum limit of

$$C^{xy}(0) = \langle \sum_i \frac{(p_i^x)^2 (p_i^y)^2}{V^2 (p_i^0)^2} \rangle \to \sum_{a=1}^{N_{spec}} \frac{g_a z_a}{30\pi^2 V} \int_0^\infty dp\ \frac{p^6}{m_a^2 + p^2} \exp\left(-\frac{\sqrt{m_a^2 + p^2}}{T} \right)\ , \tag{6}$$

where $z_a = \exp(\mu_a/T)$ is the fugacity of the species a and g_a its degeneracy factor. Since $C^{xy}(0)$ is known analytically, it is reasonable to fix it in the fit of the correlation function. Indeed, Fig. 2 shows that a fit using a floating intercept systematically undershoots the analytical value of $C^{xy}(0)$ at higher temperatures. Even though an unconstrained exponential might lead to a better fit of the original curve, a precise value for the initial value is required in our final viscosity calculation (Eq. (5)). Therefore, $C^{xy}(0)$ needs to be fixed.

Let us now turn our attention to the question of how many points to consider when fitting. We tried two different schemes to this effect. The first one is to consider a fixed interval from $t = 0$ to $t = 5$ fm/c; this interval is small enough that it always very closely fits the earliest part of the curve. The second scheme takes into account that the growth rate of the relative error of the correlation is much larger at higher temperatures, which is done using a fitting interval for which the cutoff depends on this relative error. This is illustrated by the four points which have error bars on both thick curves of figure 1: they correspond in each case to the point where the relative statistical error reaches 2%, 4%, 6% and 10%. The plotted fits correspond here in all cases to using all points up to 6% relative error.

In order to determine which of these cutoffs to use, we now compare the final shear viscosity yielded by each to an analytical calculation of the shear viscosity of the previous two simple systems using the Chapman-Enskog approximation to solve the Boltzmann equation [11]. Fig. 3 shows the effect of varying the cutoff on shear viscosity in the case of a system interacting through constant cross sections, whereas Fig. 4 shows the same for the case of energy-dependent interactions. In the second case, resonance lifetimes have been decreased to zero in SMASH, so that the comparison is carried out between comparable systems using point-like 2-to-2 interactions. All calculations use an isotropic cross-section.

As expected from looking at the correlation function on Fig. 1, the final effect of varying the cutoff is rather limited when cross-sections are isotropic. All proposed cutoff schemes and values appear to describe well the analytical calculation, with the largest deviations remaining under 8% in all cases. This value is then taken as a systematic error on the method for all further calculations (note that statistical error bars are smaller than the symbol size). If one now looks at Fig. 4, where cross-sections are energy dependent, the picture is different. It is here very clear there are cases where the sensitivity to the cutoff is large. Cutoffs at 4%, 6% and 10% relative error manage to fit the Chapman-Enskog calculation within systematic errors. While it appears that in this regime it is still possible to use exponential fits, one should keep these deviations in mind when using this ansatz, and possibly look into different ones if deviations become larger.

4. Shear viscosity of a hadron gas

As a final result, the viscosity of the full hadron gas is calculated within SMASH (Fig. 5) at $\mu_B = 0$ MeV. Once again, the computation is performed using the three previously mentioned cutoffs that were accurately describing analytical calculations. In the case of the full hadron gas, however, such a comparison is not feasible, as no analytical equivalent currently exists. We obtain in this regard results which are very similar to the previous case: although there is some variation in the actual value of the shear viscosity, the different cutoffs produce results which are consistent with each other within systematic errors. We thus conclude from this technical study that it is sufficient to use a cutoff corresponding to a relative error of 6% when fitting the correlation function. All curves being mutually consistent, this corresponds to the middle point and reproduces best the available analytical calculations.

Fig. 6 shows the final result of this procedure: the ratio of shear viscosity to entropy density. Here the entropy density is calculated according to the Gibbs formula $s = \frac{\epsilon + p - n_B \mu_B}{T}$, where ϵ is the energy density, p the pressure, and n_B and μ_B respectively the baryonic net density and chemical potential. We observe that the profile of this quantity is decreasing, which is the expected behavior when approaching a phase transition, and eventually reaches a plateau.

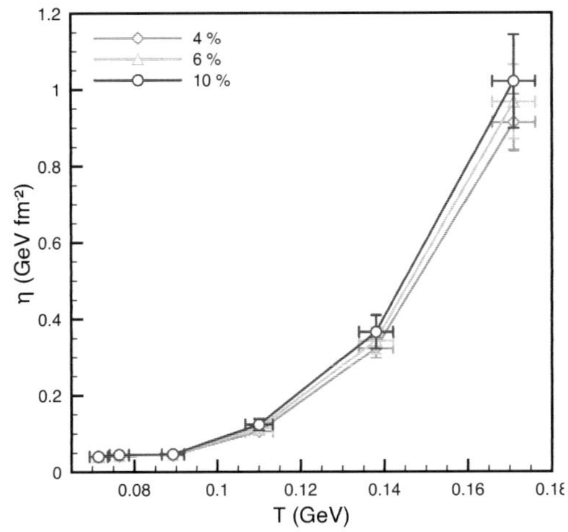

Figure 5. Shear viscosity of a full hadron gas as a function of temperature, for various fitting schemes, at $\mu_B = 0$ MeV.

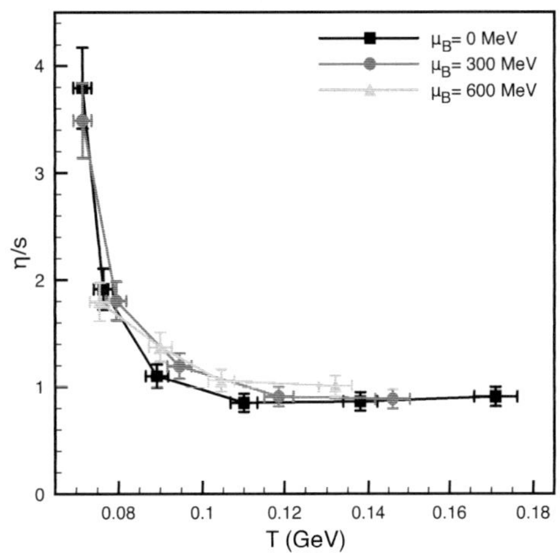

Figure 6. Shear viscosity over entropy density of a full hadron gas, for various baryochemical potentials.

This behavior is consistent with the results in Ref. [2]. We furthermore observe a very mild dependence on baryonic chemical potential, although all results are here within systematic error of each other. This could then also be consistent to an independence on chemical potential up to these values, as previously found in Ref. [12].

Acknowledgments

This work was made possible thanks to funding from the Helmholtz Young Investigator Group VH-NG-822 from the Helmholtz Association and GSI, and supported by the Helmholtz International Center for the Facility for Antiproton and Ion Research (HIC for FAIR) within the framework of the Landes-Offensive zur Entwicklung Wissenschaftlich-Oekonomischer Exzellenz (LOEWE) program launched by the State of Hesse. Computing services were provided by the Center for Scientific Computing (CSC) of the Goethe University Frankfurt.

References

[1] Muronga A, *Phys. Rev. C* **69**, 044901 (2004) [nucl-th/0309056].
[2] Demir N and Bass S A, *Phys. Rev. Lett.* **102**, 172302 (2009)
[3] Romatschke P and Pratt S, *Preprint* arXiv:1409.0010 [nucl-th].
[4] Pratt S, Baez A and Kim J, Phys. Rev. C **95**, no. 2, 024901 (2017)
[5] Rose J-B, Torres-Rincon J M, Schäfer A, Oliinychenko D R and Petersen H, arXiv:1709.03826 [nucl-th].
[6] Toda M, Kubo R and Saito N, *Statistical Physics I* , Springer-Verlag, Berlin, (1992)
[7] Zubarev R and Morozov O, *Statistical Mechanics of Non-equilibrium Processes. Vol. 2: Relaxation and Hydrodynamics Processes*, Akademie Verlag GmbH, (1996)
[8] Weil J *et al.*, *Phys. Rev. C* **94**, no. 5, 054905 (2016)
[9] Wesp C, El A, Reining F, Xu Z, Bouras I and Greiner C, *Phys. Rev. C* **84**, 054911 (2011)
[10] Plumari S, Puglisi A, Scardina F and Greco V, *Phys. Rev. C* **86**, 054902 (2012)
[11] Torres-Rincon J M, "Hadronic Transport Coefficients from Effective Field Theories," arXiv:1205.0782 [hep-ph].
[12] Itakura K, Morimatsu O and Otomo H, *Phys. Rev. D* **77**, 014014 (2008)

Quark susceptibilities in a generalized quasiparticle model

T. Steinert and W. Cassing
Institut für Theoretische Physik, Universität Giessen, Germany
E-mail: Thorsten.Steinert@theo.uni-giessen.de

Abstract. The QCD equation of state as predicted by lattice QCD calculations (lQCD) is well reproduced in terms of effective quasiparticle models. These models so far fail to describe the susceptibilties and underestimate the pressure at finite densities. We present a generalised quasiparticle model where the partonic propagators explicitly depend on the three-momentum with respect to the medium. Within this extended model we reproduce simultaneously the equation of state and the susceptibilties as provided by lQCD. We calculate the shear and bulk viscosity as well as the electric conductivity and compare to default quasiparticle models. We find a good agreement between our model and available lattice data for all transport coefficients. We use thermodynamic consistency to extend the model to finite baryon chemical potential and present predictions for the equation of state at finite baryon densities.

1. Introduction

One of the main goals of heavy-ion physics is the study of hot and dense QCD matter. Tremendous efforts have been invested from the experimental and theoretical side to investigate the QCD phase diagram. In recent years lQCD calculations were able to determine the thermodynamic properties at vanishing quark chemical potential μ_q with high accuracy and quasiparticle models with medium-dependent propagators could successfully reproduce the predicted equation of state. The properties of QCD for finite chemical potential have been studied by Taylor expansions in the chemical potential, where the first contribution is the second-order susceptibility. Default quasiparticle models can not simultaneously describe the equation of state and the susceptibility and are therefore not suited to study the QCD thermodynamics at finite baryon chemical potential μ_B. We here present an extension of the DQPM quasiparticle model (DQPM*) that incorporates momentum-dependent selfenergies and simultaneously reproduces the equation of state and the second-order susceptibility.

2. The dynamical quasiparticle model

The Dynamical QuasiParticle Model (DQPM) is an effective model for QCD that treats partons as quasiparticles with medium dependent masses and widths [1, 2, 3]:

$$M_g^2 = \frac{g^2}{6}\left(\left(N_c + \frac{1}{2}N_f\right)T^2 + \frac{N_c}{2}\sum_q \frac{\mu_q^2}{\pi^2}\right), \qquad M_{q\bar{q}}^2 = \frac{N_c^2-1}{8N_c}g^2\left(T^2 + \frac{\mu_q^2}{\pi^2}\right), \tag{1}$$

$$\gamma_g = N_c\frac{g^2 T}{24\pi}\log\left(1+\frac{2c}{g^2}\right), \qquad \gamma_{q\bar{q}} = \frac{N_c^2-1}{2N_c}\frac{g^2 T}{24\pi}\log\left(1+\frac{2c}{g^2}\right). \tag{2}$$

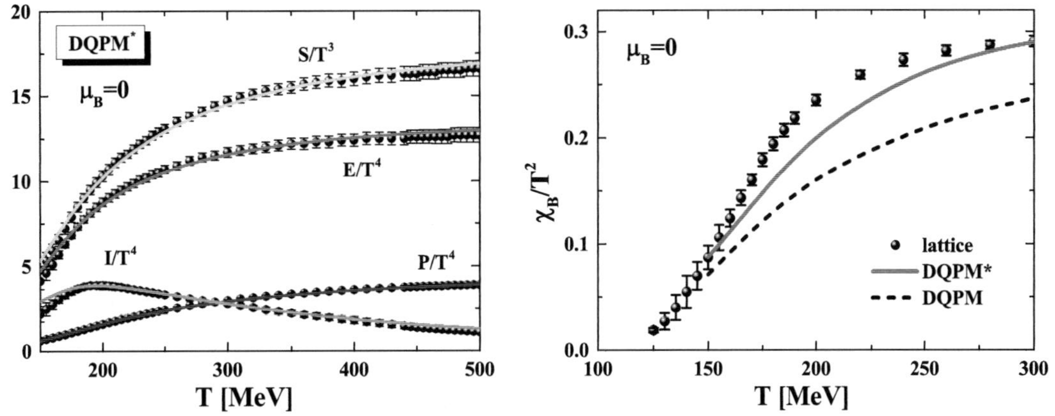

Figure 1. (Left) The DQPM* equation of state as a function of the temperature T for vanishing baryon chemical potential. The lattice results are taken from Ref. [4]. The interaction measure is $I = E - 3P$. (Right) The quasiparticle baryon number susceptibility as a function of the temperature T for vanishing baryon chemical potential $\mu_B = 0$ for DQPM and DQPM*. The lattice results are taken from Ref. [5].

The nonperturbative information of the system is contained in the effective coupling $g^2(T, \mu_B)$. The starting point of the DQPM is the quasiparticle entropy and particle density that are defined by the propagator S and the corresponding complex selfenergies Σ [6],

$$s^{dqp} = -\int \frac{d^3k}{(2\pi)^3} \frac{d\omega}{2\pi} \frac{\partial n_{B/F}}{\partial T} \left(\mathrm{Im}(\ln S^{-1}) - \mathrm{Re}(S)\mathrm{Im}(\Sigma)\right), \tag{3}$$

$$n^{dqp} = -\int \frac{d^3k}{(2\pi)^3} \frac{d\omega}{2\pi} \frac{\partial n_{B/F}}{\partial \mu} \left(\mathrm{Im}(\ln S^{-1}) - \mathrm{Re}(S)\mathrm{Im}(\Sigma)\right). \tag{4}$$

The medium dependence introduces nontrivial contributions to the thermodynamic potential that one has to determine from its derivatives, i.e. the entropy $s = \partial P/\partial T$ and the particle density $n = \partial P/\partial \mu$.
The lQCD equation of state at small baryon chemical potential μ_B is encoded in the baryon number susceptibility

$$\chi_B = \left.\frac{\partial^2 P}{\partial \mu_B^2}\right|_{\mu_B=0}. \tag{5}$$

In quasiparticle approaches χ_B is completely defined by the quasiparticle properties for $\mu_B = 0$. Quasiparticle models which are tuned to reproduce the equation of state as predicted by lQCD tend to underestimate the susceptibility [7].
The partons in the DQPM are heavy and can not reproduce the correct perturbative behavior of massless partons for large momenta. We introduce a correction factor in the DQPM selfenergies, that transfers the quasiparticle propagator into the perturbative one at large three-momenta. The momentum-dependence improves the susceptibility compared to regular quasiparticle approaches [8, 9]. Within this model, denoted by DQPM*, we can simultaneously describe the lQCD equation of state and the susceptibility from Refs. [4, 5] (cf. Fig. 1).
We calculate the shear and bulk viscosity as well as the electric conductivity in the relaxation time approximation [8, 9, 10, 11]. The DQPM has an advantage compared to default quasiparticle models, which employ only an effective mass. They have to estimate the relaxation time or compute it from parton-parton scattering. The relaxation time in the DQPM is defined

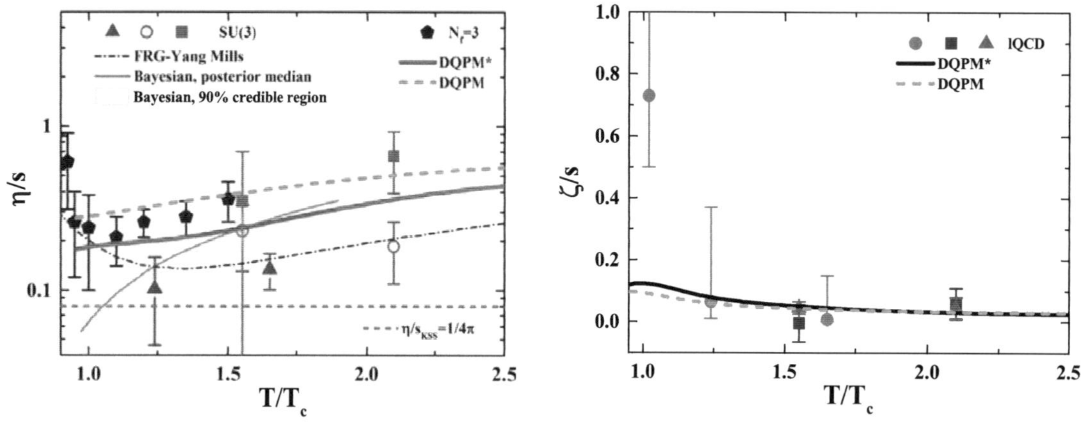

Figure 2. The shear viscosity over entropy density (left) and the bulk viscosity over entropy density (right) as a function of the normalized temperature T/T_c. The lQCD data for η/s are taken from Refs. [12], [13], [14] and [15]. The dash-dotted brown line shows η/s calculated from the functional-renomalization group. The blue line displays the estimate from a bayesian analysis of experimantal heavy-ion data, while the blue shaded area is the 90 % credible region [16, 17]. The dashed grey line indicates the KSS-limit [18]. The lQCD data for ζ/s are taken from Refs. [13] and [19].

by the quasiparticle width, $\tau = \gamma^{-1}$. We find good results for all transport coefficients in comparison to available lQCD prediction. As seen from Figs. 1, 2 and 3 the DQPM* is in fair agreement with QCD thermodynamics and correlators for $\mu_B = 0$.

3. Quasiparticle models at finite density

Quasiparticle models differ from other thermodynamic models. They can not directly access the thermodynamic pressure. Instead they calculate the entropy and particle density. This is fine for $\mu_B = 0$ where the equation of state follows from the entropy alone, but it will cause problems at finite density. The medium-dependent quasiparticle properties will break thermodynamic consistency and the pressure of the system is not unique. We here present an approach that enforces thermodynamic consistency in the whole T-μ_B plane:
If the entropy and particle density follow from the same thermodynamic potential, they have to fulfill the Maxwell relation,

$$\left.\frac{\partial s}{\partial \mu_B}\right|_T = \left.\frac{\partial n_B}{\partial T}\right|_{\mu_B}. \tag{6}$$

This equation implies a differential equation for the effective coupling g^2 [20],

$$a_T \frac{\partial g^2}{\partial T} + a_\mu \frac{\partial g^2}{\partial \mu_B} = a_0. \tag{7}$$

We solve this flow equation with the coupling at $\mu_B = 0$ as starting values and obtain the effective coupling at finite chemical potential, which defines the DQPM also at finite μ_B. The flow equation (7) ensures the validity of the Maxwell relation. We compare the particle density calculated from the quasiparticle density to the thermodynamic definition $\partial P/\partial \mu$. Only the flow equation ensures identical results within both methods (cf. Fig. 3 (right)).
One can easily generalize the flow equation to arbitrary chemical potentials. An important constraint for heavy-ion collisions is strangeness neutrality. We achieve this condition by fixing

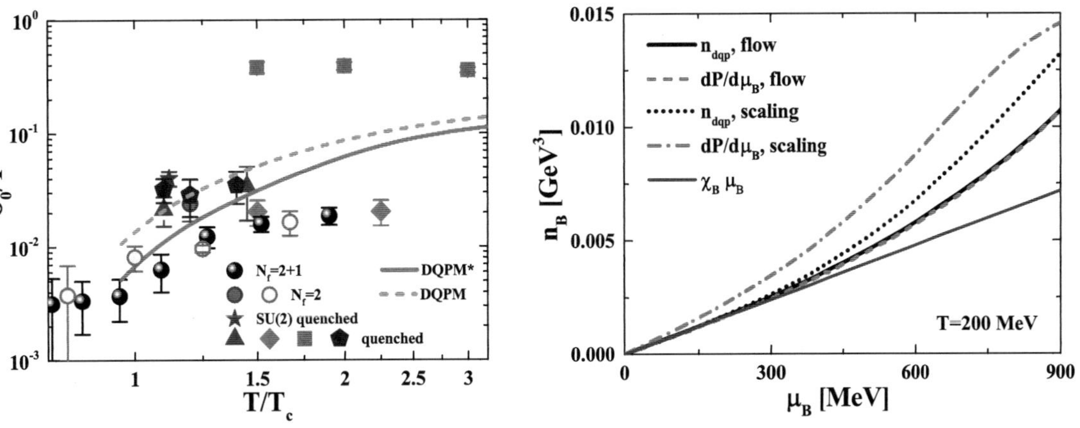

Figure 3. (Left) The electric conductivity σ_0 over temperature T as a function of the normalized temperature T/T_c. The lQCD data for σ_0/T are taken from Refs. [21], [22], [23], [24], [25], [26], [27] and [28]. (Right) The baryon number density for the DQPM* calculated from the quasiparticle density (solid and dotted line) and the thermodynamic definition (dashed and dash-dotted line) as a function of the baryon chemical potential μ_B for $T = 200$ MeV.

the strange quark chemical potential to zero in all expressions. This excludes the strange sector from entering the density n_B. However, the effective coupling is rather insensitive to strangeness neutrality and changes only marginally for this special case.

With the couplings at hand we calculate the equation of state for finite chemical potential and find that for $\mu_B < 300$ MeV the density is well described by the first Taylor coefficient, $n_B = \chi_B \cdot \mu_B$. This marks the region where one can use the lattice equation of state. At larger μ_B the behavior differs and one needs higher corrections. The equation of state for $\mu_B = 600$ MeV follows no longer from the simple Taylor expansion. Nevertheless, we see no drastic deviations in the functional form compared to the equation of state for $\mu_B = 0$ (cf. Fig. 4).

We determine also the transport coefficients for finite μ_B and find a similar scaling of the transport coefficients and the total quasiparticle densities. Both behave for small μ_B as

$$\frac{X(\mu_B)}{X(0)} = 1 + c_X \frac{\mu_B^2}{T^2}. \tag{8}$$

We find $c_\eta = 0.32$ for the shear viscosity, which is the same coefficient as for the total quark density, and for the conductivity $c_{\sigma_0} = 0.47$ which is similar to the light/charge density. The quasiparticle properties are almost unchanged at small μ_B. The transport coefficients feel only the increasing number of particles in the system. However, the bulk viscosity differs from this picture. It is sensitive to changes in the equation of state and the medium dependence of the quasiparticles. Once these effects become subleading at larger temperatures, the bulk viscosity scales with $c_\zeta = 0.43$.

Acknowledgments

The authors acknowledge valuable discussions and contributions from H. Berrehrah and thank the Helmholtz International Center for FAIR (HIC for FAIR), the Helmholtz Graduate School for Hadron and Ion Research (HGS-HIRe), the Helmholtz Research School for Quark Matter Studies in Heavy-Ion collisions (H-QM), and the Bundesministerium für Bildung und Forschung (BMBF) for support.

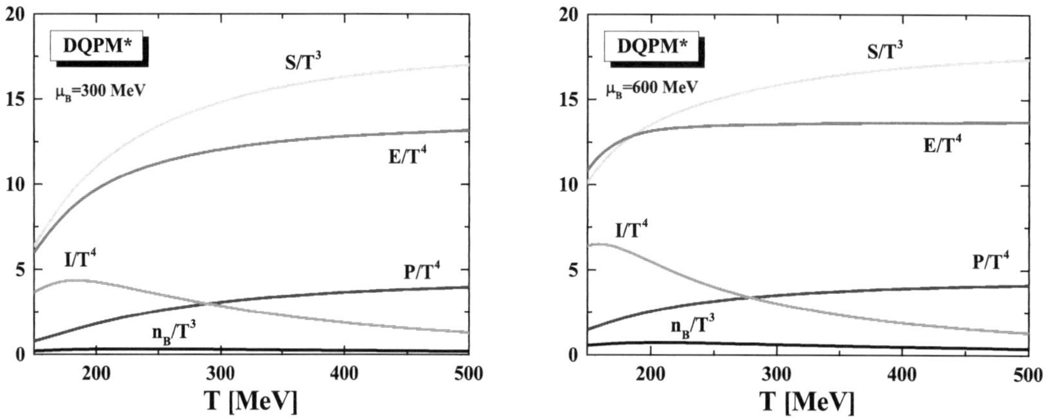

Figure 4. The dimensionless equation of state for the DQPM* as a function of the temperature T for $\mu_B = 300$ MeV and $\mu_B = 600$ MeV ($I = E - 3P$).

References

[1] Peshier A 2004 *Phys. Rev.* D **70** 034016
[2] Peshier A 2005 *J. Phys.* G **31** S371–S378
[3] Ozvenchuk V, Linnyk O, Gorenstein M I, Bratkovskaya E L and Cassing W 2013 *Phys. Rev.* C **87** 024901
[4] Borsanyi S, Fodor Z, Hoelbling C, Katz S D, Krieg S and Szabo K K 2014 *Phys. Lett.* B **730** 99–104
[5] Borsanyi S, Endrodi G, Fodor Z, Katz S D, Krieg S, Ratti C and Szabo K K 2012 *JHEP* **08** 053
[6] Blaizot J P, Iancu E and Rebhan A 2001 *Phys. Rev.* D **63** 065003
[7] Plumari S, Alberico W M, Greco V and Ratti C 2011 *Phys. Rev.* D **84** 094004
[8] Berrehrah H, Cassing W, Bratkovskaya E and Steinert T 2016 *Phys. Rev.* C **93** 044914
[9] Berrehrah H, Bratkovskaya E, Steinert T and Cassing W 2016 *Int. J. Mod. Phys.* E **25** 1642003
[10] Chakraborty P and Kapusta J I 2011 *Phys. Rev.* C **83** 014906
[11] Marty R, Bratkovskaya E, Cassing W, Aichelin J and Berrehrah H 2013 *Phys. Rev.* C **88** 045204
[12] Meyer H B 2007 *Phys. Rev.* D **76** 101701
[13] Sakai S and Nakamura A 2007 *PoS* **LAT2007** 221 [AIP Conf. Proc.893,5(2007)] (*Preprint* `0710.3625`)
[14] Nakamura A and Sakai S 2005 *Phys. Rev. Lett.* **94** 072305
[15] Astrakhantsev N, Braguta V and Kotov A 2017 *JHEP* **04** 101
[16] Bernhard J E, Moreland J S and Bass S A 2017 *26th International Conference on Ultrarelativistic Nucleus-Nucleus Collisions (Quark Matter 2017) Chicago,Illinois, USA, February 6-11, 2017* (*Preprint* `1704.04462`) URL `http://inspirehep.net/record/1591631/files/arXiv:1704.04462.pdf`
[17] Bass S A, Bernhard J E and Moreland J S 2017 *26th International Conference on Ultrarelativistic Nucleus-Nucleus Collisions (Quark Matter 2017) Chicago,Illinois, USA, February 6-11, 2017* (*Preprint* `1704.07671`) URL `http://inspirehep.net/record/1594915/files/arXiv:1704.07671.pdf`
[18] Kovtun P, Son D T and Starinets A O 2005 *Phys. Rev. Lett.* **94** 111601
[19] Meyer H B 2008 *Phys. Rev. Lett.* **100** 162001
[20] Peshier A, Kämpfer B and Soff G 2000 *Phys. Rev.* C **61** 045203
[21] Kaczmarek O, Laermann E, Müller M, Karsch F, Ding H T, Mukherjee S, Francis A and Soeldner W 2012 *PoS* **ConfinementX** 185 (*Preprint* `1301.7436`)
[22] Aarts G, Allton C, Foley J, Hands S and Kim S 2007 *Phys. Rev. Lett.* **99** 022002
[23] Gupta S 2004 *Phys. Lett.* B **597** 57–62
[24] Kaczmarek O and Müller M 2014 *PoS* **LATTICE2013** 175 (*Preprint* `1312.5609`)
[25] Buividovich P V, Chernodub M N, Kharzeev D E, Kalaydzhyan T, Luschevskaya E V and Polikarpov M I 2010 *Phys. Rev. Lett.* **105** 132001
[26] Brandt B B, Francis A, Meyer H B and Wittig H 2013 [PoSConfinementX,186(2012)] (*Preprint* `1302.0675`)
[27] Brandt B B, Francis A, Jäger B and Meyer H B 2016 *Phys. Rev.* D **93** 054510
[28] Aarts G, Allton C, Amato A, Giudice P, Hands S and Skullerud J I 2015 *JHEP* **02** 186

IOP Conf. Series: Journal of Physics: Conf. Series **1024** (2018) 012030 doi:10.1088/1742-6596/1024/1/012030

XYZ states at BESIII

Riccardo Farinelli

INFN, Sezione di Ferrara, via G. Saragat 1, 44122 Ferrara, Italy

E-mail: rfarinelli@fe.infn.it

Abstract. The BESIII experiment at the Beijing Electron Positron Collider (BEPCII) provided a significant contribution to the charmonium-like state spectroscopy thanks to the large data samples collected in the center of mass energy region between 3.8 and 4.6 GeV. New and surprising resonant states have been discovered. Their interpretation goes beyond the traditional quark and anti-quark model paradigm of the charmonium: for example, the two isospin triplets $Z_c(3900)$ and $Z_c(4020)$ that are electrically charged quarkonium-like states. In this proceeding, it will be reported recent studies on XYZ states made by BESIII in particular the radiative transition between various states, i.e. in the channel $Y(4260) \to \gamma X(3872)$.

1. Introduction

The charmonium spectrum is described by c and $\bar{c}$ quarks bound together by strong interaction and by secondary order effects depending on their quantum numbers that combine L and S to J=0, 1, 2. The potential model predicts the masses of the resonances as a function of their quantum number up to the production threshold of 2 M_D and found in a very good agreement with the experimental measurements [1]. However, above threshold there are many predicted states, but only few of them have been observed. In the last decades several experiments like BESIII, BaBar, CLEO-C, D0 and others have discovered some states that seem not to match with potential prediction. These ones have been labeled as "exotic states". First of all the X(3872), seen for the first time in 2003 by Belle [2]: due to its narrow width of few MeV, it can not be explained as an ordinary $c\bar{c}$ state. Afterward, Y states with quantum number 1^{--} such as Y(4260), Y(4660), and so on [5], have been discovered. These overpopulate the region of the vector state and do not fit the potential model. To complete the family of exotic states there are the Z states such as Z(4020) and Z(3900), discovered in the invariant mass spectrum of the decay $\pi^+ J/\psi$, which are evidently exotic since they are charged and have been interpreted as possible tetraquarks, quark molecules or hybrids [4]. A connection point between these exotic states is their probability to decay in conventional $c\bar{c}$ states through a photon or a pion emission, contrarily to the states above the open charm threshold that decay mainly in D mesons instead of $c\bar{c}$ states. The nature of this particle family is unclear and to understand them better it is necessary to extend the knowledge of their decay channels, quantum number and cross section. BESIII (BEijing Spectrometer) is a spectrometer that studies the e^+e^- collisions produced by BEPCII (Beijing Electron-Positron Collider) in the energy range between 2 and 4.6 GeV. In such reactions charmed states can be directly produced with quantum numbers 1^{--} such as J/ψ and other $c\bar{c}$ states or the Y states. The mail goals of the experiment is to investigate the charm region and the XYZ states.

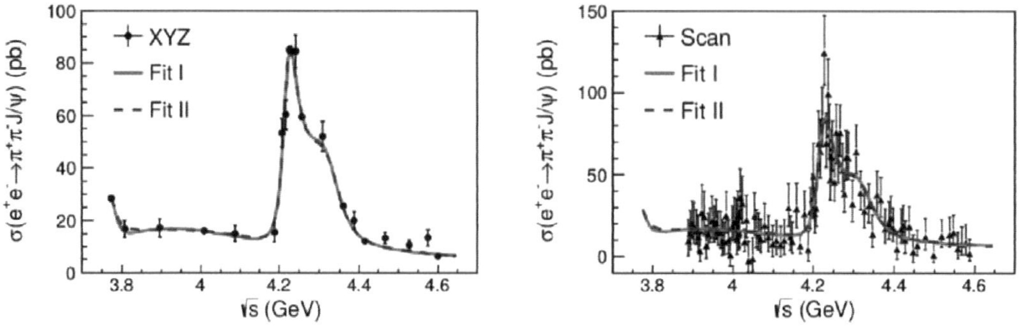

Figure 1. Measured cross section $\sigma(e^+e^- \to \pi^+\pi^- J/\psi)$ and simultaneous fit to two data samples (left and right panels) with the coherent sum of three Breit-Wigner functions (red solid curve) and the coherent sum of an exponential continuum and two Breit-Wigner functions (blue dashed curve). The dots with the error bars are the data. [7]

2. Y states

The Y states have been observed for the first time by BaBar in the initial state radiation process $\pi^+\pi^- J/\psi$: Y(4260) is a resonance with quantum number 1^{--} confirmed by Belle, CLEO-C and BESIII [3, 6]. In this proceeding are presented the results found in the channels $e^+e^- \to \pi^+\pi^- J/\psi$, $e^+e^- \to \pi^+\pi^- h_c$ and $e^+e^- \to \omega\chi_{cJ}$. The Y(4260) is a resonance studied at BESIII with 9.05 fb^{-1} data collected. This Y state suggests the existence of two resonances on the decay channel $\pi\pi J/\psi$ [7], one compatible with the one discovered by Belle and BaBar Y(4260) [3, 6]: M = 4222 ± 3.1 ± 1.4 MeV/c^2 and Γ = 44.1 ± 4.3 ± 2.0 MeV; plus another resonance with M = 4320 ± 10.4 ± 7.0 MeV/c^2 and Γ = 101.4 ± 25.3 ± 10.2 MeV compatible with the Y(4360) seen by BaBar [5] in the spectrum of $M(\pi^+\pi^-\psi(2S))$. The study has been performed with two different data sets and two different fits: one uses three Breit-Wigner functions to search the Y(4008) [6] discovered by Belle and the other two Breit-Wigner and an exponential function for the background. The result does not confirm the resonance Y(4008) seen by Belle. The lineshape of the Born cross section as function of the energy has been measured and it is shown in Fig. 1.

In the decay channel $e^+e^- \to \pi^+\pi^- h_c$, with $h_c \to \gamma\eta_c$ and η_c decays in 16 exclusive channels, a cross section compatible with $e^+e^- \to \pi^+\pi^- J/\psi$ and $e^+e^- \to \pi^+\pi^-\psi(2S)$ has been measured. Two resonances have been measured: M = 4218.4 ± 5.5 ± 0.9 MeV/c^2 and Γ = 66.0 ± 12.3 ± 0.4 MeV, M = 4391 ± 6.3 ± 1.0 MeV/c^2 and Γ = 139.5 ± 16.2 ± 0.6 MeV [8]. The first one is compatible with the previous in $\pi\pi J/\psi$. The second one is a new resonance and it is incompatible with the others discovered in this region (see Fig. 2).

Another proposed Y state is the one observed for the first time in the process $e^+e^- \to \omega\chi_{c0}$ in the data samples collected at a $\sqrt{s}$ between 4.21 and 4.42 GeV with $\omega \to \pi^+\pi^-\pi^0$, $\chi_{c0} \to \pi\pi, KK\chi_{c_0}$. The measured Born cross section is (55.4 ± 6.0 ± 5.9) pb at $\sqrt{s} = 4.23$ GeV and (23.7 ± 5.3 ± 3.5) pb at 4.26 GeV. By assuming the $\omega\chi_{c0}$ signals, a resonance has been measured with M = 4230 ± 8 ± 6 MeV/c^2 and Γ = 38 ± 12 ± 2 MeV. No resonances show up in the decay with J=1,2 states.

3. X states

Concerning the X states, in this proceeding only $e^+e^- \to \gamma X(3872)$ and $X(3872) \to \pi\pi J/\psi$ is presented, in the center of mass from 4.009 to 4.420 MeV, where the process has been observed with a significance of 6.3σ. The measured mass of the X(3872) is (3871.9 ± 0.7 ± 0.2) MeV/c^2

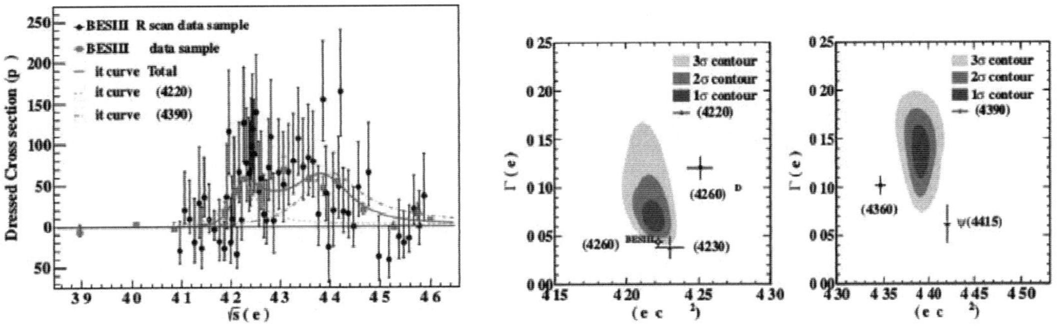

Figure 2. Right: Fit to the dressed cross section $e^+e^- \to \pi^+\pi^- h_c$ with coherent sum of two Breit-Wigner functions (solid curve). The dash (dash-dot) curve shows the contribution from the two structures Y (4220) [Y (4390)]. The dots with the errors are data. Left: The likelihood contours in the mass and width planes for Y(4220) and Y(4390).

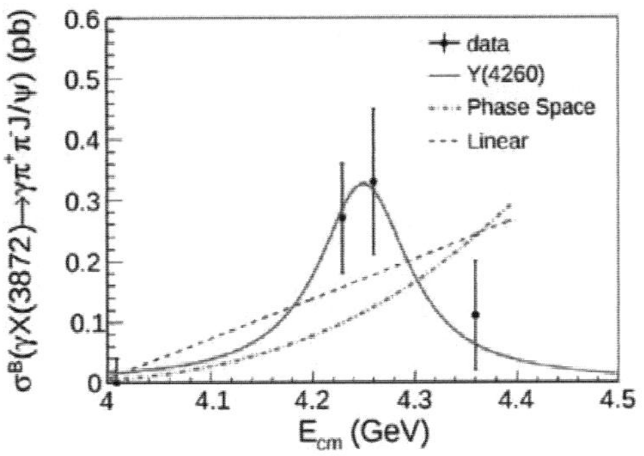

Figure 3. Fit to the cross section $\sigma[e^+e^- \to \gamma X(3872)] \times B[X(3872) \to \pi^+\pi^- J/\psi]$ with three hypotheses: a Y(4260) resonance (red solid curve), a linear continuum (blue dashed curve), or a E1-transition phase space term (red dotted-dashed curve).

[10]. The cross section of this process has been measured for four different energies and its shape shows an increment at about 4260 MeV. The lineshape of the cross section has been described with three hypotheses: the phase space of E_1 transition, a linear continuum or a resonant Y(4260), see Fig. 3. The last one is the fit with the best χ^2 and these observations strongly support the existence of the radiative transition process $Y(4260) \to \gamma X(3872)$.

4. Z states

The Z states are the one clearly exotic. They can decay in a $c\bar{c}$ state plus a charged particle such as a π^+, thus they are composed by four quarks and carry charge. We consider here the following Z decay channels: $e^+e^- \to \pi^{\pm,0}(\pi^{\mp,0}J/\psi)$ [11, 12], $e^+e^- \to \pi^{\pm,0}(\pi^{\mp,0}h_c)$ [15, 16], $e^+e^- \to \pi^{\pm,0}(D\bar{D}^*)^{\mp,0}$ [13, 14] and $e^+e^- \to \pi^{\pm,0}(D^*\bar{D}^*)^{\mp,0}$ [17, 18]. By the study of the mass spectrum $\pi^\pm J/\psi$ in the channel $\pi^+\pi^- J/\psi$ at the energy of 4.26 GeV a cross section of $62.9 \pm 1.9 \pm 3.7$ pb has been measured with $L = 525$ pb^{-1} and a mass M $= 3899.0 \pm 3.6 \pm 4.9$ MeV/c^2 and $\Gamma = 46 \pm 10 \pm 20$. The same resonance has been measured in $\pi^0\pi^0 J/\psi$ in the invariant mass spectrum of $\pi^0 J/\psi$. This is the neutral partner of the $Z^\pm(3900)$ and these three states

IOP Conf. Series: Journal of Physics: Conf. Series **1024** (2018) 012030 doi:10.1088/1742-6596/1024/1/012030

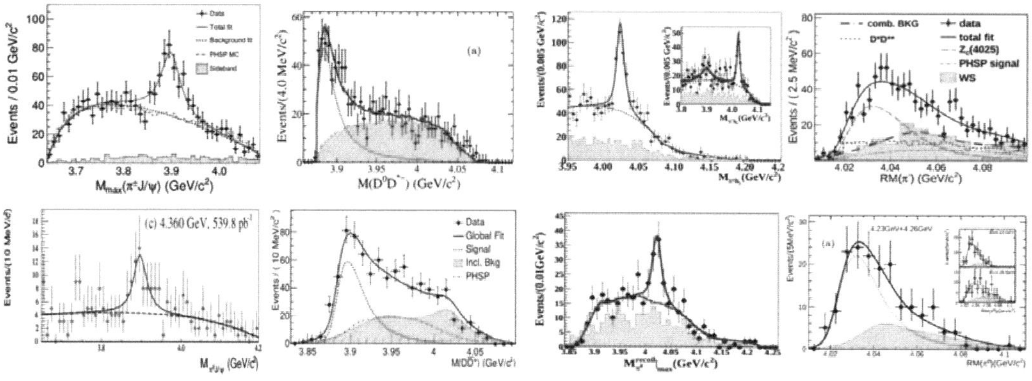

Figure 4. Summary of the Z states observed in the proposed eight channels. The four plots on the top row show the invariant mass (recoil mass) of the charged Z (the pion), while in the bottom part there is the neutral partner of the Z's isospin triplet. From left to right: $Z_c(3900)^{\pm,0} \to \pi^{\pm,0}J/\psi$, $Z_c(3885)^{\pm,0} \to (D\bar{D}^*)^{\pm,0}$, $Z_c(4020)^{\pm,0} \to \pi^{\pm,0}h_c$, $Z_c(4025)^{\pm,0} \to (D^*\bar{D}^*)^{\pm,0}$

determine an isospin triplet and light up the mystery of the Z states. A similar analysis has been done in $e^+e^- \to \pi^\pm(DD^*)^\mp$ and they have determined another Z triplet with M = 3881.7 ± 1.6 ± 1.6 MeV/c^2 and Γ = 26.2 ± 2.0 ± 2.1 MeV. These Z states are compatible with the ones in the channel $\pi J/\psi$ and they could be the same resonance that decays in two different channels. Same analogy can be done in $e^+e^- \to \pi^{\pm,0}(\pi^{\mp,0}h_c)$ and $e^+e^- \to \pi^{\pm,0}(D^*D^*)^{\mp,0}$ channels that define another Z labeled Z(4020). The Z states decay from $c\bar{c}$ states with quantum numbers 1^{--} and similarly to the connection between the X and the Y, here we can assume that at the energy of 4.26 GeV a certain amount of Y states are produced and these decay in a Z plus a pion. A summary of these Z states is shown in Fig. 4.

5. Conclusion

The nature of the XYZ is still unknown but possible connections in this multitude of exotic states show up in the recent discoveries by BESIII in the τ-charm region, such as the presence of two close but separated Y states, the isospin triplets for Z(3900) and Z(4020) discovered in several channels and the possible decay of the Y into X(3872) (radiative) or Z (hadronic).

References

[1] T. Barnes et al. Higher charmonia - Phys. Rev. D **72**, (2005) 054026
[2] G. Pakhlova *et al.* [Belle Collaboration], Phys. Rev. Lett. **26**, no.26
[3] B. Aubert *et al.* [BaBar Collaboration] Phys. Rev. Lett. **95**, 142001 (2005)
[4] M. Ablikim *et al.* [BESIII Collaboration] Phys. Rev. Lett. **110**, 252001 (2013)
[5] B. Aubert *et al.* [BaBar Collaboration] Phys. Rev. D **89**, 111103(R)
[6] G. Pakhlova *et al.* [Belle Collaboration] Phys. Rev. Lett. **110**,, 252002 (2013)
[7] M. Ablikim *et al.* [BESIII Collaboration] Phys. Rev. Lett. **118**, 092001 (2017)
[8] M. Ablikim *et al.* [BESIII Collaboration] Phys. Rev. Lett. **118**, 092002 (2017)
[9] M. Ablikim *et al.* [BESIII Collaboration] Phys. Rev. Lett. **114**, 092003 (2015)
[10] M. Ablikim *et al.* [BESIII Collaboration] Phys. Rev. Lett. **112**, 092001 (2014)
[11] M. Ablikim *et al.* [BESIII Collaboration] Phys. Rev. Lett. **110**, 252001 (2013)
[12] M. Ablikim *et al.* [BESIII Collaboration] Phys. Rev. Lett. **115**, 112003 (2015)
[13] M. Ablikim *et al.* [BESIII Collaboration]Phys. Rev. Lett. **112**, 022001 (2014)
[14] M. Ablikim *et al.* [BESIII Collaboration] Phys. Rev. Lett. **115**, 222002 (2015)
[15] M. Ablikim *et al.* [BESIII Collaboration] Phys. Rev. Lett. **111**, 242001 (2013)
[16] M. Ablikim *et al.* [BESIII Collaboration] Phys. Rev. Lett. **113**, 212002 (2014)

[17] M. Ablikim *et al.* [BESIII Collaboration] Phys. Rev. Lett. **112**, 132001 (2013)
[18] M. Ablikim *et al.* [BESIII Collaboration] Phys. Rev. Lett. **115**, 182002 (2015)

Precision Laser Spectroscopy Technique for Exotic Radioactive Beams at CERN-ISOLDE

Xiaofei Yang and the COLLAPS and CRIS collaboration
KU Leuven, Instituut voor Kern- en Stralingsfysica, B-3001 Leuven, Belgium
School of Physics and State Key Laboratory of Nuclear Physics and Technology, Peking University, Beijing 100871, China

E-mail: xiaofei.yang@pku.edu.cn

Abstract. The combination of high-resolution laser spectroscopy techniques and radioactive ion beams has provided major input for nuclear physics studies in different regions of the nuclear chart by accessing nuclear intrinsic properties, such as spins, magnetic and quadrupole moments and charge radii of exotic nuclei. As more and more exotic radioactive beams become available, laser spectroscopy and its relevant techniques start to be renascent with many advanced experimental setups being established and planned at radioactive-beam facilities world-wide. Meanwhile, a further exploration to expand the laser spectroscopy of radioactive beams into not only nuclear physics but also multiple interdisciplinary fields is being in progress. In this contribution, by using the relevant laser spectroscopy setups at ISOLDE as examples, the details of high precision laser spectroscopy experimental techniques applied to radioactive beams will be presented, together with a few highlights and achievements during the past years.

1. Introduction

It is well-known nowadays that the laser spectroscopy technique can be applied in various fields of science and technology. It was more than 30 years ago that this technique was firstly applied to nuclear physics studies [1, 2]. By probing the atomic hyperfine (HF) splitting and isotope shift (IS) using a high-precision laser spectroscopy method, multiple nuclear properties of ground/isomeric states of exotic nuclei far from stability, such as nuclear spins, nuclear magnetic and quadrupole moments and charge radii can be accessed simultaneously in a model independent way. With continuous development of radioactive ion (RI) beam facilities and production of bunched ions beam using the rf-quadrupole ion cooler and trap [3], the sensitivity of the laser spectroscopy technique has been significantly enhanced by several orders of magnitude in the last 10 years. The more and more exotic isotopes far from stability became accessible for laser spectroscopy, yielding many major inputs for nuclear structure study and providing important benchmark for the development of state-of-art nuclear theory [4, 5, 6, 7].

On another hand, considerable efforts have also been dedicated on the development of the laser spectroscopy technique with higher sensitivity and precision by a combination of many advantages offered by the conventional techniques [8, 9, 10] and by initiating many novel techniques [11, 12, 13]. These techniques have been established or are under construction or planned at different RI beam facilities across the globe and now can take advantages of RI beams produced from not only isotope separation on line (ISOL) but also in-flight project fragment (PF) facilities, as the The BEam COoler and LAser spectroscopy (BECOLA) [14] setup by using the RI beam at NSCL of MSU, US, and PALIS gas cell laser spectroscopy [15] developed to use

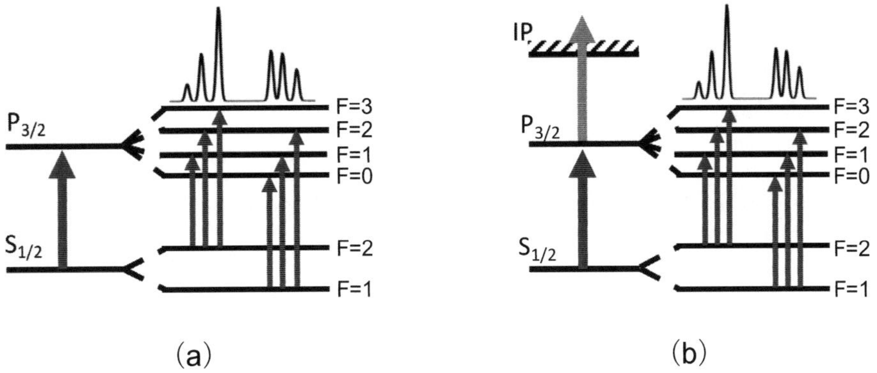

Figure 1. Example for HF structure of a given atomic transition from $^2S_{1/2}$ to $^2P_{3/2}$ with a nuclear spin of $I = 3/2$. It also presented the basic principle for (a) a collinear laser spectroscopy based on photo detection and (b) a typical collinear resonance ionisation spectroscopy by using ion detection

the RI beam produced at bigRIPS at RIBF of RIKEN, JAPAN. Therefore, by now, the nuclear spins, moments and charge radii of ground and isomeric states from more than 1900 isotopes have been investigated by various laser spectroscopy techniques, as has been summarized in Ref. [16]. Furthermore, the advantages offered by laser spectroscopy techniques can also be applied to interdisciplinary fields for solid state physics, fundamental physics, biophysics studies and so on, as pioneered at TRIUMF [17] and now also initiated at ISOLDE by polarized RI beam setup VITO (Versatile Ion-polarized Techniques On-line) [18, 19]. As new-generation RI ion beam facilities are foreseen, a growing development of advanced higher precision and high sensitivity laser spectroscopy techniques are urgently needed to fully exploit the nuclear physics opportunities.

The on-line laser spectroscopy method was firstly used for RI beams at ISOLDE-CERN. Until today, ISOLDE is still a well-known venue to produce most of the measurements of basic properties of RI beams by two complementary laser spectroscopy techniques, the conventional collinear laser spectroscopy technique (using the COLLAPS set-up) and the newly developed collinear resonance ionization spectroscopy (CRIS) by using the cooled and bunched ion beams. In this contribution, both COLLAPS and CRIS laser spectroscopy techniques currently active at ISOLDE will be introduced in detail as well as a few main inputs for nuclear structure studies in the last few years.

2. Precision lasers spectroscopy at ISOLDE

Figure 1 shows a HF splitting of atomic energy levels ($^2S_{1/2}$ and $^2P_{3/2}$) for a given atomic transition with a nuclear spin of $I = 3/2$, which is in a typical energy scaler of $10^{-7}eV$ or MHz and results in 6 allowed transition. This HF splitting is due to the interaction of the nucleus and electrons, and therefore provide directly several observables: nuclear spin (I), magnetic dipole (A) and electric quadrupole (B) HF parameters, and IS, as described in detail in Ref. [20] (Section 3). Therefore, by obtaining these parameters of a given isotope, fundamental properties (spins, moments and charge radii) of the nucleus can be extracted in a nuclear model independent way. Laser spectroscopy has acted as a powerful experimental technique for nuclear physics studies of RI beams over the past two decades. Certainly, several techniques can be applied for such measurement, for example in-source resonance ionization spectroscopy (RIS), collinear laser spectroscopy and spectroscopy of trapped atoms. The merit and demerit of different method has been evaluated in several review publication [16, 20], mainly regarding to three aspects: resolution, efficiency, accessible elements. Currently, the precision laser spectroscopy technique at ISOLDE is a compromise or balance between experimental efficiency and resolution by probing the ions or atoms with the low energy of 30-60 keV in a collinear geometry with the laser beam.

This technique has been employed at several RI beam facilities world-wide, and contributed the major part of the summarized nuclear chart for ground state properties measurements of RI beams (Figure 1 of Ref. [16]).

2.1. Collinear laser spectroscopy (COLLAPS)
COLLAPS is one of the collinear laser spectroscopic setups devoted to the study of radioactive isotopes, located at the isotope separator on-line facility ISOLDE at CERN. A detail overview about the past, present and future of COLLAPS setup can be found in a recent review publication [21]. As shown in Figure 1, this technique is based on the photon detection emitted from the laser resonance exited atoms or ions.

The different parts of the basic setup can be shortly described as (see FIG. 2 of Ref [22] for the schematic representation of the setup): The bunched ions beam delivered from ISOLDE is overlapped with the laser light from a cw Ti:sapphire or Dye laser (or together with a frequency doubling device), which is usually locked to a fixed frequency to match a Doppler shifted atomic/ionic transition and stabilized with an external stabilization system to avoid systematic errors and instability of the laser frequency. A charge exchange cell (CEC) filled with K or Na vapor (depends on the studied element) is used when an atomic transition from a certain atomic state was used for the HF structure measurement. This atomic state is populated when the RI beam passes thought the high-density vapor filled CEC. For the measurement process, instead of tuning the laser frequency, a variable voltage is applied to the light collection region (or CEC) to tune the velocity of the beam onto resonance with the laser beam. The photons emitted from the laser excited atoms or ions are recorded by four photomultiplier tubes (PMTs) as a function of the tunable voltage to obtain the HF spectra. Note that by using the bunched beam from ISOLDE (typical every $\sim$ 100 ms, depends on the studied isotopes), the collection time for the PMTs can be gated with the bunch width of the beams (typical $\sim$ 5 μs) in which the atom bunches passed through the light collection system. In this way, the background arising from the scattered laser light was reduced by a factors of $\sim 10^4$ ($\sim$ 5 μs /100 ms) and therefore the overall sensitivity of the experiment has been significantly improved, as have been demonstrated for many experiments in the last few years [4, 6, 23].

Another noteworthy point about COLLAPS is regarding to its versatility. Namely, the basic setup of COLLAPS can be modified or extended to a higher sensitivity measurement. For example, by a combination of a basic COLLAPS and ultra-sensitive collisional ionization detection [24], the exotic Kr isotopes were investigated, reaching a sensitivity of 10^3 atoms/s with continuous beam from ISOLDE; combing the optical pumping and β-NMR technique, COLLAPS has been used to studied exotic Mg isotopes [25], providing key input for the study of isotopes in the region of 'island of inversion', as has been reviewed in Ref. [26]; In-cooler optical pumping has been applied to Mn isotopes recently to populate a metastable state, enabling the measurement of the quadrupole moments of $^{53-63}$Mn and showing a clear onset of collectivity towards N = 40 [27]; a dedicated setup with ultra high sensitivity has been extended from the basic COLLAPS setup, aiming to measure the ground state properties of $^{53-54}$Ca with a production yield less than 10 ion/s [28].

2.2. Collinear resonance ionization spectroscopy
It is generally known that the production of exotic isotopes usually drops when go to the neutron-rich or proton-rich side, while the typical sensitivity for the COLLAPS technique is around 10^{3-4} ions/s. In-source spectroscopy using the efficient and selective resonant laser ionization method is another technique applied for rare isotope studies with very high efficiency [29]. Resonant laser ionization is also used in laser ion sources at many on-line facilities around the world [30]. However, due to pressure and Doppler broadening as well as the linewidth of commonly used pulsed lasers, the achieved resolution for the HF spectra from this technique is typically larger

than 1GHz, which prevents a precision measurement of ground state properties of light and medium mass isotopes.

CRIS is proposed to combine the high resolution achieved in collinear spectroscopy and the high sensitivity of resonance laser ionization spectroscopy. With a continuous effort for the development of high-resolution pulsed laser, CRIS has achieved a resolution of less than 100 MHz (in some cases even 20 MHz [10]) and sensitivity of 20 ions/s for Cu isotopes recently [31]. As detailed in Fig. 1 of Ref. [9], the delivery of the ISOLDE ion beam, the collinear geometry of the beam line, and the CEC system used in the CRIS setup are comparable to those in the COLLAPS setup. Specifically, instead of detecting the laser induced fluorescence, CRIS detects the laser resonantly re-ionized ions, which can be deflected from the remaining background atoms in the beam. Thus the sensitivity limitation of photon detection is overcome and also the background photon detection due to scattered laser light is avoided, resulting in a higher overall sensitivity [8]. This is realized by multiple-step laser excitation until the ionization potential of the studied element is reached. The process occurs in an interaction region with ultra-high vacuum (UHV: currently 10^{-9} mbar) to suppress the production of background ions from collisional ionization. This UHV region after the CEC is achieved by a differential pumping section [9]. The laser ionized ions are deflected onto an ion detector and recored. Note that a high resolution laser system with a linewidth of at most a few 10 MHz is required for the first step excitation to reach high resolution HF spectra, and also this laser is required to be pulsed or chopped to eliminate the possible low sensitivity resulting from optical pumping processes for certain elements [10, 31]. Therefore, it is the combination of bunched ion beams, the resonance ionization process, ion detection, high-resolution pulsed laser system and the UHV to reduce non-resonant excitation that makes CRIS a high resolution and high sensitivity laser spectroscopy method for the measurement of ground state properties of rare isotopes with low production yields down to few 10 ions/s.

Furthermore, different long-lived states of a studied isotope can be selected by the multiple-step laser ionization procedure and send to a decay spectroscopy station for laser-assisted nuclear decay spectroscopy on a pure isomeric beam. This method has been successfully used for alpha-tagging of HF components and for the alpha decay of ground and isomeric states of neutron-deficient Fr isotopes [9].

2.3. Recent highlights from laser spectroscopy measurements at ISOLDE

As laser spectroscopy provides four nuclear parameters at the same time, it can offer different aspects of nuclear structure information. For example, the nuclear spins and magnetic moments can be used to study the shell evolution of the single particle states of nucleons [32]; the magnetic and quadrupole moments can be used to check the configuration of the wavefunction for a given state with the help of the shell model calculation [33, 27]; the quadrupole moments and mean square charge radii can probe the collectivity and deformation of the nucleus [27, 34]; charge radii itself can also be used to verify the halo structure observed in light nuclei [35]; a combination of spins, moments and charge radii of ground and isomeric states also signal the existence of shape coexistence in a given nucleus [7].

Recently, many efforts for laser spectroscopy measurements at ISOLDE have been applied to the isotopes near the double magic nuclei such as ^{52,54}Ca, ^{78}Ni, ^{100,132}Sn, resulting in many fruitful optical spectra results, providing key inputs for nuclear structure studies. Around ^{52,54}Ca, HF spectra of Ca isotopes have been measured between $N = 20$ and $N = 32$ at COLLAPS, yielding the nuclear spins, moments and charge radii up to ^{52}Ca. The spins and moments of Ca isotopes show the single particle nature of the $^{41-49}$Ca, but with some mixing in the wave function of ^{51}Ca due to excitations of neutrons across N =32 [6]. The unexpected increase in the charge radii up to $N = 32$ challenges the earlier suggested magicity of neutron number 32 and tested the most advanced nuclear theory from both phenomenological and microscopic interactions [36]. Neutron rich Cu isotopes up to $N = 49$ have been studied by the CRIS experiment, and achieved the highest efficiency until now for a high-resolution HF spectra measurement. With one proton added to the proton closed shell $Z = 28$, the magnetic and quadrupole moments of $^{76-78}$Cu are therefore very sensitive to the cross shell excitation of proton and neutron [31]. Assisted by theoretical calculations from state-of-the-art shell model, this measurement provides further evidence for the robustness of $Z = 28$ shell closure near ^{78}Ni because of a subsequent decrease of proton and neutron excitations across $Z = 28$ and $N = 50$ when approach to $N = 50$. Experimental results in the Sn region from both laser spectroscopy techniques at ISOLDE are ongoing and under analysis, which will certainly provide important information on the nuclear structure around the doubly-magic isotopes $^{100}_{50}$Sn and $^{132}_{50}$Sn. Further study of more exotic isotopes near these so called double magic nucleus are still on going.

3. Summary

Laser spectroscopy is a powerful tool for a systematic measurement of nuclear ground state properties of exotic isotopes. Here, we introduced two complementary precision laser spectroscopy techniques at ISOLDE. The measurements achieived from these setups have provided massive nuclear data on the fundamental properties (nuclear spins, moments and charge radii) of ground and long-lived isomeric states of exotic isotopes produced at ISOLDE, and contributed a lot to the nuclear structure studies in the different regions of nuclear chart.

References

[1] R. Neugart, Nucl. Instrum. Methods Phys. Res. **186** 165 (1981).
[2] A. C. Mueller et al., Nucl. Phys. A **403**, 234 (1983).
[3] H. Franberg et al., Nucl. Instrum. Meth. B. **266**, 4502 (2008).
[4] J. Papuga et al., Phys. Rev. Lett. **110**, 172503 (2013).
[5] M. L. Bissell et al., Phys. Rev. Lett. **113**, 052502 (2014).
[6] R. F. Garcia Ruiz et al., Phys. Rev. C **91**, 041304 (2015).
[7] X. F. Yang et al., Phys. Rev. Lett. **116** , 182502 (2016).
[8] K. T. Flanagan et al., Phys. Rev. Lett. **111**, 212501(2013).
[9] K. M. Lynch et al., Phys. Rev. X **4**, 011055 (2014).
[10] R. P. de Groote et al., Phys. Rev. Lett. **115**, 132501 (2015).
[11] X. F. Yang et al., Phys. Rev. A. **90**, 052516 (2014).
[12] A. Takamine et al., Phys. Rev. Lett. **112**, 162502 (2014).
[13] R. Ferrer et al., Nature Communications **8**, 14520 (2017).
[14] K. Minamisono et al., Phys. Rev. Lett. **117**, 252501 (2016).
[15] T. Sonoda et al., Hyperfine Interactions **216**, 103 (2013)
[16] P. Campbell et al., Prog. Part. Nucl. Phys. **86**, 127 (2016).
[17] R.F. Kiefl et al., Nucl. Instrum. Meth. B, **204**, 682 (2003).
[18] R. F. Garcia Ruiz et al., EPJ Web Conf. **93**, 07004 (2015).
[19] M. Kowalska et al., J. Phys. G **44**, 084005 (2017) .
[20] B. Cheal and K. T. Flanagan. Phys. G: Nucl. Part. Phys. **37**,113101 (2010).
[21] R. Neugart et al., J. Phys. G: Nucl. Part. Phys. **44**, 064002 (2017).
[22] J. Papuga et al., Phys. Rev. C **90**, 034321(2014).
[23] C. Wraith et al., Phys. Lett. B **771**, 385 (2017).
[24] M. Keim et al., Nuclear Physics A **586**, 219 (1995).
[25] G. Neyens et al., Phys. Rev. Lett. **94**, 022501 (2005).
[26] G. Neyens, J. Phys. G **43**, 024007 (2016).
[27] C. Babcock et al., Phys. Lett. B 760 387(2016).
[28] R. F. Garcia Ruiz et al., J. Phys. G: Nucl. Part. Phys. **44**, 044003 (2017).
[29] T. E. Cocolios et al., Phys. Rev. Lett. **106**, 052503(2011).
[30] V. Fedosseev et al., J. Phys. G: Nucl. Part. Phys. **44**, 084006 (2017).
[31] R.P. de Groote et al., Phys. Rev. C. **96**, 041302 (R) (2017).
[32] K. T. Flanagan et al., Phys. Rev. Lett. **103**, 142501 (2009).
[33] C. Babcock et al., Phys. Lett. B **750**, 176 (2015).
[34] H. Heylen et al., Phys. Rev. C **94**, 054321 (2016).
[35] W. Nörtershäuser et al., Phys. Rev. Lett **102**, 062503 (2009).
[36] R. F. Garcia Ruiz et al. Nature Physics **12**, 594 (2016).

IOP Conf. Series: Journal of Physics: Conf. Series **1024** (2018) 012032 doi:10.1088/1742-6596/1024/1/012032

Electromagnetic decays of the neutral pion investigated in the Dyson-Schwinger formalism

Esther Weil[1], Gernot Eichmann[2], Christian S. Fischer[1], Richard Williams[1]

[1] Institut für Theoretische Physik, Justus-Liebig Universität Gießen, 35392 Gießen, Germany
[2] CFTP, Instituto Superior Técnico, Universidade de Lisboa, 1049-001 Lisboa, Portugal

E-mail: Esther.D.Weil@physik.uni-giessen.de

Abstract. We summarize recent work in determining the transition form factor (TFF) of the neutral pion ($\pi^0 \to \gamma^*\gamma^*$), by solving the non-perturbative Dyson-Schwinger and Bethe-Salpeter equations. We first study the transition form factor, followed by the rare decay $\pi^0 \to e^+e^-$, which requires the TFF as input. In addition to the aspects of truncation required to compute the solution, we discuss unexpected behavior in the large Q^2 regime (with Q^2 the photon virtuality), and also touch on a path deformation in the complex plane to access the total decay rate for the rare decay.

1. Introduction

Meson transition form factors are intermediate states of many decay and collision processes. As experiments continue to reduce statistical uncertainties on such observables, it is important to match the effort on the theoretical side. To calculate these non-perturbative quantities we are using the framework of the Dyson-Schwinger and Bethe-Saltpeter equations. We summarize the calculations described in Refs. [1, 2]. These two publications are concerned with the calculation of the pion transition form factor $\pi^0 \to \gamma^*\gamma^*$ and further leptonic decays of the pion, such as the rare decay $\pi^0 \to e^+e^-$, which poses an interesting case due to a discrepancy between theoretical and experimental data.

2. Transition form factor $\pi^0 \to \gamma\gamma$

The $\pi \to \gamma^*\gamma^*$ transition matrix element is given by the tensor

$$\Lambda^{\mu\nu}(Q, Q') = e^2 \, \frac{F(Q^2, Q'^2)}{4\pi^2 f_\pi} \, \varepsilon^{\mu\nu\alpha\beta} Q'^\alpha Q^\beta \,, \tag{1}$$

with incoming and outgoing photon momenta Q' and Q, the pion's electroweak decay constant $f_\pi \approx 92$ MeV and the electromagnetic charge e. The transition is described by a single scalar function, called a transition form factor (TFF), depending on virtualities of the two photon momenta $F(Q^2, Q'^2)$, and the convention of prefactors is chosen such that $F(0, 0) = 1$ in the chiral limit, constrained by the Abelian anomaly. Besides the on-shell point the form factor is evaluated in different kinematic regions, see Fig 1 (b). The so called singly virtual or asymmetric limit (one photon on-shell $Q'^2 = 0$ and one off-shell $Q^2 \neq 0$) is the quantity measured in most

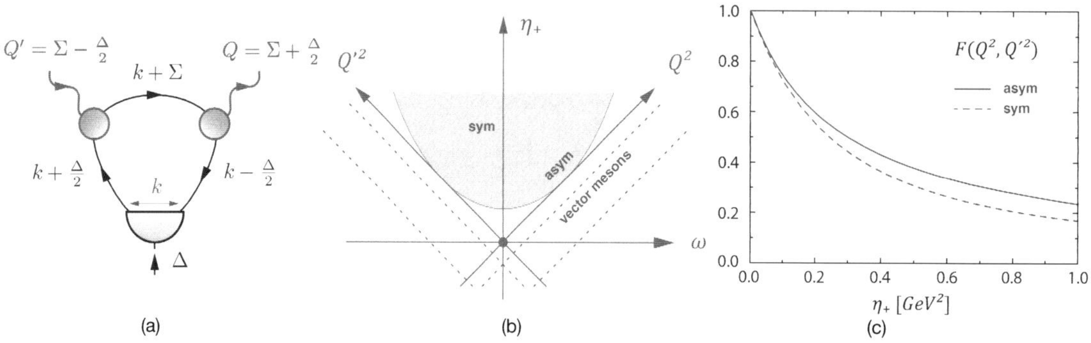

Figure 1. Transition form factor ($\pi^0 \to \gamma^*\gamma^*$) : (a) transition matrix element, (b) kinematic domains in Q^2 and Q'^2 including the symmetric and asymmetric limit. The dotted lines indicate the vector pole locations. (c) Result for the on-shell transition form factor in the symmetric and asymmetric limit. Here we use $\omega = (Q^2 - Q'^2)/2$ and $\eta_+ = (Q^2 + Q'^2)/2$.

experiments. The symmetric limit ($Q^2 = Q'^2$) is the kinematic region needed as an input for the decay into a di-lepton pair $\pi^0 \to e^+e^-$. These are both contained in the space-like domain ($Q^2 > 0$, $Q'^2 > 0$). The time-like region ($Q^2 < 0$, $Q'^2 < 0$) is probed e.g when evaluating the pion Dalitz decays $\pi \to \gamma e^+e^-$, $\pi^0 \to 2e^+2e^-$. This region contains the physical singularities such as vector meson poles which make the evaluation of the form factor in this region challenging.

2.1. The triangle diagram

In the impulse approximation, the transition form factor $\pi^0 \to \gamma^*\gamma^*$ is given by, Fig. 1 (a)

$$\Lambda^{\mu\nu} = 2e^2 \,\mathrm{Tr} \int \frac{d^4k}{(2\pi)^4} S(k_+)\, \Gamma_\pi(k, \Delta)\, S(k_-)\Gamma^\mu(r_-, -Q)\, S(k + \Sigma)\, \Gamma^\nu(r_+, Q')\,, \tag{2}$$

where $\Sigma = (Q^2 + Q'^2)/2$ is the average photon momentum, $\Delta = Q - Q'$ the pion momentum, k is the loop momentum, $k_\pm = k \pm \Sigma/2$ and $r_\pm = k + \Sigma/2 \pm \Delta/4$. The trace in Eq. 2 is over Dirac indices only and the factor 2 in front of the diagrams is a symmetry factor. The ingredients of the triangle are obtained by numerically solving Dyson-Schwinger (DSE) and Bethe-Salpeter equations (BSE).

The internal lines in the triangle diagram Fig. 1(a) are quark propagators S. Those include effects of dynamical mass generation due to dynamical breaking of chiral symmetry and we calculate them solving their corresponding DSE. The pion bound-state composed from two quarks is described by the pseudoscalar Bethe-Salpeter amplitude Γ_π (yellow half circle). Both elements are described by scalar functions times their tensor structure in Dirac, color and flavor space. The quark propagator is described by two tensor structures, while the pion has four. The quark-photon vertex Γ^μ fulfills an inhomogeneous Bethe-Salpeter equation and is decomposed into twelve tensor structures. These can be split into a transverse and non-transverse part, the so-called Ball-Chiu vertex [3]. The transverse part of the vertex contains vector meson states. Thus by properly solving the BSE for the vertex, vector-meson-dominance effects are automatically incorporated into our approach.

Furthermore it should be mentioned that DSE/BSE's can not be solved without truncating the equations. In context of the form factor calculation we used the so called rainbow-ladder truncation and the Maris-Tandy model [4] with parameters $\Lambda = 0.74$ GeV and $\eta = 1.85 \pm 0.2$.

IOP Conf. Series: Journal of Physics: Conf. Series **1024** (2018) 012032 doi:10.1088/1742-6596/1024/1/012032

Furthermore the input quark mass is $m_q = 3.57$ MeV at a renormalization point of $\mu = 19$ GeV. The pion mass and decay constant we obtain are $m_{\pi^0} = 135.0(2)$MeV and $f_{\pi^0} = 92.4(2)$MeV.

2.2. *Results*

With all the building blocks as described in the previous section we were able to determine the form factor in the space-like domain as well as in a small region into the time-like area ($Q^2 > -m_\pi^2$). The result for the form factor in the asymmetric and symmetric limit are shown in Fig. 1 (c). At the on-shell point we obtain a value of $F(0,0) = 0.996$, which is close to the anomaly constraint of $F(0,0) = 1$ in the chiral limit. A direct calculation is however only possible in a restricted kinematic region, because of singularities in the quark propagator. For on-shell pion momentum we are limited to values $Q^2 \lesssim 4$ GeV^2 in the asymmetric limit. To calculate the form factor in the whole space-like plane we applied a technique briefly touched up on in section 4. With this we were able to provide a fit for the form factor in the whole space-like domain, which is given in Ref. [2], Eq. (12).

3. Rare decay

The two-body decay of the neutral pion into a di-lepton pair ($\pi \to e^+e^-$) poses an interesting puzzle, as the current theoretical estimates show a discrepancy with the experimental result from the KTeV E799-II experiment at Fermilab [5,8] at the order of 2σ. The experimental value for the branching ratio is $B(\pi^0 \to e^+e^-) = (6.87 \pm 0.36) \times 10^{-8}$, after reanalysis subtracting radiative corrections [6, 7].

To lowest order in QED the process is described by the one-loop Feynman diagram, Fig. 2(b), which includes the transition form factor $F(Q^2, Q'^2)$ as the only non-perturbative input (indicated in yellow). The corresponding normalized branching ratio is given by

$$R = \frac{B(\pi^0 \to e^+e^-)}{B(\pi^0 \to \gamma\gamma)} = 2\left(\frac{m\,\alpha_{\text{em}}}{\pi m_\pi}\right)^2 \beta(t_0)\,|\mathcal{A}(t_0)|^2\,, \tag{3}$$

with $t = \Delta^2/4$, $\beta(t) = \sqrt{1 + m^2/t}$ stems from the two-body phase-space integration and $B(\pi^0 \to \gamma\gamma) = 0.988$. The scalar amplitude $\mathcal{A}(t)$ is the averaged and spin-summed matrix element, which must be evaluated at the on-shell pion point $t_0 = -m_\pi^2/4$.

After taking traces the result reduces to

$$\mathcal{A}(t) = \frac{1}{2\pi^2 t}\int d^4\Sigma\, \frac{(\Sigma\cdot\Delta)^2 - \Sigma^2\Delta^2}{(p+\Sigma)^2 + m^2}\,\frac{F(Q^2, Q'^2)}{Q^2\,Q'^2}\,. \tag{4}$$

This integral has poles in the integration domain and thus cannot be naively integrated. A common technique to work around problems of such kinds is the use of dispersive methods (see e.g. [8, 9, 12]). Another way to circumvent the problem is path deformation in the complex

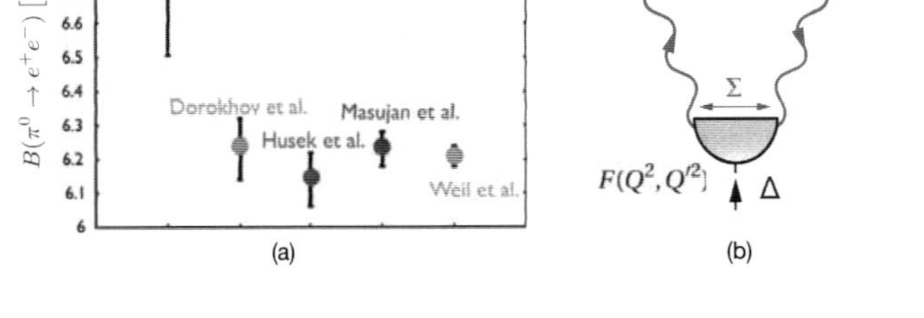

Figure 2. Rare decay of the neutral pion $\pi^0 \to e^+e^-$: (a) Comparison of the latest experimental results [5, 6] and theoretical calculations [5–11]. (b) Feynman diagram to Eq. 4, the yellow half-circle denotes the pion transition form factor $F(Q^2, Q'^2)$, as only non-perturbative input.

IOP Conf. Series: Journal of Physics: Conf. Series **1024** (2018) 012032 doi:10.1088/1742-6596/1024/1/012032

plane. For a chosen value of t one knows the locations of poles and branch cuts in the complex plane and can thus find a path to navigate around these infinities, such that the Euclidean integration is possible and not ill-defined. The detailed pole analysis for Eq. 4 and the intricate integration path can be found in [1], Section IV B.

We calculated the branching ratio using both methods. For the direct calculation we obtain a branching fraction of $B(\pi^0 \to e^+e^-) = 6.22(3) \times 10^{-8}$ which compares very nicely to our result reached when using dispersion relation $B(\pi^0 \to e^+e^-) = 6.21(3) \times 10^{-8}$. We are furthermore in agreement with previous theoretical predictions, see overview Fig. 2(a) and thus a 2σ discrepancy to the experimental value remains. With this we are able to provide a valuable cross check for previous calculations using dispersive tools [8–11].

4. Large Q^2-behavior

The latest two experimental measurements for the singly-virtual form factor [15,16] have stirred considerable interest in the large-momentum behavior [17–37], with the result from the BaBar collaboration suggesting a deviating from scaling. Within the DSE/BSE framework, the large-momentum behavior of the TFF has been studied previously in Refs. [18,35,36]. We employed a different technique to calculate the form factor in the whole space-like domain and also take into account physical singularities stemming from vector meson dominance, a genuinely non-perturbative effect [1].

For off-shell kinematics the form factor is a function of three Lorentz invariants, such as $\{Q^2, Q'^2, Q \cdot Q'\}$, and can be calculated up to arbitrary high space-like values in Q^2, Q'^2 without crossing any singularity in the integrand. We thus calculate the off-shell form factor and exploit a fit to the physical point constrained by the vector-meson-pole, see [1] for details. The result for the form factor up the experimentally measured momentum range is displayed in the left panel of Fig. 3 and describes the experimental data well. The right panel shows the singly-virtual form factor at asymptotically large momenta beyond the region where experimental data exists. Here we observe a continued rise until the curve plateaus to its asymptotic behavior at around $10^2 - 10^3$ GeV^2. The error-band comes from a conservative variation of the fit parameters.

As compared with the well-know Efremov-Radyushkin-Brodsky-Lepage (ERBL) scaling limit [38,39], which is an asymptotic-QCD prediction, our result agrees in case of the symmetric limit ($Q^2 = Q'^2$), but for the asymmetric case exceeds their prediction. We have checked

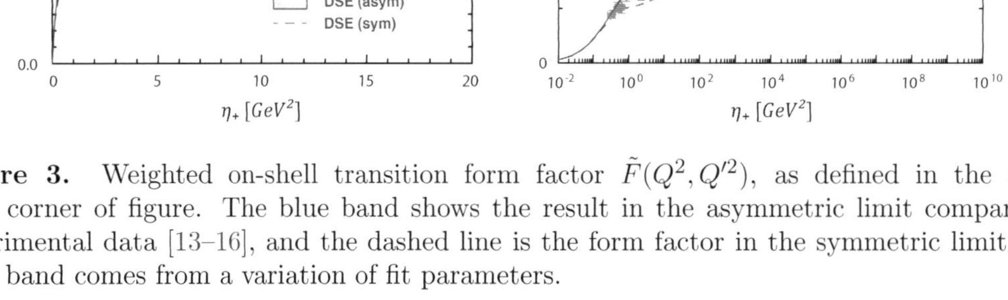

Figure 3. Weighted on-shell transition form factor $\tilde{F}(Q^2, Q'^2)$, as defined in the upper right corner of figure. The blue band shows the result in the asymmetric limit compared to experimental data [13–16], and the dashed line is the form factor in the symmetric limit. The error band comes from a variation of fit parameters.

explicitly that discarding the vector-meson pole constraints leads to a recovery of the ERLB limit. The discrepancy still has to be understood.

5. Conclusion

We have summarized recent calculation of the neutral pion TFF using the Dyson-Schwinger and Bethe-Salpeter equations. Generally the results show good consistency with existing theoretical and experimental determinations.

For the decay into a di-lepton pair ($\pi \to e^+e^-$) we confirm the previous theoretical calculation with a branching ratio of $B(\pi^0 \to e^+e^-) = 6.22(3) \times 10^{-8}$, leaving a 2σ discrepancy to the experimental value. We were able to provide a cross-check for previous theoretical calculation through using a different method to solve the integral.

References

[1] Eichmann G, Fischer C, Weil E and Williams R 2017 (*Preprint* 1704.05774)
[2] Weil E, Eichmann G, Fischer C S and Williams R 2017 *Phys. Rev.* **D96** 014021 (*Preprint* 1704.06046)
[3] Eichmann G, Sanchis-Alepuz H, Williams R, Alkofer R and Fischer C S 2016 *Prog. Part. Nucl. Phys.* **91** 1–100 (*Preprint* 1606.09602)
[4] Maris P and Tandy P C 1999 *Phys. Rev.* **C60** 055214 (*Preprint* nucl-th/9905056)
[5] Abouzaid E *et al.* (KTeV) 2007 *Phys. Rev.* **D75** 012004 (*Preprint* hep-ex/0610072)
[6] Husek T, Kampf K and Novotny J 2014 *Eur. Phys. J.* **C74** 3010 (*Preprint* 1405.6927)
[7] Vasko P and Novotny J 2011 *JHEP* **10** 122 (*Preprint* 1106.5956)
[8] Dorokhov A E and Ivanov M A 2007 *Phys. Rev.* **D75** 114007 (*Preprint* 0704.3498)
[9] Dorokhov A E 2010 *JETP Lett.* **91** 163–169 (*Preprint* 0912.5278)
[10] Husek T and Leupold S 2015 *Eur. Phys. J.* **C75** 586 (*Preprint* 1507.00478)
[11] Masjuan P and Sanchez-Puertas P 2015 (*Preprint* 1504.07001)
[12] Bergstrom L, Masso E, Ametller L and Bramon A 1983 *Phys. Lett.* **B126** 117–121
[13] Behrend H J *et al.* (CELLO) 1991 *Z. Phys.* **C49** 401–410
[14] Gronberg J *et al.* (CLEO) 1998 *Phys. Rev.* **D57** 33–54 (*Preprint* hep-ex/9707031)
[15] Aubert B *et al.* (BaBar) 2009 *Phys. Rev.* **D80** 052002 (*Preprint* 0905.4778)
[16] Uehara S *et al.* (Belle) 2012 *Phys. Rev.* **D86** 092007 (*Preprint* 1205.3249)
[17] Khodjamirian A 1999 *Eur. Phys. J.* **C6** 477–484 (*Preprint* hep-ph/9712451)
[18] Anikin I V, Dorokhov A E and Tomio L 2000 *Phys. Lett.* **B475** 361–371 (*Preprint* hep-ph/9909368)
[19] Melic B, Mueller D and Passek-Kumericki K 2003 *Phys. Rev.* **D68** 014013 (*Preprint* hep-ph/0212346)
[20] Radyushkin A V 2009 *Phys. Rev.* **D80** 094009 (*Preprint* 0906.0323)
[21] Polyakov M V 2009 *JETP Lett.* **90** 228–231 (*Preprint* 0906.0538)
[22] Agaev S S, Braun V M, Offen N and Porkert F A 2011 *Phys. Rev.* **D83** 054020 (*Preprint* 1012.4671)
[23] Agaev S S, Braun V M, Offen N and Porkert F A 2012 *Phys. Rev.* **D86** 077504 (*Preprint* 1206.3968)
[24] Ruiz Arriola E and Broniowski W 2010 *Phys. Rev.* **D81** 094021 (*Preprint* 1004.0837)
[25] Kroll P 2011 *Eur. Phys. J.* **C71** 1623 (*Preprint* 1012.3542)
[26] Gorchtein M, Guo P and Szczepaniak A P 2012 *Phys. Rev.* **C86** 015205 (*Preprint* 1102.5558)
[27] Brodsky S J, Cao F G and de Teramond G F 2011 *Phys. Rev.* **D84** 033001 (*Preprint* 1104.3364)
[28] Brodsky S J, Cao F G and de Teramond G F 2011 *Phys. Rev.* **D84** 075012 (*Preprint* 1105.3999)
[29] Noguera S and Vento V 2012 *Eur. Phys. J.* **A48** 143 (*Preprint* 1205.4598)
[30] El-Bennich B, de Melo J P B C and Frederico T 2013 *Few Body Syst.* **54** 1851–1863 (*Preprint* 1211.2829)
[31] Dorokhov A E and Kuraev E A 2013 *Phys. Rev.* **D88** 014038 (*Preprint* 1305.0888)
[32] Dorokhov A E 2010 *JETP Lett.* **92** 707–719
[33] Maris P and Tandy P C 2002 *Phys. Rev.* **C65** 045211 (*Preprint* nucl-th/0201017)
[34] Holl A, Krassnigg A, Maris P, Roberts C D and Wright S V 2005 *Phys. Rev.* **C71** 065204 (*Preprint* nucl-th/0503043)
[35] Raya K, Chang L, Bashir A, Cobos-Martinez J J, Gutierrez-Guerrero L X, Roberts C D and Tandy P C 2016 *Phys. Rev.* **D93** 074017 (*Preprint* 1510.02799)
[36] Raya K, Ding M, Bashir A, Chang L and Roberts C D 2016 (*Preprint* 1610.06575)
[37] Mikhailov S V, Pimikov A V and Stefanis N G 2016 *Phys. Rev.* **D93** 114018 (*Preprint* 1604.06391)
[38] Lepage G P and Brodsky S J 1980 *Phys. Rev.* **D22** 2157
[39] Efremov A V and Radyushkin A V 1980 *Phys. Lett.* **94B** 245–250

$\Lambda N \to NN$ EFT potentials and hypertriton non-mesonic weak decay

Axel Pérez-Obiol[1], David R. Entem[2], Andreas Nogga[3]

[1] Nuclear Physics Institute, 25068 Řež, Czech Republic
[2] Institute for Advanced Simulation, Institut für Kernphysik, and Jülich Center for Hadron Physics, Forschungszentrum Jülich, D-52425 Jülich, Germany
[3] Grupo de Física Nuclear and Instituto Universitario de Física Fundamental y Matemáticas (IUFFyM) Universidad de Salamanca, E-37008 Salamanca, Spain

E-mail: perez-obiol@ujf.cas.cz

Abstract. The potential for the $\Lambda N \to NN$ weak transition, the main responsible for the non-mesonic weak decay of hypernuclei, has been developed within the framework of effective field theory (EFT) up to next-to-leading order (NLO). The leading order (LO) and NLO contributions have been calculated in both momentum and coordinate space, and have been organised into the different operators which mediate the $\Lambda N \to NN$ transition. We compare the ranges of the one-meson and two-pion exchanges for each operator. The non-mesonic weak decay of the hypertriton has been computed within the plane-wave approximation using the LO weak potential and modern strong EFT NN potentials. Formally, two methods to calculate the final state interactions among the decay products are presented. We briefly comment on the calculation of the $^3_\Lambda H \to ^3 He + \pi^-$ mesonic weak decay.

1. Introduction

The hypertriton, a bound system of a proton, a neutron and a Λ, is the lightest known hypernucleus, and therefore also the simplest to study theoretically. Because it contains only three particles, it is possible to describe its wave function, its weak decay modes, and the corresponding final state interactions with exact few-body techniques and precise strong and weak potentials.

Hypernuclei decay weakly in mainly two modes, mesonically, resembling the weak decay of the Λ in free space, $\Lambda \to N\pi$, and non-mesonically, in which the Λ is converted into a nucleon by interacting with a nucleon (or nucleons) from the medium, $\Lambda N \to NN$, $\Lambda NN \to NNN$, etc. The non-mesonic weak decay mode constitutes a very small part of the total decay rate for the lightest hypernuclei, —the hypertriton, $^4_\Lambda H$ and $^4_\Lambda He$—. However, these decay rates can in principle be measured independently of the mesonic ones, and can offer useful information to constrain the $\Lambda N \to NN$ interaction. In particular, the experiment E22 in J-PARC, plans to measure the non-mesonic weak decay partial rates for $^4_\Lambda H$ and $^4_\Lambda He$ [1].

The hypertriton non-mesonic and mesonic weak decays were studied previously using one-meson-exchange potentials for the strong and weak forces and the Faddeev few-body scheme to implement all the interactions [2], [3]. In this work we focus on studying the non-mesonic weak decay of the hypertriton by implementing the same few-body framework and using modern strong and weak EFT potentials for the initial and final wave functions. We put special emphasis in the non-mesonic weak decay transition, $\Lambda N \to NN$, for which we have developed the EFT potentials up to NLO in momentum and coordinate space. For the strong interactions NLO NN and YN EFT potentials are used [4]. At the end, we briefly comment on how to compute the mesonic partial decay rate $\Gamma(^3_\Lambda H \to ^3 He + \pi^-)$, which might be relevant in understanding the specially short lifetime of the hypertriton [5]. Note that one should, in principle, also consider the role of the Σ baryon in the weak decays, as it contributes to the hypertriton wave function due to $\Lambda - \Sigma$ mixing. In our work we have focused in the Λ decay modes since they represent the most dominant contribution.

2. $\Lambda N \to NN$ potential with EFT

The weak transition $\Lambda N \to NN$, driving the non-mesonic weak decay of the hypertriton, has been described within the EFT framework up to next-to-leading order (NLO). As in the EFT description of the NN and YN strong interactions, the idea is to model this transition in a systematic, general way through a separation of the relevant scales. The soft and hard scales of a baryon-baryon low-energy interaction are the three-momenta (p) and the masses (M) of the baryons, respectively. The EFT considers only the degrees of freedom with masses (m) of the order of the low-energy scale, $m \sim p$, and a potential expanded up to a certain order $(p/M)^n$, $(m/M)^n$. At each order n, the EFT potential consists in all the terms of order $\mathcal{O}(p/M)^n$ that are allowed by these degrees of freedom and the symmetry constraints of the interaction.

In contrast to the NN and YN cases, the $\Lambda N \to NN$ reaction is weak and exothermic — there is a variation in strangeness and a mass difference between the initial ΛN pair and the final NN one. Therefore, in our case we must allow for PV contributions, and consider that the soft scale is determined not by the typical momenta of the hypertriton constituents, but by the higher momenta of the two emerging nucleons, of the order of $p \sim 400$ MeV. The pion mass is well below p, and since the kaon mass is of the same order we consider both mesons as explicit degrees of freedom at leading order (LO). At NLO, for simplicity and to not add more uncertainties coming from the unknown kaon couplings, we have only considered the two-pion exchange.

The EFT potential is thus divided in the different orders in the expansion, each order containing the corresponding meson exchanges and contact interactions. The contact terms consist in all the possible four-point operators of order $\mathcal{O}(p/M)^n$. They describe the unresolved short range dynamics and their coefficients, or so called low-energy constants (LECs), must be fixed to the data. In momentum space, the contact part of the potential up to $\mathcal{O}(q^2/M^2)$, consisting of all the PC and PV operational structures, is shown in Tab. 1.

$\mathcal{O}(q^0)$	$1,\ \vec{\sigma}_1\cdot\vec{\sigma}_2$
$\mathcal{O}(q)$	$\vec{\sigma}_1\cdot\vec{q},\ \vec{\sigma}_2\cdot\vec{q},\ \vec{\sigma}_1\cdot\vec{p},\ \vec{\sigma}_2\cdot\vec{p},\ (\vec{\sigma}_1\times\vec{\sigma}_2)\cdot\vec{q},\ (\vec{\sigma}_1\times\vec{\sigma}_2)\cdot\vec{p},$
$\mathcal{O}(q^2)$	$\vec{q}^{\,2},\ \vec{p}^{\,2},\ \vec{q}\cdot\vec{p},\ (\vec{\sigma}_1\cdot\vec{\sigma}_2)\vec{q}^{\,2},\ (\vec{\sigma}_1\cdot\vec{\sigma}_2)\vec{p}^{\,2},\ (\vec{\sigma}_1\cdot\vec{\sigma}_2)\vec{q}\cdot\vec{p}$ $\vec{\sigma}_1\cdot(\vec{q}\times\vec{p}),\ \vec{\sigma}_2\cdot(\vec{q}\times\vec{p}),\ (\vec{\sigma}_1\cdot\vec{p})(\vec{\sigma}_2\cdot\vec{p}),$ $(\vec{\sigma}_1\cdot\vec{q})(\vec{\sigma}_2\cdot\vec{p}),\ (\vec{\sigma}_1\cdot\vec{p})(\vec{\sigma}_2\cdot\vec{q}),\ (\vec{\sigma}_1\cdot\vec{q})(\vec{\sigma}_2\cdot\vec{q}).$

Table 1. All contact operational structures up to next-to-leading order. $\vec{q}$, $\vec{p}$ are the average initial momentum and the transferred momentum, respectively, and $\vec{\sigma}_1$ and $\vec{\sigma}_2$ are the Pauli matrices connecting the spins of the two pairs of particles.

The LO part of the potential was first studied in [6], [7]. Its mesonic exchange contribution consists in the one-pion and one-kaon exchanges at LO, and all the possible two-pion exchanges at NLO. The strong vertices appearing in the diagrams are taken from the strong chiral Lagrangian. The $YN\pi$ and NNk weak vertices are phenomenological, while the ΛN and $\Lambda N\pi\pi$ ones, entering at NLO diagrams, are given by the weak SU(3)$_F$ Lagrangian [8].

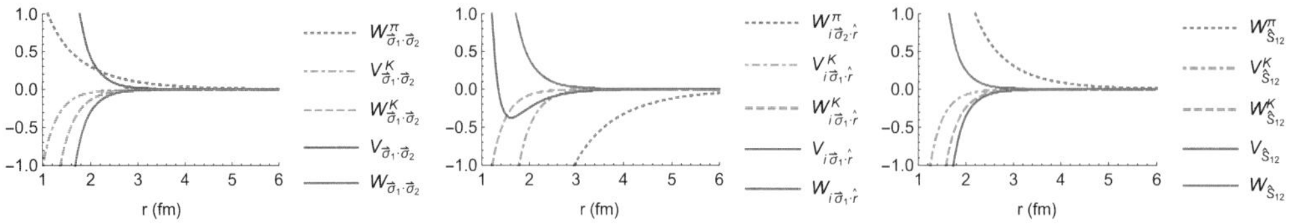

Figure 1. LO one-pion and one-kaon exchange potentials in coordinate space (dashed lines) organised into the three spin operators to which they contribute, $\vec{\sigma}_1 \cdot \vec{\sigma}_2$, $\vec{\sigma}_2 \cdot \hat{r}$, and the tensor term $\hat{S}_{12} \equiv 3(\vec{\sigma}_1 \cdot \hat{r})(\vec{\sigma}_2 \cdot \hat{r}) - (\vec{\sigma}_1 \cdot \vec{\sigma}_2)$, and the two isospin ones, 1 and $\vec{\tau}_1 \cdot \vec{\tau}_2$, labeled with V and W, respectively. Together are plotted the NLO contributions to the same operators, coming from the two-pion exchange (solid lines).

The potentials have been first calculated in momentum space and then they have been Fourier-transformed to coordinate space. In both cases, the explicit expressions are shown in terms of four basic integrals that are computed numerically and can be found in [9] and [10], respectively. The NLO potentials contain many more operational structures than the LO ones, and therefore allow for many more possible spin transitions. In Fig. 1 we compare the LO and NLO potentials in coordinate space for the six operators to which the LO diagrams contribute. It is shown that the one-pion and one-kaon exchanges dominate in the long range, and get surpassed by the two-pion exchanges at ~ 2 and ~ 3 fm, respectively. The NLO thus plays a relevant role, both quantitatively and qualitatively. It contains, however, a large number of unknown coefficients which, at the moment, cannot be fixed with the available data on hypernuclear non-mesonic weak decay.

3. Non-mesonic weak decay rate

The calculation of the non-mesonic weak decay rate of the hypertriton consists in a matrix element driving the transition between the hypertriton and the final state of three baryons, an average over initial hypertriton spin projections, and a sum over all possible final states (momenta, spins, and isospins). The rates for d+n and 3N can be written as

$$\Gamma^{d+n} = \frac{32}{9}\pi^2 M_N p_3^{(d+n)} \sum_{m_{J_3} m_{j_{12}} m_3} \left| \langle \Psi_{m_{j_{12}} m_3} | (1+P) V_{12}^w | \psi_{^3_\Lambda H} \rangle + \langle \Psi_{m_{j_{12}} m_3} | P \, | U \rangle \right|^2 , \tag{1}$$

$$\Gamma^{3N} = \frac{128}{9}\pi^3 M_N \sum_{\substack{m_{J_3} m_1 m_2 m_3 \\ m_{t_1} m_{t_2} m_{t_3}}} \int p_{12}^2 dp_{12} p_3^{(3N)} \sin(\theta) d\theta \left| \langle \Psi_{\substack{m_1 m_2 m_3 \\ m_{t_1} m_{t_2} m_{t_3}}} | (1+P) \left(V_{12}^w | \psi_{^3_\Lambda H} \rangle + | U \rangle \right) \right|^2 . \tag{2}$$

m_{J_3} and $m_{j_{12}}$ are the spin projections of the hypertriton and the deuteron, and m_i and m_{t_i} label the spin and isospin projections of the three nucleons ($i = 1, 2, 3$). The conservation of energy and momentum has been used to fix the momenta $p_3^{(d+n)} \equiv \sqrt{4M_N/3(\Delta M + \epsilon - \epsilon_d)}$ and $p_3^{(3N)} \equiv \sqrt{4M_N/3\,(\Delta M + \epsilon - \vec{p}_{12}^{\,2}/M_N)}$, where $\Delta M \equiv M_\Lambda - M_N$, with M_N and M_Λ the masses of the nucleon and the Λ. ϵ and ϵ_d are the binding energies of the hypertriton and the deuteron, and $P \equiv P_{12}P_{23} + P_{13}P_{23}$ is an antisymmetry operator, where P_{ij} exchanges the quantum numbers and coordinates of particles i and j. V_{12}^w is the weak $\Lambda N \to NN$ potential and $\psi_{^3_\Lambda H}$, $\Psi^{(-)}_{m_d m_n}$ and $\Psi^{(-)}_{\substack{m_1 m_2 m_3 \\ m_{t_1} m_{t_2} m_{t_3}}}$ are the wave functions for the hypertriton, the outgoing deuteron and neutron, and the three outgoing nucleons.

Implementing the LO EFT potential into Eqs. (1) and (2) and setting $|U\rangle = 0$ we obtain the decay rates to d+n and 3N within the plane-wave approximation. The decay rates obtained by considering only the OPE in the weak transition are $\Gamma_{d+n} = 0.16 \cdot 10^8\ s^{-1}$ and $\Gamma_{3N} = 0.17 \cdot 10^9\ s^{-1}$. Including both the OPE and the OKE, the decay rates get reduced by about a factor three, $\Gamma_{d+n} = 0.45 \cdot 10^7\ s^{-1}$ and $\Gamma_{3N} = 0.54 \cdot 10^8\ s^{-1}$. The low-energy constants appearing in the LO $\Lambda N \to NN$ potential were constrained in [11] by fitting them to the most recent data on hypernuclear non-mesonic weak decay of ${}^5_\Lambda He$, ${}^{11}_\Lambda B$, and ${}^{12}_\Lambda C$. However, the fitting did not give precise enough values, and the LECs were found to depend on the model used for the $SU(3)_F$ couplings appearing in the weak transition and the final state interactions. We have computed the decay rates for a reasonable range of the values of the LECs to study how they depend on the two constants. We observe that both Γ_{d+n} and Γ_{3N} increase (decrease) when the LECs grow with opposite (same) signs.

The final state interactions are being computed in the present with two different methods. The first one consists in solving the rescattering state $|U\rangle$ in Eqs. (1) and (2). This is done by solving iteratively the following Lippmann-Schwinger equation,

$$|U\rangle \equiv V_{12} G\,(1+P) V_{12}^w |\psi_{{}^3_\Lambda H}\rangle = V_{12} G_0\,(1+P) V_{12}^w |\psi_{{}^3_\Lambda H}\rangle + V_{12} G_0\,(1+P)\,|U\rangle\,, \tag{3}$$

where $G = \frac{1}{E+i\epsilon-H}$, $G_0 = \frac{1}{E+i\epsilon-H_0}$, H (H_0) being the strong (free) Hamiltonian and E the energy of the final state, and V_{12} the two-body strong force.

Alternatively, we use the Stieltjes transformation to compute the total non-mesonic decay rate. In this method, we add a spurious degree of freedom to the decay rate, w, which can be thought of as an extra mass of the hypertriton. Labelling $|f\rangle$ a general final state, the total decay rate with this new degree of freedom can be written as

$$\Gamma(w) = \frac{1}{2} \sum_m \int df\; 2\pi \delta(M_{{}^3_\Lambda H} - E_f + w) |\langle f|\hat{O}|\psi_{{}^3_\Lambda H} m\rangle|^2. \tag{4}$$

Applying the Stieltjes transformation to $\Gamma(w)$, $\phi(\sigma) = \int_{w_0}^\infty dw \frac{\Gamma(w)}{w+\sigma}$, and using completeness, one obtains

$$\phi(\sigma) = -\frac{1}{2} \sum_m 2\pi \langle {}^3_\Lambda H m|\hat{O} \left[\frac{1}{\Delta M + E_{{}^3_\Lambda H} - \sigma - H} + |{}^3 H\rangle \frac{1}{E_t - \Delta M - E_{{}^3_\Lambda H} + \sigma} \langle {}^3 H| \right] \hat{O}|{}^3_\Lambda H m\rangle. \tag{5}$$

The decay rate in the transformed space, $\phi(\sigma)$, can be obtained by solving the iterative equation

$$|\phi\rangle = \frac{1}{\Delta M + E_{{}^3_\Lambda H} - \sigma - H} \left(1 - \langle {}^3_\Lambda H|\hat{O}|{}^3_\Lambda H m\rangle \right) \hat{O}|{}^3_\Lambda H\rangle + V|\phi\rangle. \tag{6}$$

In order to obtain the total non-mesonic decay rate, $\Gamma(0)$, we make the inverse Stieltjes transformation through the singular value decomposition method and set $w = 0$. The advantage of this method is that for large enough σ, we avoid the pole of the deuteron in solving the iterative equation. The drawback is that we cannot distinguish between both partial decay rates.

4. On the mesonic weak decay ${}^3_\Lambda H \to {}^3 He + \pi^-$

Recent experimental results on the lifetime of the hypertriton ($\tau_{{}^3_\Lambda H} = 216^{+19}_{-16}$ ps) [5], seem to indicate that it decays considerably faster than the Λ in free space ($\tau_\Lambda = 263 \pm 2$ ps). In order to understand the relatively fast decay rate of the hypertriton, one should study its mesonic decay mode, since it is the dominant one in the lightest hypernucleus. Previous works have studied this decay, but none of the results seem to be compatible with the short hypertriton lifetime.

Once the non-mesonic weak decay formalism is established, one can apply the same few-body scheme to compute the mesonic decay rates. In this case, the weak one-body transition, $\Lambda \to p\pi^-$, is implemented by the operator $V_3^w = i\sqrt{2}G_F m_\pi^2(A + B\vec{\sigma}\cdot\vec{q})$, where the couplings $A = 1.05$ and $B = -7.15$ are known from the weak decay of the Λ in free space. $G_F m_\pi^2 = 2.21 \times 10^{-7}$, $\vec{q}$ is the momentum of the pion, and the factor $\sqrt{2}$ appears due to isospin. For the strong part the same EFT potentials can be used. The possible final decay products are not only the three free nucleons or the deuteron and the nucleon, but also ^3H or ^3He. This is because the emerging nucleon in the mesonic decay does not carry as much energy as the outgoing nucleons in the non-mesonic one, allowing the three final nucleons to form a bound state. Note that the decays with an emerging π^- are twice as the ones with an outgoing π^0 due to isospin, e.g. $\Gamma(^3_\Lambda H \to ^3 He + \pi^-) = 2\Gamma(^3_\Lambda H \to ^3 H + \pi^0)$. Therefore, there are only three independent channels.

As an example, the rate for the decay to $^3_\Lambda H \to ^3 He + \pi^-$ is, integrating the deltas of energy and momentum conservation,

$$\Gamma^{He} = \frac{1}{12\pi}\frac{M_{He}}{M_{He}+E_\pi} q_{He} \sum_{m_{J_3}} \sum_{m_{He}} \left| \langle \Psi_{m_{He}} | (1+P) V_3^w | \psi_{^3_\Lambda H} \rangle \right|^2 , \tag{7}$$

where q_{He} is the modulus of the momentum of the outgoing π^-,

$$q_{He} \equiv \sqrt{2M_{He}} \sqrt{M_{^3_\Lambda H} - \sqrt{m^2 + 2M_{He}M_{^3_\Lambda H} - M_{He}^2}}, \tag{8}$$

and $E_\pi = \sqrt{m_\pi^2 + q_{He}^2}$ its energy, M_{He} the Helium-3 mass, and $\Psi_{m_{He}}$ its wave function.

This decay was computed previously in [3] using OME potentials for both the strong and weak part and not considering strong interactions between the pion and the rest of the nucleons. The authors of this work obtained, in combination with the non-mesonic decay rate, a value for the hypertriton lifetime of $\tau_{^3_\Lambda H} = 256$ ps, about 3 % smaller than τ_Λ, still far from the 11 – 24 % difference from [5]. As in the non-mesonic case, our aim is to compute it within the same Faddeev scheme but with more modern EFT potentials. Note that in Γ^{He}, once the wave functions of the hypertriton and Helium-3 are known, the only final strong interactions to be computed are the ones between the pion and the nucleons. In order to study its effect, specially on the hypertriton lifetime, the plane-wave of the pion can be replaced by a distorted wave described through optical potentials. In contrast to the non-mesonic weak decay, which contains LECs that are still not well constrained, the weak part of the mesonic transition is fixed, and in principle one should be able to study, already on the Helium-3 mode, how the more modern EFT wave functions and the final state interactions affect the hypertriton lifetime.

This work was partly supported by the GAČR Grant No. 15-04301S, by the NSFC Grant No. 11261130311, and by the MEC Contract No. FPA2013-47443-C2-2-P.

References

[1] Ajimura S. et al. 2006 J-PARC proposal P22 http://j-parc.jp/researcher/Hadron/en/Proposal_e.html
[2] Golak J et al. 1997 Phys. Rev. C **55** 2196
[3] Kamada H et al. 1998 Phys. Rev. C **57** 1595
[4] Polinder H, Haidenbauer J, and Meiner U G 2006 Nucl. Phys. A **779** 244
[5] Rappold C et al. 2014 Phys. Lett. B **728** 543
[6] Jun J H 2001 Phys. Rev. C **63** 044012
[7] Parreño A, Bennhold C, and Holstein B R 2004 Phys. Rev. C **70** 051601, 2005 Nucl. Phys. A **754** 127c
[8] Donoghue J F, Golovich E, and Holstein B R 1992 *Dynamics of the Standard Model* (Cambridge University Press, Cambridge)
[9] Pérez-Obiol A et al. 2013 Phys. Rev. C **87** 044614
[10] Pérez-Obiol A et al. 2016 Nucl. Phys. A **914** 199
[11] Pérez-Obiol A, Parreño A, and Juliá-Díaz B 2011 Phys. Rev. C **84** 024606

Λ_c^+ physics at BESIII

Weiping Wang (on behalf of the BESIII collaboration)

University of Science and Technology of China, No.96, JinZhai Road Baohe District, Hefei, Anhui, 230026, P.R.China.

E-mail: cloud13@mail.ustc.edu.cn

Abstract. Based on the data sets collected by the BESIII detector near the $\Lambda_c^+\bar{\Lambda}_c^-$ production threshold, *i.e.* at $\sqrt{s}$ = 4574.5, 4580.0, 4590.0, and 4599.5 MeV, we report the preliminary study of the production behaviour of $e^+e^- \to \Lambda_c^+\bar{\Lambda}_c^-$ process, including the Born cross section and electromagnetic form factor ratios. Using the large statistic data at $\sqrt{s}$ = 4599.5 MeV, we measured the absolute branching fractions of Cabibbo-favored hadronic decays of Λ_c^+ baryon with a double-tag technique. The branching fractions for 12 hadronic decay modes are significantly improved. We also report the model-independent measurement of the branching fraction of $\Lambda_c^+ \to \Lambda e^+\nu_e$ and $\Lambda_c^+ \to \Lambda\mu^+\nu_\mu$ semi-leptonic decays .

1. Introduction

The electromagnetic structure of hadrons, parameterized in terms of electromagnetic form factors (EMFFs) provides a key to understand the strong interaction. Assuming that one-photon exchange dominates the production of spin 1/2 baryons B, the Born cross section of the process $e^+e^- \to B\bar{B}$ can be parameterized in terms of EMFFs, *i.e.* G_E and G_M, in the following way [1]:

$$\sigma_{B\bar{B}}(s) = \frac{4\pi\alpha^2 C\beta}{3s}|G_M(s)|^2[1 + \frac{2m_B^2c^4}{s}|\frac{G_E(s)}{G_M(s)}|^2]. \tag{1}$$

Here α is the fine-structure constant $\beta{=}\sqrt{1 - 4m_B^2c^4/s}$ the velocity of the baryon, s the square of center-of-mass (CM) energy and m_B is the mass of the baryon. The Coulomb factor C parameterizes the electromagnetic interaction between the outgoing baryon-antibaryon pair. For neutral baryons the Coulomb factor is equal to 1, while for point-like charged fermions it is equal to $C = \varepsilon R$ [2, 3], where $\varepsilon = \pi\alpha/\beta$ is an enhancement factor resulting in a nonzero cross section at threshold, and $R = \sqrt{1-\beta^2}/(1 - e^{-\pi\alpha/\beta})$ is the Sommerfeld resummation factor. In the $e^+e^- \to p\bar{p}$ process, the BaBar collaboration observed a rapid rise of the cross section near threshold, followed by a plateau around 200 MeV above threshold [4]. The BESIII collaboration also observed the cross section enhancement [5]. The non-vanishing cross section near threshold as well as the wide-range plateau have led to various theoretical interpretations [6, 7, 8]. Recently, the BESIII collaboration has observed the non-zero cross section near threshold in the $e^+e^- \to \Lambda\bar{\Lambda}$ process. Naturally, it is also interesting to explore the production behaviour of Λ_c^+, the lightest baryon containing the charm quark. Previously, the Belle collaboration measured the cross section of $e^+e^- \to \Lambda_c^+\bar{\Lambda}_c^-$ using initial-state radiation (ISR) technique [9], but the results suffer from significant uncertainties in CM energy and cross section. Therefore, near $\Lambda_c^+\bar{\Lambda}_c^-$ threshold, precise measurements of the production Born cross section and EMFF ratios are highly needed.

IOP Conf. Series: Journal of Physics: Conf. Series **1024** (2018) 012034 doi:10.1088/1742-6596/1024/1/012034

The decays of charm baryon provide crucial information for the study of both strong and weak interactions. The hadronic decays of Λ_c^+ provide important input to Λ_b physics, while the semi-leptonic (SL) decays of Λ_c^+ provide a stringent test on non-perturbative theoretical models. The $\Lambda_c^+ \to \Lambda l^+ \nu_l$ decay is dominated by the Cabibbo-favored transition $c \to s l^+ \nu_l$, which occurs, to a good approximation, independently on the spin-zero spectator ud di-quark. In addition, theoretical calculations are proved to be quite challenging for lattice quantum chromodynamics (LQCD) due to the complexity of form factors in $\Lambda_c^+ \to \Lambda l^+ \nu_l$ [10]. Consequently, the model-independent measurement of hadronic and SL decays with better precision is a key ingredient in theoretical predictions and LQCD calculation, which in turn will play an important role in understanding different Λ_c^+ decays.

2. The production of Λ_c^+

The Born cross section of $e^+e^- \to \Lambda_c^+\bar{\Lambda}_c^-$ reaction is measured at four CM energies: $\sqrt{s} = 4574.5$, 4580.0, 4590.0, and 4599.5 MeV. At each CM energy, 10 Cabibbo-favored hadronic decay modes: $\Lambda_c^+ \to pK^-\pi^+$, pK_S^0, $\Lambda\pi^+$, $pK^-\pi^+\pi^0$, $pK_S^0\pi^0$, $\Lambda\pi^+\pi^0$, $pK_S^0\pi^+\pi^-$, $\Lambda\pi^+\pi^+\pi^-$, $\Sigma^0\pi^+$, and $\Sigma^+\pi^+\pi^-$, as well as the 10 corresponding charge-conjugate modes, are independently reconstructed. Each mode will produce one measurement of the Born cross section, and the total cross section is obtained from weighted average over the 20 individual measurements [11]. The resulting cross sections at four CM energies are listed in Table 1 and shown in Figure 1 together with the Belle data [9] for comparison.

Table 1. The average Born cross section of $e^+e^- \to \Lambda_c^+\bar{\Lambda}_c^-$ measured at each CM energy, where the uncertainties are statistical and systematic, respectively (BESIII preliminary results). The $\mathcal{L}_{\rm int}$ denotes the integrated luminosity, and the $f_{\rm ISR}$ represents for the ISR correction factor.

$\sqrt{s}$ (MeV)	$\mathcal{L}_{\rm int}$ (pb^{-1})	$f_{\rm ISR}$	σ (pb)
4574.5	47.67	0.45	$236 \pm 11 \pm 46$
4580.0	8.54	0.66	$207 \pm 17 \pm 13$
4590.0	8.16	0.71	$245 \pm 19 \pm 16$
4599.5	566.93	0.74	$237 \pm 3 \pm 15$

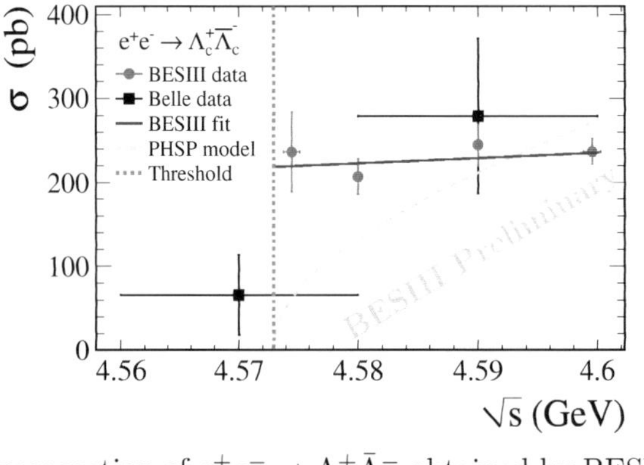

Figure 1. The Born cross section of $e^+e^- \to \Lambda_c^+\bar{\Lambda}_c^-$ obtained by BESIII (this work) and Belle. The blue solid curve represents the input line-shape for KKMC, the Monte Carlo (MC) generator inside the BESIII framework, when determining the $f_{\rm ISR}$. The dash-dot cyan curve denotes the prediction of the phase space (PHSP) model.

The data collected at $\sqrt{s} = 4574.5$ and 4599.5 MeV are large enough to perform a detailed study in the CM frame of the Λ_c polar angle θ_{Λ_c}, which is defined as the angle between the Λ_c momentum and the beam direction. The data fulfilling all selection criteria are divided into 10 bins in $\cos\theta_{\Lambda_c^+}$. In each $\cos\theta_{\Lambda_c^+}$ bin, the total yield is obtained by summing yields of all the 10 tagged modes. The one-dimensional bin-by-bin efficiency corrections are applied to these total yields. The same procedure is performed by tagging $\bar{\Lambda}_c^-$ decay channels. The total yields of Λ_c^+ and $\bar{\Lambda}_c^-$ are combined bin-by-bin, and the shape function $f(\theta) \propto (1 + \alpha_{\Lambda_c}\cos^2\theta)$ is fitted to the combined data, as shown in Figure 2.

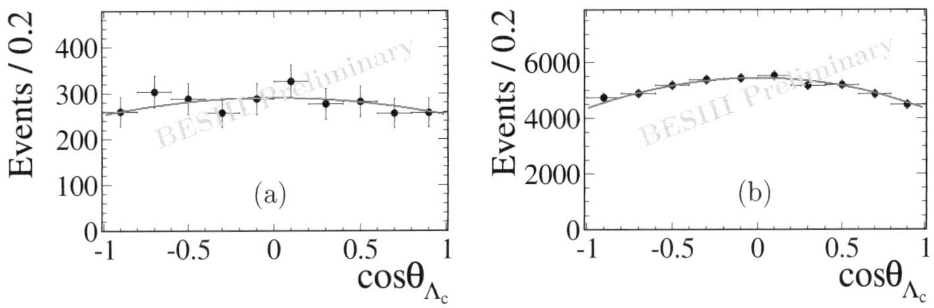

Figure 2. The angular distribution and corresponding fit results in data at $\sqrt{s} = 4574.5$ MeV (a) and 4599.5 MeV (b).

Table 2 lists the resulting α_{Λ_c} parameters obtained from the fits, as well as the $|G_E/G_M|$ ratios extracted using the equation:

$$|G_E/G_M|^2(1-\beta^2) = (1-\alpha_{\Lambda_c})/(1+\alpha_{\Lambda_c}). \tag{2}$$

Table 2. Shape parameters of the angular distribution and $|G_E/G_M|$ ratios at $\sqrt{s} = 4574.5$ and 4599.5 MeV. The uncertainties are statistical and systematic (BESIII preliminary results).

$\sqrt{s}$ (MeV)	α_{Λ_c}	$\lvert G_E/G_M\rvert$
4574.5	$-0.13 \pm 0.12 \pm 0.08$	$1.14 \pm 0.14 \pm 0.07$
4599.5	$-0.20 \pm 0.04 \pm 0.02$	$1.23 \pm 0.05 \pm 0.03$

3. Λ_c^+ hadronic decay

Based on the data sample with an integrated luminosity of 566.9 pb^{-1} collected with the BESIII detector [12] at $\sqrt{s} = 4599.5$ MeV, we apply a tagged technique pioneered by the MARK-III Collaboration [13] to determine the branching fraction (BF) of Λ_c^+ hadronic decay. To identify the $\Lambda_c^+\bar{\Lambda}_c^-$ signal candidates, we firstly reconstruct one $\bar{\Lambda}_c^-$ baryon [called single tag (ST)] through the final states of any of the singly tagged modes. For a given decay mode j, the ST yields is

$$N_j^{\rm ST} = N_{\Lambda_c^+\bar{\Lambda}_c^-} \cdot \mathcal{B}_j \cdot \varepsilon_j, \tag{3}$$

where $N_{\Lambda_c^+\bar{\Lambda}_c^-}$ is the total number of produced $\Lambda_c^+\bar{\Lambda}_c^-$ pairs, and ε_j is the corresponding efficiency. Then we define double-tag (DT) events as those where the partner Λ_c^+ recoiling against the $\bar{\Lambda}_c^-$ is reconstructed in one of the signal modes. That is, in DT events, both the Λ_c^+ and $\bar{\Lambda}_c^-$ are reconstructed. The DT yield with $\Lambda_c^+ \to i$ (signal mode) and $\bar{\Lambda}_c^- \to j$ (tagging mode) is

$$N_{ij}^{\rm DT} = N_{\Lambda_c^+\bar{\Lambda}_c^-} \cdot \mathcal{B}_i \cdot \mathcal{B}_j \cdot \varepsilon_{ij}, \tag{4}$$

Table 3. Comparison of the measured BFs in this work [14] with previous results from PDG [15]. For our results, the first uncertainties are statistical and the second are systematic.

Mode	This work (%)	PDG (%)
pK_S^0	$1.52 \pm 0.08 \pm 0.03$	1.15 ± 0.30
$pK^-\pi^+$	$5.84 \pm 0.27 \pm 0.23$	5.0 ± 1.3
$pK_S^0\pi^0$	$1.87 \pm 0.13 \pm 0.05$	1.65 ± 0.50
$pK_S^0\pi^+\pi^-$	$1.53 \pm 0.11 \pm 0.09$	1.30 ± 0.35
$pK^-\pi^+\pi^0$	$4.53 \pm 0.23 \pm 0.30$	3.4 ± 1.0
$\Lambda\pi^+$	$1.24 \pm 0.07 \pm 0.03$	1.07 ± 0.28
$\Lambda\pi^+\pi^0$	$7.01 \pm 0.37 \pm 0.19$	3.6 ± 1.3
$\Lambda\pi^+\pi^-\pi^+$	$3.81 \pm 0.24 \pm 0.18$	2.6 ± 0.7
$\Sigma^0\pi^+$	$1.27 \pm 0.08 \pm 0.03$	1.05 ± 0.28
$\Sigma^+\pi^0$	$1.18 \pm 0.10 \pm 0.03$	1.00 ± 0.34
$\Sigma^+\pi^+\pi^-$	$4.25 \pm 0.24 \pm 0.20$	3.6 ± 1.0
$\Sigma^+\omega$	$1.56 \pm 0.20 \pm 0.07$	2.7 ± 1.0

where ε_{ij} is the efficiency for simultaneously reconstructing modes i and j. Hence, the ratio of the DT yield (N_{ij}^{DT}) and ST yield (N_j^{ST}) provides an absolute measurement of the BF:

$$\mathcal{B}_i = \frac{N_{ij}^{\mathrm{DT}}}{N_j^{\mathrm{ST}}}\frac{\varepsilon_j}{\varepsilon_{ij}}. \tag{5}$$

Because of the large acceptance of the BESIII detector and the low multiplicities of Λ_c hadronic decays, $\varepsilon_{ij} \approx \varepsilon_i\varepsilon_j$. Hence, the ratio $\varepsilon_j/\varepsilon_{ij}$ is insensitive to most systematic effects associated with the decay mode j, and a signal BF $\mathcal{B}_i$ obtained using this procedure is nearly independent on the efficiency of the tagging mode. Therefore $\mathcal{B}_i$ is sensitive to the signal mode efficiency (ε_i), whose uncertainties dominate the contribution to the systematic error from the efficiencies. We use a least-squares fitter, which considers statistical and systematic correlations among the different hadronic modes, to obtain the BFs of the 12 Λ_c^+ decay modes globally. In total, there are 13 free parameters (12 $\mathcal{B}_i$ and 1 $N_{\Lambda_c^+\bar{\Lambda}_c^-}$) to be estimated. The extracted BFs of Λ_c^+ are listed in Table 3. The total number of $\Lambda_c^+\bar{\Lambda}_c^-$ pairs produced is obtained to be $N_{\Lambda_c^+\bar{\Lambda}_c^-} = (105.9 \pm 4.8 \pm 0.5) \times 10^3$. The goodness of fit is evaluated as $\chi^2/ndf = 9.9/(24-13) = 0.9$.

4. Λ_c^+ semi-leptonic decay

Using the similar strategy in hadronic decay measurements, we select the data sample of $\bar{\Lambda}_c^-$ baryons by reconstructing exclusive hadronic decays. The ST $\bar{\Lambda}_c^-$ are reconstructed using 11 hadronic decay modes: $\bar{\Lambda}_c^- \to \bar{p}K_S^0$, $\bar{p}K^+\pi^-$, $\bar{p}K_S^0\pi^0$, $\bar{p}K^+\pi^-\pi^0$, $\bar{p}K_S^0\pi^+\pi^-$, $\bar{\Lambda}\pi^-$, $\bar{\Lambda}\pi^-\pi^0$, $\bar{\Lambda}\pi^-\pi^+\pi^-$, $\bar{\Sigma}^0\pi^-$, $\bar{\Sigma}^-\pi^0$ and $\bar{\Sigma}^-\pi^+\pi^-$, where the intermediate particles K_S^0, $\bar{\Lambda}$, $\bar{\Sigma}^0$, $\bar{\Sigma}^-$ and π^0 are reconstructed by their decays into $K_S^0 \to \pi^+\pi^-$, $\bar{\Lambda} \to \bar{p}\pi^+$, $\bar{\Sigma}^0 \to \gamma\bar{\Lambda}$ and $\bar{\Lambda} \to \bar{p}\pi^+$, $\bar{\Sigma}^- \to \bar{p}\pi^0$ and $\pi^0 \to \gamma\gamma$, respectively. The total observed events for the 11 ST modes is $N_{\Lambda_c^{tag}} = (14415 \pm 159)$. The signal candidates for $\Lambda_c^+ \to \Lambda l^+\nu_l$ are selected from the remaining tracks recoiling against the ST $\bar{\Lambda}_c^-$ candidates. As the neutrino is the missing particle of the event, we employ a variable

$$U_{\mathrm{miss}} = E_{\mathrm{miss}} - c|\vec{p}_{\mathrm{miss}}|$$

to obtain information of the neutrino, where E_{miss} and $\vec{p}_{\mathrm{miss}}$ are the missing energy and momentum carried by the neutrino, respectively. They are calculated by $E_{\mathrm{miss}} = E_{\mathrm{beam}} -$

$E_\Lambda - E_{e^+}$, and $\vec{p}_{\rm miss} = \vec{p}_{\Lambda_c^+} - \vec{p}_\Lambda - \vec{p}_{e^+}$, where $\vec{p}_{\Lambda_c^+}$ is the momentum of Λ_c^+ baryon, $E_\Lambda(\vec{p}_\Lambda)$ and E_{e^+} $(\vec{p}_{e^+})$ are the energies (momenta) of the Λ and the positron, respectively. Here, the $\vec{p}_{\Lambda_c^+}$ is given by $\vec{p}_{\Lambda_c^+} = -\hat{p}_{\rm tag}\sqrt{E^2_{\rm beam} - m^2_{\bar{\Lambda}_c^-}}$, where $\hat{p}_{\rm tag}$ is the momentum direction of ST $\bar{\Lambda}_c^-$ and $m_{\bar{\Lambda}_c^-}$ is the nominal $\bar{\Lambda}_c^-$ mass. For signal events, $U_{\rm miss}$ is expected to peak around zero.

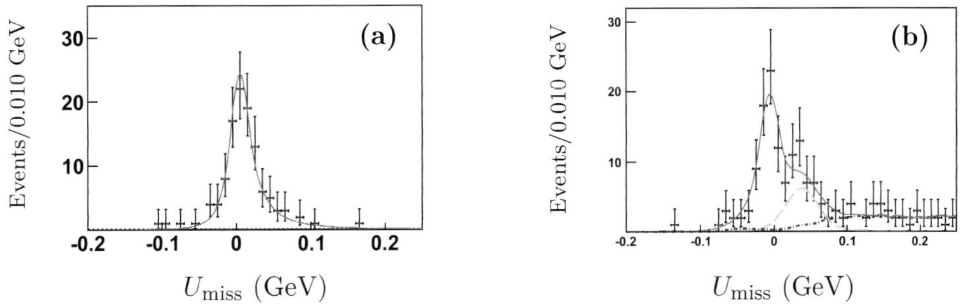

Figure 3. (a) Fit to the $U_{\rm miss}$ distribution of process $\Lambda_c^+ \to \Lambda e^+\nu_e$ [16]. (b) Fit to the $U_{\rm miss}$ distribution of process $\Lambda_c^+ \to \Lambda\mu^+\nu_\mu$ [17]. The points with error bars are data, the (red) solid curve shows the total fit and the (blue) dashed curve is the background shape. The green-dashed line in the right subfigure denotes the MC-driven background shapes which is supposed to simulate the remaining background.

Figure 3(a) shows the fit result of the $U_{\rm miss}$ distribution for $\Lambda_c^+ \to \Lambda e^+\nu_e$. From the fit, we obtain the number of SL signals to be (109.4 ± 10.9). After subtracting all the background, we determine the net number of $\Lambda_c^+ \to \Lambda e^+\nu_e$ to be $N_{\rm semi} = (103.5 \pm 10.9)$, where the uncertainty is statistical. The absolute BF for $\Lambda_c^+ \to \Lambda e^+\nu_e$ is determined by

$$\mathcal{B}(\Lambda_c^+ \to \Lambda e^+\nu_e) = \frac{N_{\rm semi}}{N^{\rm tot}_{\bar{\Lambda}_c^-} \times \varepsilon_{\rm semi} \times \mathcal{B}(\Lambda \to p\pi^-)}, \tag{6}$$

where $\varepsilon_{\rm semi} = (30.92 \pm 0.26)\%$ is the overall efficiency for detecting the $\Lambda_c^+ \to \Lambda e^+\nu_e$ decay in ST events, weighted by the ST yields of data for each tag. Inserting the values of $N_{\rm semi}$, $N^{\rm tot}_{\bar{\Lambda}_c^-}$, $\epsilon_{\rm semi}$ and $\mathcal{B}(\Lambda \to p\pi^-)$ in Eq. (6), we get $\mathcal{B}(\Lambda_c^+ \to \Lambda e^+\nu_e) = (3.63 \pm 0.38 \pm 0.20)\%$, where the first error is statistical and the second systematic.

Using an analogous technique to that of $\Lambda_c^+ \to \Lambda e^+\nu_e$, in which the $U_{\rm miss}$ is used as the final signal variable, the process $\Lambda_c^+ \to \Lambda\mu^+\nu_\mu$ is also studied. The fitting model includes MC-driven background shapes to simulate the remaining background, shown in Fig. 3(b). Accordingly, the $\mathcal{B}(\Lambda_c^+ \to \Lambda\mu^+\nu_\mu)$ is determined to be $(3.49 \pm 0.46 \pm 0.27)\%$. With the result of $\mathcal{B}(\Lambda_c^+ \to \Lambda e^+\nu_e)$ in hand, the ratio $\mathcal{B}(\Lambda_c^+ \to \Lambda\mu^+\nu_\mu)/\mathcal{B}(\Lambda_c^+ \to \Lambda\mu^+\nu_\mu)$ is obtained to be $(0.96 \pm 0.16({\rm stat.}) \pm 0.04({\rm sys.}))\%$, which verified the lepton universality in baryon decays.

5. Summary

In summary, based on the data sets collected by BESIII detector near the $\Lambda_c^+\bar{\Lambda}_c^-$ production threshold, we report the preliminary study of the production behaviour of $e^+e^- \to \Lambda_c^+\bar{\Lambda}_c^-$ process, including the Born cross section and electromagnetic form factor ratios. Using the data at $\sqrt{s}$ = 4599.5 MeV, BESIII performed the first measurement of the absolute hadronic branching fractions of 12 Cabibbo-favored decays of Λ_c^+. BESIII also presented the first model-independent measurement of the branching fractions of the semi-leptonic decay $\Lambda_c^+ \to \Lambda e^+\nu_e$ and $\Lambda_c^+ \to \Lambda\mu^+\nu_\mu$.

IOP Conf. Series: Journal of Physics: Conf. Series **1024** (2018) 012034 doi:10.1088/1742-6596/1024/1/012034

6. Reference

[1] Cabibbo N and Gatto R, *Phys. Rev.* **124** 1577
[2] Sommerfeld A, *Ann. Phys.* **403** 257
[3] Sakharov A D, *Sov. Phys. Usp.* **34** 375
[4] Lees J P *et al.* (BaBar Collaboration), *Phys. Rev.* D **87** 092005; Lees J P *et al.* (BaBar Collaboration), *Phys. Rev.* D **88** 072009
[5] Ablikim M *et al.* (BESIII Collaboration), *Phys. Rev.* D **91** 112004
[6] Dalkarov O D, Khakhulin P A and Voronin A Yu, *Nucl. Phys.* A **833** 104
[7] El-Bennich B, Lacombe M, Loiseau B and Wycech S, *Phys. Rev.* C **79** 054001; Haidenbauer J, Hammer H W, Messener U G and Sibirtsev A, *Phys. Lett.* B **643** 29
[8] Baldini R, Pacetti S, Zallo A and Zichichi A, *Eur. Phys. J.* A **39** 315; Baldini R, Pacetti S and Zallo A, *Eur. Phys. J.* A **48** 33
[9] Pakhlova G *et al.* (Belle Collaboration), *Phys. Rev. Lett.* **101** 172001
[10] Rosner J L, *Phys. Rev.* D **86** 014017
[11] Schmelling M, *Phys. Scripta* **51** 676
[12] Ablikim M *et al.* (BESIII Collaboration), *Nucl. Instrum. Meth.* A **614** 345
[13] Baltrusaitis R M *et. al.* (MARK-III Collaboration), *Phys. Rev. Lett.* **56** 2140; Adler J *et. al.* (MARK-III Collaboration), *Phys. Rev. Lett.* **60** 89
[14] Ablikim M *et al.* (BESIII Collaboration), *Phys. Rev. Lett.* **116** 052001
[15] Olive K A *et al.* (Particle Data Group), *Chin. Phys.* C **38** 090001
[16] Ablikim M *et al.* (BESIII Collaboration), *Phys. Rev. Lett.* **115** 221805
[17] Ablikim M *et al.* (BESIII Collaboration), *Phys. Lett.* B **767** 42

IOP Conf. Series: Journal of Physics: Conf. Series **1024** (2018) 012035 doi:10.1088/1742-6596/1024/1/012035

Light tetraquarks in a Dyson-Schwinger/Bethe-Salpeter approach

Paul C. Wallbott[1], Gernot Eichmann[2] and Christian S. Fischer[1]

[1] Institut für Theoretische Physik, Justus-Liebig Universität Gießen, 35392 Gießen, Germany
[2] CFTP, Instituto Superior Técnico, Universidade de Lisboa, 1049-001 Lisboa, Portugal

E-mail: paul.wallbott@physik.uni-giessen.de, gernot.eichmann@tecnico.ulisboa.pt, christian.fischer@theo.physik.uni-giessen.de

Abstract. A number of bound states of the strong interaction with potential non-$q\bar{q}$ nature are waiting to be understood within QCD. A prominent example are the light scalars with $J^{PC} = 0^{++}$, which have been interpreted as tetraquarks already long ago. Recent four-body calculations in the Dyson-Schwinger/Bethe-Salpeter approach support the tetraquark hypothesis and find the σ meson to have a strong pion-pion component.

1. Are the light scalar mesons tetraquarks?

The perhaps oldest tetraquark candidates are the light scalar mesons with $J^P = 0^{++}$, namely, the $\sigma/f_0(500)$ and its multiplet partners $\kappa(800)$, $a_0(980)$ and $f_0(980)$, which have been subject to intense debate for many decades [1]. Combining new data with dispersive approaches to $\pi\pi$ scattering using Roy equations [2, 3], the σ is now again listed in the PDG with a pole position $(400\ldots550) - i(200\ldots350)$ MeV [4]. Theoretically, the situation is complicated by the fact that the quark content of the σ and its partners cannot be unambiguously identified, because in principle they can be mixtures of $q\bar{q}$, $qq\bar{q}\bar{q}$ and other configurations. Phenomenological evidence for their non-$q\bar{q}$ nature comes from the comparison with other mesons, cf. Fig. 1: in the non-relativistic quark model the light scalars are p waves with orbital angular momentum $L = 1$, which should put them in a similar mass range as the axialvector and tensor mesons. This is clearly not the case in the experimental spectrum, where the members of the 'first radially excited' nonet in the 0^{++} channel appear to be better candidates for the actual $q\bar{q}$ ground states. But what are then the light scalars? Even more striking is the mass ordering within the 0^{++} multiplet. If they were $q\bar{q}$ states, the isotriplet a_0 and isosinglet σ should be made of light quarks and mass-degenerate, followed by the κ and the $f_0(980)$ as the $s\bar{s}$ state. Instead, the a_0 is approximately degenerate with the f_0, which seems rather nonsensical in a $q\bar{q}$ picture. Their decay widths are quite different as well: the a_0 and f_0 have masses just at or below $K\bar{K}$ threshold with widths below 100 MeV, whereas the σ and κ are broad states above the $\pi\pi$ and $K\pi$ thresholds, respectively.

These problems disappear within a tetraquark picture. In the simplest case advocated by Jaffe [5], one combines a (color and flavor antitriplet) scalar diquark with a scalar antidiquark, which produces again a flavor nonet of color singlets. However, in this case the σ would be the lightest state with four light quarks, followed by the κ with one strange quark and the mass-degenerate a_0 and f_0 with two strange quarks. This naturally explains the mass ordering

IOP Conf. Series: Journal of Physics: Conf. Series **1024** (2018) 012035 doi:10.1088/1742-6596/1024/1/012035

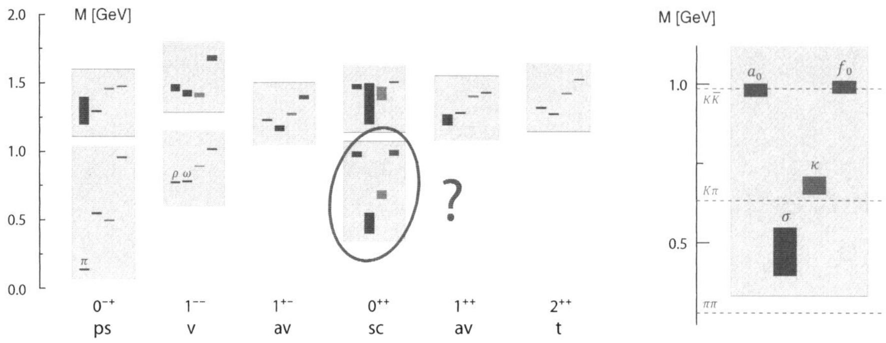

Figure 1. Light meson spectrum from the PDG [4]; the bars show the experimental mass range. The light scalar mesons are magnified on the right.

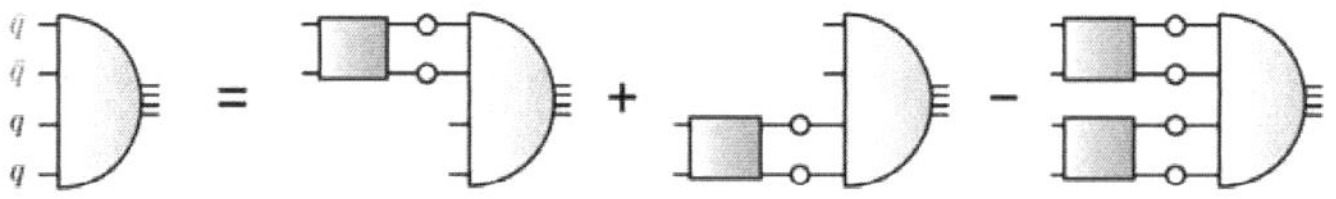

Figure 2. Four-body equation for tetraquarks. The lines with circles stand for a dressed quark propagator and the yellow blobs for the tetraquark Bethe-Salpeter amplitude. Note that the figure only shows the $(qq)(\bar{q}\bar{q})$ topology; there are two further $(q\bar{q})(q\bar{q})$ topologies which all add up to the final equation.

as well as the decay widths: the dominant a_0 decay into $\pi\eta$ and its coupling to $K\bar{K}$ support an internal $s\bar{s}$ pair, whereas the σ and κ can simply fall apart into $\pi\pi$ and $K\pi$ through the 'OZI-superallowed' mechanism, which does not require the exchange of gluons. The non-$q\bar{q}$ interpretation of the light scalars is supported by various approaches such as QCD sum rules [6], unitarized ChPT [7,8], quark models [9], and the extended linear σ model [10]. What remains to be understood is the precise binding mechanism: are such tetraquarks strongly bound diquark-antidiquark states, loosely bound meson molecules, or a bit of both? If they are made of diquarks and antidiquarks, then why are their masses so unexpectedly low? And is it possible to distinguish between these scenarios without any prior knowledge of mesons or diquarks, i.e., when starting directly from four current quarks in QCD?

2. Tetraquarks in the Dyson-Schwinger/Bethe-Salpeter framework

To address these questions, we utilize the framework of Dyson-Schwinger and Bethe-Salpeter equations (DSEs and BSEs) [11–13]. It has been successfully applied to a range of meson and baryon properties (see e.g. [14] for a review) and first investigations of glueballs [15] and light tetraquarks [16, 17] have recently been pioneered as well. In [17], the 0^{++} tetraquarks were calculated from a genuine four-quark equation and the resulting masses were found to qualitatively reproduce the experimentally observed 0^{++} spectrum. In the following we will summarize the methods used in that work and discuss the corresponding results. The tetraquark BSE is shown in Fig. 2. It describes a $q\bar{q}q\bar{q}$ system that is bound by two-body interactions between quarks and antiquarks, whereas three- and four-body interactions are neglected at this point. For the two-body interactions we employ the rainbow-ladder truncation where the kernels reduce to gluon exchanges, see e.g. [18, 19] for details. Combined with the solution for the quark propagator, this setup ensures chiral symmetry and its dynamical breaking. Systematic ways

IOP Conf. Series: Journal of Physics: Conf. Series **1024** (2018) 012035 doi:10.1088/1742-6596/1024/1/012035

of improving the truncation have been successfully applied to the light meson sector lately [20]. The four-body Bethe-Salpeter amplitude is a complicated object which contains a color (C), a flavor (F), and a Dirac (D) part:

$$\Gamma = \Gamma^D \otimes \Gamma^C \otimes \Gamma^F, \qquad \Gamma^D = \sum_{i=1}^{256} \tau_i(p,q,k,P)\, f_i(\Omega), \qquad K(M)\, f(M) = \lambda(M)\, f(M). \quad (1)$$

Here p, q, and k are the relative momenta between the (anti-) quarks and P is the total momentum of the tetraquark. The τ_i are the basis elements in Dirac space, the f_i are the scalar dressing functions of the tetraquark amplitude, and Ω is the set of nine Lorentz invariants that can be built from p, q, k, P with $P^2 = -M^2$ onshell. The last (eigenvalue) equation follows after taking traces and discretizing the momentum variables. Here $K(M)$ is the Bethe-Salpeter kernel matrix, $f(M)$ is the eigenvector corresponding to the tetraquark amplitude, and $\lambda(M)$ a mass-dependent eigenvalue which satisfies $\lambda(M) = 1$ at the physical bound-state mass. The practical difficulty with this equation is the enormous amount of Lorentz invariants on which the $f_i(\Omega)$ depend: with two color tensors and N grid points in each direction, the dimension of the kernel matrix is $2 \times 256 \times N^9$, which impedes a straightforward numerical solution.

The solution of this equation was made possible by two strategies. One is the restriction of the 256 Dirac basis tensors to s waves only, i.e., the Fierz-complete set of 16 tensors which depend on the total momentum P but not any relative momentum. In the analogous three-body equation for baryons (where the full set of tensors is implemented) the same approximation is reliable up to $\sim 10\%$, which is sufficient for our present purposes. The second strategy is the extensive use of the equation's symmetries in terms of S_4 variables [21]. It turns out to be useful to group the Lorentz invariants in the set Ω into S_4 multiplets: a singlet $\mathcal{S}_0$, a doublet $\mathcal{D}$ and two triplets $\mathcal{T}_0$ and $\mathcal{T}_1$. For example, the singlet and doublet read:

$$\mathcal{S}_0 = \frac{p^2+q^2+k^2}{4}, \qquad \mathcal{D} = \frac{1}{4\mathcal{S}_0}\begin{pmatrix} \sqrt{3}\,(q^2-p^2) \\ p^2+q^2-2k^2 \end{pmatrix}. \quad (2)$$

The singlet $\mathcal{S}_0$ carries the scale dependence of the dressing functions f_i above, whereas the phase space of the multiplets ($\mathcal{D}$, $\mathcal{T}_0$, $\mathcal{T}_1$) is restricted to geometrical objects (triangle, tetrahedron, sphere). In this way, groups of momentum variables can be switched on and off at once during a calculation without destroying the permutation symmetries of the system, thus dramatically reducing the size of the kernel matrix in Eq. (1).

3. Results

The resulting inverse eigenvalues $1/\lambda(M)$ of the tetraquark BSE, Eq. (1), for different multiplet combinations are shown in Fig. 3. If only the momentum dependence in $\mathcal{S}_0$ is retained in the equation, the resulting tetraquark mass is $\sim$ 1500 MeV. Naively, this is what one expects from combining four current quarks with a dynamical mass function: the result is roughly 'four times the constituent-quark mass'. Switching on the triplet $\mathcal{T}_1$ changes almost nothing, whereas the triplet $\mathcal{T}_0$ reduces the mass by a few hundert MeV. A much more drastic effect happens when the doublet $\mathcal{D}$ is switched on, because in that case the tetraquark mass drops to $\sim$ 350 MeV. This is because the equation dynamically develops meson and diquark poles, which appear in the Mandelstam plane formed by the doublet triangle in Eq. (2). Each side of the triangle corresponds to one of the three topologies in the BSE – 'diquark-antidiquark', (12)(34), or 'meson-meson', (13)(24) and (14)(23). Because the diquark mass scales are much larger than the pion mass (about 800 MeV for a scalar diquark), the diquark poles have a much weaker effect on the dressing functions than the two pion poles sitting to the left and the right. Hence, the light mass of the σ is driven by the pion poles, which leads to a predominant 'meson molecule' interpretation of the tetraquark.

IOP Conf. Series: Journal of Physics: Conf. Series **1024** (2018) 012035 doi:10.1088/1742-6596/1024/1/012035

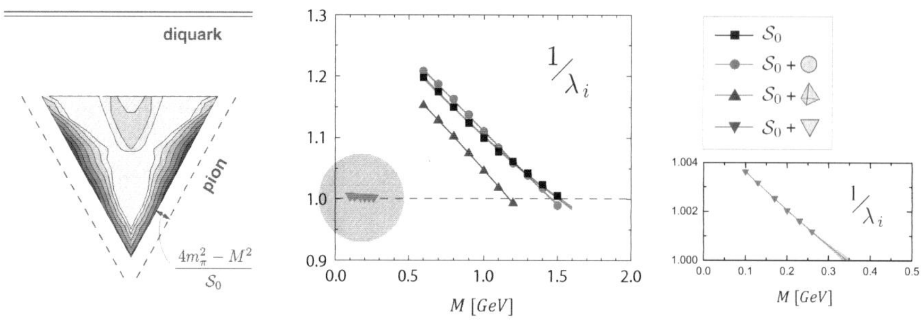

Figure 3. *Left panel:* dominant dressing function f_1 of Eq. (1) within the doublet triangle in the case of the σ, together with the pole positions of the dynamically generated pions and diquarks. *Right panels:* inverse eigenvalues of Eq. (1) for different S_4 multiplets.

The results for the 0^{++} multiplet are shown in the left panel of Fig. 4 as a function of the current-quark mass. One can see that they are in qualitative agreement with the 0^{++} spectrum from the PDG. The problems of the mass ordering within the multiplet and the alignment in comparison to other multiplets discussed in Sec. 1 do not appear. The dynamical features of the σ repeat themselves in the multiplet partners κ, a_0 and f_0, which are dominated by internal $K\pi$ and $\eta\pi$ / $K\bar{K}$ poles, respectively. For tetraquark masses $M > 2m_\pi$ the pion poles eventually move inside the integration domain and the σ becomes a resonance. So far we performed an extrapolation to $\lambda = 1$ to estimate the tetraquark mass, whereas a proper implementation of the pole structure requires contour-deformation methods – see [22] for an analogous treatment of the $\pi^0 \to e^+e^-$ decay. In any case, for large current-quark masses the resulting σ mass goes below threshold and thus becomes a genuine bound state. This is analogous to the case of the ρ meson shown in the right panel of Fig. 4. At large current masses (large m_π^2), $m_\rho < 2m_\pi$ and the system is a bound state; at low current masses, chiral dynamics becomes important and the ρ becomes a resonance. The figure compares the rainbow-ladder result for m_ρ with contemporary lattice results. In contrast to the tetraquarks, the ρ meson in this case is *always* a bound state because chiral effects are missing in rainbow-ladder. The comparison with lattice and experiment suggests, however, that the effects on the real part of the mass are rather modest. In this sense an eigenvalue extrapolation beyond the threshold should also provide a reasonable first estimate in the four-body case, where the resonance dynamics is already built into the system.

To summarize, the genuine four-quark ($q\bar{q}q\bar{q}$) equation, where the quarks interact by gluon exchange in a chiral-symmetry preserving way, can qualitatively reproduce the mass pattern of the light scalar mesons. This does not happen in the corresponding two-body ($q\bar{q}$) equation, where the resulting a_0 and σ are mass-degenerate and one would require large unquenching effects to establish an (accidental) mass degeneracy between a_0 and f_0. Therefore, it provides further support for the tetraquark interpretation of the light scalar mesons. The four-body equation dynamically produces meson poles in the Bethe-Salpeter amplitude, which drive the system and are responsible for the low masses of the 0^{++} multiplet. In this sense the light scalar mesons are conceptually closer to 'meson molecules' than 'diquark-antidiquark' states.

4. Outlook

Recent experimental findings at BES III, Belle and LHCb point towards a number of tetraquark candidates in the charm region. Furthermore, PANDA will provide more detailed information about channels other than 1^{--}. Therefore, in the future we would like to extend our calculations to other quantum numbers to help understand the nature of some of these states.

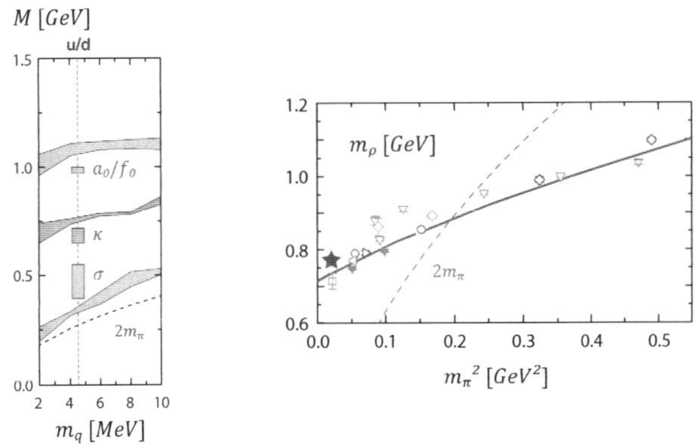

Figure 4. *Left:* masses of the 0^{++} multiplet members around the light u/d quark mass and compared to the PDG. The error bands correspond to different extrapolation schemes. *Right:* ρ-meson mass calculated in rainbow-ladder (solid, blue) compared to lattice results (see [14] for references). The star is the PDG value and the dashed line is the $\pi\pi$ threshold.

5. Acknowledgement

This work has been supported by DFG, FI 970/11-1.

[1] Pelaez J R 2016 *Phys. Rept.* **658** 1 (*Preprint* 1510.00653)
[2] Caprini I, Colangelo G and Leutwyler H 2006 *Phys. Rev. Lett.* **96** 132001 (*Preprint* hep-ph/0512364)
[3] Garcia-Martin R, Kaminski R, Pelaez J R and Ruiz de Elvira J 2011 *Phys. Rev. Lett.* **107** 072001 (*Preprint* 1107.1635)
[4] Patrignani C *et al.* (Particle Data Group) 2016 *Chin. Phys.* **C40** 100001
[5] Jaffe R L 1977 *Phys. Rev.* **D15** 267
[6] Chen H X, Hosaka A and Zhu S L 2007 *Phys. Rev.* **D76** 094025 (*Preprint* 0707.4586)
[7] Pelaez J R 2004 *Mod. Phys. Lett.* **A19** 2879–2894 (*Preprint* hep-ph/0411107)
[8] Ruiz de Elvira J, Pelaez J R, Pennington M R and Wilson D J 2011 *Phys. Rev.* **D84** 096006 (*Preprint* 1009.6204)
[9] Ebert D, Faustov R N and Galkin V O 2009 *Eur. Phys. J.* **C60** 273–278 (*Preprint* 0812.2116)
[10] Parganlija D, Giacosa F and Rischke D H 2010 *Phys. Rev.* **D82** 054024 (*Preprint* 1003.4934)
[11] Roberts C D and Williams A G 1994 *Prog. Part. Nucl. Phys.* **33** 477–575 (*Preprint* hep-ph/9403224)
[12] Alkofer R and von Smekal L 2001 *Phys. Rept.* **353** 281 (*Preprint* hep-ph/0007355)
[13] Fischer C S 2006 *J. Phys.* **G32** R253–R291 (*Preprint* hep-ph/0605173)
[14] Eichmann G, Sanchis-Alepuz H, Williams R, Alkofer R and Fischer C S 2016 *Prog. Part. Nucl. Phys.* **91** 1–100 (*Preprint* 1606.09602)
[15] Sanchis-Alepuz H, Fischer C S, Kellermann C and von Smekal L 2015 *Phys. Rev.* **D92** 034001 (*Preprint* 1503.06051)
[16] Heupel W, Eichmann G and Fischer C S 2012 *Phys. Lett.* **B718** 545–549 (*Preprint* 1206.5129)
[17] Eichmann G, Fischer C S and Heupel W 2016 *Phys. Lett.* **B753** 282–287 (*Preprint* 1508.07178)
[18] Maris P and Tandy P C 1999 *Phys. Rev.* **C60** 055214 (*Preprint* nucl-th/9905056)
[19] Eichmann G 2009 *Hadron Properties from QCD Bound-State Equations* Ph.D. thesis Graz U. (*Preprint* 0909.0703)
[20] Williams R, Fischer C S and Heupel W 2016 *Phys. Rev.* **D93** 034026 (*Preprint* 1512.00455)
[21] Eichmann G, Fischer C S and Heupel W 2015 *Phys. Rev.* **D92** 056006 (*Preprint* 1505.06336)
[22] Weil E, Eichmann G, Fischer C S and Williams R 2017 *Phys. Rev.* **D96** 014021 (*Preprint* 1704.06046)

IOP Conf. Series: Journal of Physics: Conf. Series **1024** (2018) 012036 doi:10.1088/1742-6596/1024/1/012036

Two-pole structure of the $D_0^*(2400)$

P. Fernandez-Soler[1], M. Albaladejo[1,2], F.-K. Guo[3,4], J. Nieves[1]

[1] Instituto de Física Corpuscular (IFIC), Centro Mixto CSIC-Universidad de Valencia, Institutos de Investigación de Paterna, Aptd. 22085, E-46071 Valencia, Spain

[2] Departamento de Física, Facultad de Química, Universidad de Murcia, E-30071, Murcia, Spain

[3] CAS Key Laboratory of Theoretical Physics, Institute of Theoretical Physics, Chinese Academy of Sciences, Beijing 100190, China

[4] School of Physical Sciences, University of Chinese Academy of Sciences, Beijing 100049, China

E-mail: pedro.fernandez@ific.uv.es

Abstract. We study the interaction of the $D\pi$, $D\eta$ and $D_s\bar{K}$ channels in $J^P = 0^+$ in the framework of unitarized chiral perturbation theory in finite volume, predicting the volume dependence of the energy levels. The results succesfully describe the energy levels provided in a recent lattice QCD calculation of the same system. We find two poles in the energy region of the $D_0^*(2400)$ resonance, with masses 2105^{+6}_{-8}, 2451^{+36}_{-26} MeV and half-widths 102^{+10}_{-12}, 134^{+7}_{-8} MeV respectively. We study their evolution in the light-flavor $SU(3)$ limit, which allows us to identify the $D_{s0}^*(2317)$ resonance and the pole with lower mass as partners of the same $SU(3)$ multiplet. The mass of the higher pole is obtained between the $D\eta$ and $D_s\bar{K}$ threshold energies, strongly coupled to the latter channel, thus we expect it to influence the $D_s\bar{K}$ invariant mass distribution.

1. Introduction

The $D_0^*(2400)$ and $D_{s0}^*(2317)$ resonances are the lowest lying scalar states in the open charm sector. Since their discovery in 2003 [1, 2], they have attracted special attention because they are key pieces in the understanding of the hadron spectroscopy in the charm sector. With an associated $c\bar{u}/\bar{d}$ and $c\bar{s}$ quark content respectively, their existence was early postulated by Godfrey *et al* in Ref. [3] as quark model states. Later, they showed up in the experiments with unexpected masses, specially in the case of the $D_{s0}^*(2317)$, whose mass was found around 150 MeV smaller than the theoretical expectations. Concerning the $D_0^*(2400)$ state in which we are interested here, the experimental mass and total width are shown in Figure 1. In that figure we can see how the early measurements provided a mass of ~ 2.4 GeV, which is in agreement with, *e.g.*, the value predicted by Godfrey in Ref. [3]. On the other hand, some experiments have found a mass of the order of 2.3 GeV for the neutral $D_0^*(2400)$, which lies close to the $D_{s0}^*(2317)$ mass. This is an interesting point since, according to their constituent quark content, one would expect the $D_{s0}^*(2317)$ to be heavier than the $D_0^*(2400)$, which is the opposite to what the experiments reveal. In the study of these resonances the two-meson thresholds with energies close to the resonance mass play an important role, as it was suggested in Ref. [4] and by lattice QCD (LQCD) studies [5, 6]. In Figure 1 we can find the values of the $D\eta$ and $D_s\bar{K}$ meson threshold energies, which lie relatively close to the $D_0^*(2400)$ mass. These two channels together with the $D\pi$ system, when coupled in S-wave and isospin $I = 1/2$, will have the same quantum numbers as the $D_0^*(2400)$ resonance. The authors of Ref. [7] performed for the first time a LQCD simulation of the charmed-strangeness sector, $(S, I) = (0, 1/2)$, considering the three two-meson channels already mentioned. We will describe the scattering of the $D\pi$, $D\eta$ and $D_s\bar{K}$ mesons in coupled channels in a finite volume, in order to compare with the results of Ref. [7]. After that, we will show some of the predictions of the model. The contents of the present manuscript are based on the work of Ref. [8].

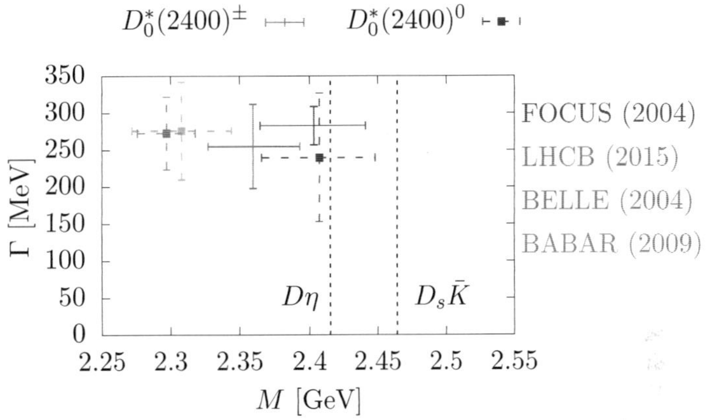

Figure 1. Experimental mass M and total width Γ for the charged and neutral $D_0^*(2400)$. The values and uncertainties shown are those provided by different experimental facilities see Ref. [9] and references therein. The vertical dashed lines show the energies of the $D\eta$ and $D_s\bar{K}$ thresholds, which coupled in S-wave have the same quatum numbers as $D_0^*(2400)$.

2. Formalism

2.1. Unitarized S-wave scattering amplitudes

We describe the $D\pi$, $D\eta$ and $D_s\bar{K}$ S-wave scattering in coupled channels. Let us label each of the channels by $H_j\phi_j$, $(j = 1, 2, 3)$, H_j and ϕ_j denote the charmed and light meson inside each channel. The unitarized scattering amplitudes are obtained solving the on-shell Bethe-Salpeter equation in its factorized version,

$$T(s) = \left(V^{-1}(s) - \mathcal{G}(s)\right)^{-1}. \tag{1}$$

The unitary matrix $T_{i,j}$ describes the scattering in the coupled channels space $H_i(p_1)\phi_i(p_2) \rightarrow H_j(p_3)\phi_j(p_4)$, where $p_{1,\ldots,4}$ is the four-momentum of the particles and s is the Mandelstam variable. The V matrix is the kernel, its matrix elements being the S-wave scattering amplitudes obtained from a next-to-leading order (NLO) chiral Lagrangian, see Ref. [10] for details. The NLO amplitudes depend on six low energy constants (LECs), denoted $h_{0,\ldots,5}$, that are taken from a previous work [11], where they were fitted to several LQCD scattering lengths in different channels and (S, I) sectors. It is worth mentioning that these sectors did not include the $(S, I) = (0, 1/2)$, which is the sector of interest here. $\mathcal{G}$ is a diagonal matrix containing the loop functions of the different channels, $\mathcal{G}_{jj}(s) = G(s, m_{H_j}, m_{\phi_j})$, see Eq. (16) in Ref. [11]. It is regularized by a substraction constant with its value, together with the already mentioned LECs, constrained in the latter reference. The interaction of the mesons may produce bound states and resonances, as well as virtual states, which will manifest as poles in the complex s-plane of T in the different Riemann Sheets (RS). Let us denote the different RS by (ξ_1, ξ_2, ξ_3), where $\xi_j = 0, 1$. The different RS are reached by analytical continuations of the loop function $\mathcal{G}_{jj} \rightarrow \mathcal{G}_{jj} + i\xi_j k_j/4\pi\sqrt{s}$, where k_j is the modulus of the three-momenta of the mesons in channel j, in the center of mass frame. For a pole located at $s = s_p$, the coupling g_j to channel j is given by $g_i g_j = \mathrm{Res}\left\{T_{ij}\right\}|_{s=s_p}$.

2.2. Finite volume

The LQCD simulation of Ref. [7] provides the energy levels for the $D\pi$, $D\eta$ and $D_s\bar{K}$ interaction in a finite cubic box of size L^3 with periodic boundary conditions, for several values of L. When we restrict the interaction to a finite volume imposing periodic boundary conditions on the fields, we get a quantization condition for the three-momentum, $\vec{q} = 2\pi\vec{n}/L$ with $\vec{n} \in \mathbb{Z}^3$. Therefore, the three-momentum integrals in the meson loops of Eq. (1) will become discrete sums over all the possible three-momentum values: $\mathcal{G}(s) \to \tilde{\mathcal{G}}_{jj}(s, L)$. Further details on the finite volume loop function $\tilde{\mathcal{G}}_{jj}(s, L)$ can be found in Refs. [8, 12, 13]. Besides, the lattice conditions involve the generation of mesons with unphysical masses and, consequently, we adopt the masses obtained in Ref. [7] in order to compare with their results. The energy levels of our meson interaction will be the poles of a finite volume version of Eq. (1) [13],

$$\tilde{T}(s, L) = \left(V^{-1}(s) - \tilde{\mathcal{G}}(s, L)\right)^{-1}. \tag{2}$$

In Eq. (2), up to the order we are working here, the kernel V does not receive any volume correction, and the volume dependence of the interaction is introduced through the loop functions. By solving $\tilde{T}^{-1}(E^2, L) = 0$ for different values of the box length L, we get the volume dependence of the energy levels $E(L)$. In the non-interacting limit, $V_{i,j} \to 0$, the energy levels are nothing but the free energies of the mesons in the finite volume, which can be labelled by their three momentum state $E_j^{\text{free}}(\vec{q}, L) = \sqrt{m_{H_j}^2 + \vec{q}^{\,2}} + \sqrt{m_{\phi_j}^2 + \vec{q}^{\,2}}$.

3. Results

3.1. Energy levels

In Figure 2 we show the results of our model in finite volume. The volume dependence of the energy levels obtained solving Eq. (2) are the red solid lines and bands. We find that our predictions are in agreement with the energy levels provided by the LQCD simulation of Moir *et al* [7], represented as black points in Figure 2. The results in the region above 2.7 GeV are beyond the range of validity of our model, this region is shown as a shaded area. Finally, the free energies are represented with dashed lines.

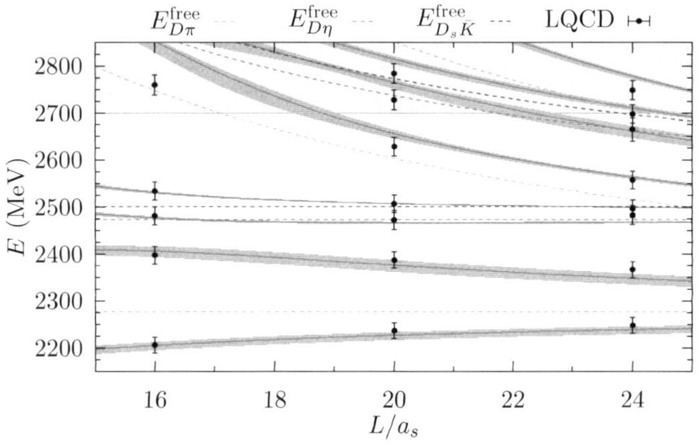

Figure 2. Volume dependence of the energy levels of Ref. [7] (black points) compared with our predictions (red coloured bands). The bands are the 1σ uncertainties propagated from the free parameters of the model, that were fitted in Ref. [11]. The dashed lines are the free energies. The length L is shown in units of the lattice spacing used in Ref. [7], $a_s = 0.12$ fm.

From the deviation of the energy levels with respect to the free energies we can extract information about the interaction. The first energy level, the one below the $D\pi$ threshold energy, is shifted to lower energies, and in Ref. [7] is interpreted as a $D\pi$ bound state. As we shall see in the next section, when we adopt the lattice masses and search for poles in Eq. (1),

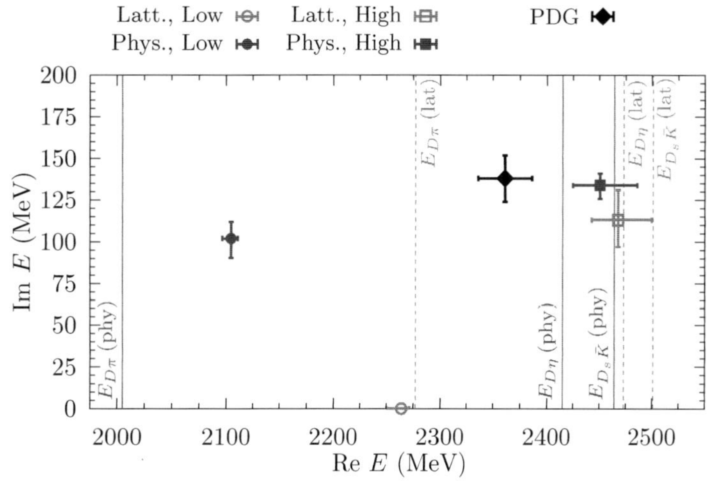

Figure 3. Pole positions in the complex energy plane. As can be seen a two-pole pattern is obtained in the case of using the unphysical masses of Ref. [7] (red points) and physical masses (blue points). The black diamond is the PDG average [9] for the $D_0^*(2400)$. We get a two-pole structure in the $D_0^*(2400)$ sector, where both poles are resonances with a similar width and masses above and below the PDG value.

we also get a bound state (see red circle in Figure 3), it must be compared with the vertical lines representing the thresholds, red-dashed type in the case of lattice and blue-solid type for physical masses. The second level in Figure 2, lying between the $D\pi$ and $D\eta$ thresholds, is also very shifted with respect to them, revealing that the interaction is strong. In fact, another pole appears right below the $D\eta$ threshold, see the red square in Figure 3.

3.2. Poles

As we have commented above, when we use the lattice meson masses of Ref. [7] and search for poles in Eq. (1), we find two poles which are the red points in Figure 3. The pole with lower mass is a bound state, mostly coupled to $D\pi$. On the other hand, when we change to physical masses, due to the heavy pion mass obtained in Ref. [7] the $D\pi$ threshold energy changes dramatically, as can be seen in the vertical lines of Figure 3. The lower pole is no longer a bound state and it is shifted to a lower mass becoming a resonance changing of RS, and being still mostly coupled to $D\pi$. The higher pole also shows a similar pattern of couplings using lattice and physical masses, it does not change of RS but it moves slightly to lower masses staying between the $D\eta$ and $D_s\bar{K}$ thresholds, and thus, becoming also a resonance. We end up with two resonance candidates in the $D_0^*(2400)$ sector, and we succesfully describe the LQCD information from [7] as a prediction of our model. In the following we will study the evolution of the poles in the light $SU(3)$ flavour limit.

3.3. Light-flavour $SU(3)$ limit

The two-pole structure has been previously reported in Refs. [14, 15, 16] and is partially rooted in the $SU(3)$ structure of the interaction of Ref. [10]. To explore it, we change the masses of the mesons recovering the light-flavor $SU(3)$ limit: $m_{H_j} \to M$, $m_{\phi_j} \to m$ [14]. In practice, this is achieved by introducing an $SU(3)$ breaking parameter, denoted x, $m_{\phi_j}(x) = m_{\phi_j} + x(m - m_{\phi_j})$ (similar for m_{H_j}). In the $SU(3)$ limit the interaction decomposes into irreducible representations (irreps), $\overline{\mathbf{3}} \otimes \mathbf{8} = \overline{\mathbf{15}} \oplus \mathbf{6} \oplus \overline{\mathbf{3}}$, and it can be diagonalized $V \to V_d = \mathrm{diag}(\mathrm{V}_{\overline{15}}, \mathrm{V}_6, \mathrm{V}_{\overline{3}})$. This fact, together with the choice of a common substraction constant for all the channels, also makes the T-matrix diagonalizable. Thus, for a particular choice of m and M, we are able to look for poles in each $SU(3)$ irrep as three independent one channel problems, see Ref. [8] for further details. It turns out that the $V_{\overline{15}}$ is repulsive, and therefore there are no poles associated to this irrep. The rest of the irreps are attractive, being the $\overline{\mathbf{3}}$ the most. This representation shows a bound state and the **6**, whose interaction is weaker, a virtual state. With the model of Ref. [10] one can

Figure 4. (Left panel) Pole paths on the complex energy plane when recovering the $SU(3)$ limit. (Right panel) Mass evolution of the different poles with the parameter x. Besides the two $(0, 1/2)$ poles, denoted as high and low (dark blue dashed and green solid lines), we show the evolution of the $(1, 0)$ bound state, which is identified with the $D^*_{s0}(2317)$ resonance (orange dot-dashed).

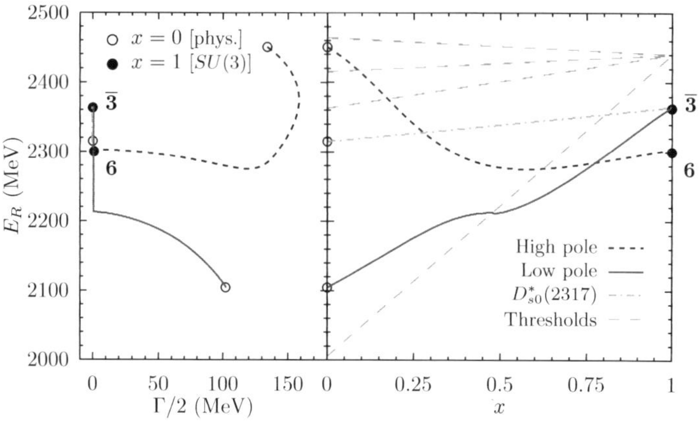

describe all the possible (S, I) sectors, not only the $(0, 1/2)$ studied here. We have also looked for poles in the $(1, 0)$ sector, the corresponding to the $D^*_{s0}(2317)$ resonance. The model predicts a bound state located at 2315^{+18}_{-28} MeV, not far from the experimental value quoted in Ref. [9]. As it happens in the $(0, 1/2)$ sector, the $(1, 0)$ sector was not considered to constrain the LECs in Ref. [11] so this bound state is another predicion. Besides, the authors have also explored the pole spectroscopy in all the (S, I) $J^P = 0^+, 1^+$ combinations, in the charm and bottom sector. A similar $(0, 1/2)$, $(1, 0)$ pole pattern is found, while for $(-1, 0)$ we find a virtual state, see Ref. [8]. We have followed the pole positions in the complex energy plane when varying x from 0 to 1, tracing the trajectories in Figure 4. The $(0, 1/2)$ pole with a ~ 2.1 GeV mass, denoted as lower pole, and the $(1, 0)$ bound state, identified with $D^*_{s0}(2317)$, have the same $SU(3)$ origin in the $\overline{\mathbf{3}}$ representation. The apparent puzzle mentioned in the introduction (namely, $D^*_0(2400)$ heavier that $D^*_{s0}(2317)$) is solved in this way, the lower pole of the two $D^*_0(2400)$ states we find is actually lighter than $D^*_{s0}(2317)$. On the other hand, the higher pole is originated from the virtual state in the **6** irrep, which is less attractive than the $\overline{\mathbf{3}}$. Given its strong coupling to the $D_s\bar{K}$ channel, we expect it to influence the invariant mass distribution of that channel in B decays, something that could be tested in future experiments.

References

[1] Aubert B *et al.* (BaBar) 2003 *Phys. Rev. Lett.* **90** 242001 (*Preprint* `hep-ex/0304021`)
[2] Abe K *et al.* (Belle) 2004 *Phys. Rev.* **D69** 112002 (*Preprint* `hep-ex/0307021`)
[3] Godfrey S and Isgur N 1985 *Phys. Rev.* **D32** 189–231
[4] van Beveren E and Rupp G 2003 *Physical Review Letters*
[5] Mohler D, Prelovsek S and Woloshyn R M 2013 *Phys. Rev.* **D87** 34501 (*Preprint* `1208.4059`)
[6] Mohler D, Lang C B, Leskovec L, Prelovsek S and Woloshyn R M 2013 *Physical Review Letters* **111** 1–5 ISSN 00319007 (*Preprint* `arXiv:1308.3175v3`)
[7] Moir G, Peardon M, Ryan S M, Thomas C E and Wilson D J 2016 *JHEP* **10** 011 (*Preprint* `1607.07093`)
[8] Albaladejo M, Fernandez-Soler P, Guo F K and Nieves J 2017 *Phys. Lett.* **B767** 465–469 (*Preprint* `1610.06727`)
[9] Patrignani C *et al.* (Particle Data Group) 2016 *Chin. Phys.* **C40** 100001
[10] Guo F K, Hanhart C, Krewald S and Meissner U G 2009 *Acta Physica Polonica B, Proceedings Supplement* **2** 209–214 ISSN 18992358 (*Preprint* `0901.1597`)
[11] Liu L, Orginos K, Guo F K, Hanhart C and Meissner U G 2013 *Phys. Rev.* **D87** 14508 (*Preprint* `1208.4535`)
[12] Martinez Torres A, Dai L R, Koren C, Jido D and Oset E 2012 *Phys. Rev.* **D85** 014027 (*Preprint* `1109.0396`)
[13] Doring M, Meissner U G, Oset E and Rusetsky A 2011 *Eur. Phys. J.* **A47** 139 (*Preprint* `1107.3988`)
[14] Kolomeitsev E E and Lutz M F M 2004 *Phys. Lett.* **B582** 39–48 (*Preprint* `hep-ph/0307133`)
[15] Guo F K, Shen P N, Chiang H C, Ping R G and Zou B S 2006 *Phys. Lett.* **B641** 278–285 (*Preprint* `hep-ph/0603072`)
[16] Guo F K, Hanhart C and Meissner U G 2009 *Eur. Phys. J.* **A40** 171–179 (*Preprint* `0901.1597`)

Charm physics at BESIII

Rupan Guo, on behalf of the BESIII Collaboration
Institute of High Energy Physics, Beijing 100049, People's Republic of China
E-mail: guorp@ihep.ac.cn

Abstract. By analyzing data samples taken at the center-of-mass energies $\sqrt{s} = 3.773$ GeV and $\sqrt{s} = 4.009$ GeV with the Beijing Spectrometer (BESIII) detector at the Beijing Electron Positron Collider (BEPCII), we determined the $D^+_{(s)}$ decay constants, the form factors of semileptonic decays of D meson, the CKM matrix elements $|V_{cs(d)}|$, the $D^+_{(s)}$ hadronic decays, and we search for charm rare decays. In this article we briefly report the results.

1. Introduction

The leptonic and semileptonic charm decays provide an ideal window to explore the strong and week interactions. The Cabibbo-Kobayashi-Maskawa (CKM) matrix elements $|V_{cs(d)}|$ could be well determined via precision measurements of these decays, which can test the CKM matrix unitary. On the other hand with precision measurements of the decay constants $f_{D^+_{(s)}}$ and form factor, we are able to calibrate the Lattice QCD calculations Hadronic decays of charmed mesons provide important information on flavor mixing, CP violation, and strong-interaction effects. Precision measurements of the hadronic decay rates also strengthen the understanding of U-spin and SU(3)-flavor symmetry breaking effects in D meson decays. In addition, the search for charm rare decays can provide plentiful information for the new physics (NP).

In this paper, the samples for the recent studies of the leptonic, semileptonic, hadronic and rare charm decays described in the following were recorded with the BESIII detector at the $D^0\bar{D}^0/D^+D^-$ threshold ($\sqrt{s} = 3.773$ GeV) and at the $D_s^+D_s^-$ threshold ($\sqrt{s} = 4.009$ GeV) with integrated luminosities of 2.93 fb^{-1} and 482 pb^{-1}, respectively. Throughout the proceeding, charge conjugate is implied.

2. BESIII and data analysis method

The BESIII is a general purpose magnetic detector working at a double-ring e^+e^- collider operating in center-of-mass energy between 2.0 GeV and 4.6 GeV, located at the Institute of High Energy Physics (IHEP), Beijing. The maximum luminosity of the BEPCII at 3.773 GeV is 1^{33}cm^{-2}s^{-1}. In data analysis, taking D_s^+ meson as an example, at $\sqrt{s} = 4.009$ GeV, the D_s^+ and D_s^- mesons are produced in e^+e^- annihilation. We first selected the D_s^- meson sample from its hadronic decay modes, which is called singly tagged D_s^- meson or single D_s^- tag, and then selected the leptonic or semileptonic D_s^+ decays in the system recoiling against the singly tagged D_s^- mesons. To select the leptonic and semileptonic D_s^+ decays with missing neutrino, we calculated $U_{\rm miss} \equiv E_{\rm miss} - |\vec{p}_{\rm miss}|$ or $M^2_{\rm miss} \equiv E^2_{\rm miss} - |\vec{p}^{\,2}_{\rm miss}|$, where $E_{\rm miss}$ and $\vec{p}_{\rm miss}$ are the missing energy and missing momentum of the event, respectively. From the $U_{\rm miss}$ or $M^2_{\rm miss}$

distributions we can get the number of signal candidates for the leptonic or semileptonic D_s^+ decays, and then measure their decay branching fractions (BFs).

3. Pure leptonic $D_{(s)}$ decays

The pure leptonic decay of charged $D_{(s)}^+$ mesons proceeds via the annihilation of the c quark and the anti-d (anti-s) quark to a virtual $W^\pm$ boson and its decay to $\ell^+\nu_\ell$. The decay rate can be parametrized as:

$$(D_{(s)}^+ \to \ell^+\nu_\ell) = \frac{G_F^2}{8\pi} f_{D_{(s)}}^2 m_\ell^2 m_{D_{(s)}} (1 - \frac{m_\ell^2}{m_{D_{(s)}}^2})^2 |V_{cd(s)}|^2, \quad (1)$$

where G_F is the Fermi coupling constant, m_ℓ and $m_{D_{(s)}}$ are the masses of the lepton and the $D_{(s)}$ meson in the final state, respectively. Hence the leptonic decays of the charmed meson provides us an opportunity to measure the decay constants $f_{D_{(s)}}$ and the magnitudes of CKM element $|V_{cd(s)}|$.

3.1. $D^+ \to \mu^+\nu_\mu$

By analyzing the data sample at $\sqrt{s} = 3.773$ GeV, we studied the $D^+ \to \mu^+\nu_\mu$ decay. Nine hadronic D^- decay modes are used to accumulate the single tagged D^-. The $D^+ \to \mu^+\nu_\mu$

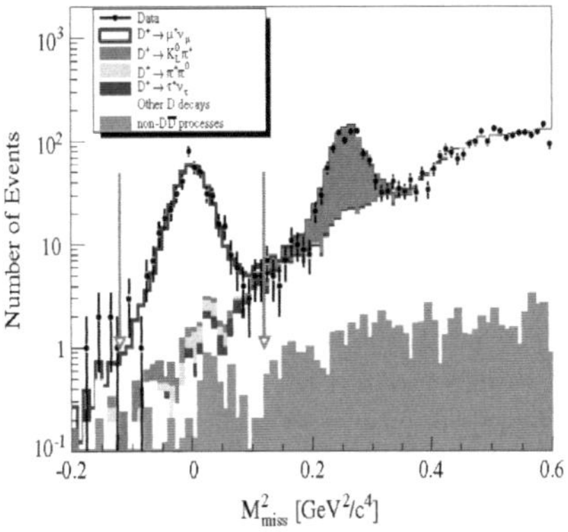

Figure 1. The $M^2_{\rm miss}$ distributions of the $D^+ \to \mu^+\nu_\mu$ candidates, where dots with error bars indicate the data, the opened histogram is for Monte Carlo simulated signal events of $D^+ \to \mu^+\nu_\mu$ decays, and the hatched histograms are for the simulated backgrounds from $D^+ \to K_L^0\pi^+$ (red), $D^+ \to \pi^0\pi^+$ (green), $D^+ \to \tau^+\nu_\tau$ (blue), all other D-meson decays (yellow), and non-$D\bar{D}$ processes (pink).

decay is selected in the recoil side of the single tagged D^-. Figure 1 shows the distribution of $M^2_{\rm miss}$. After subtracting the backgrounds originating from various D^+ decay channels, 409.0 ± 21.2 signal events are retained and the BF is measured to be $\mathcal{B}(D^+ \to \mu^+\nu_\mu) = (3.71 \pm 0.19_{\rm stat} \pm 0.06_{\rm syst}) \times 10^{-4}$, which is the most precise measurement to date. Combining this BF measurement and the Particle Data Group (PDG) values of D^+ lifetime, m_{D^+}, m_μ and magnitude of $|V_{cd}|$ determined from the global Standard Model (SM) fit, the decay constant is determined to be $f_D = (203.2 \pm 5.3_{\rm stat} \pm 1.8_{\rm syst})$ MeV. Alternately, the magnitude of CKM matrix element V_{cd} is extracted to be $V_{cd} = 0.2210 \pm 0.0058_{\rm stat} \pm 0.0047_{\rm syst}$ [1].

3.2. $D_s^+ \to \mu^+\nu_\mu$ *and* $D_s^+ \to \tau^+\nu_\tau$

By analyzing the data taken at 4.009 GeV, we also studied the leptonic D_s^+ decays. From nine D_s^- hadronic decay modes, 15127 ± 321 singly tagged D_s^- mesons were accumulated. In the system recoiling against the single tagged D_s^-, the signal events of $D_s^+ \to \ell^+\nu_\ell$ ($\ell = \mu$, or τ) decays were selected. The distribution of $M^2_{\rm miss}$ is shown in Fig. 2, we obtained 69.3 ± 9.3 $D_s^+ \to \mu^+\nu_\mu$ and 32.5 ± 4.3 $D_s^+ \to \tau^+\nu_\tau$ events. The results of the BFs are determined to be $\mathcal{B}(D_s^+ \to \mu^+\nu_\mu) = (0.495 \pm 0.067_{\rm stat} \pm 0.026_{\rm syst})\%$, $\mathcal{B}(D_s^+ \to \tau^+\nu_\tau) = (4.83 \pm 0.65_{\rm stat} \pm 0.26_{\rm syst})\%$, and the decay constant $f_{D_s} = (241.0 \pm 16.3_{\rm stat} \pm 6.6_{\rm syst})$ MeV [2].

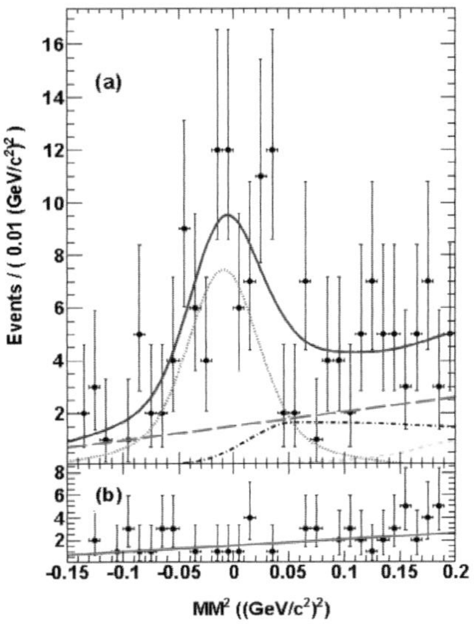

Figure 2. The $M^2_{\rm miss}$ distributions of the signal (a), and the sideband region (b). Data are shown as the points with error bars. The red dotted curve shows the $\mu^+\nu_\mu$ signal and the black dot-dashed curve shows the $\tau^+\nu_\tau$ signal. The purple longdashed line shows the non-D_s^+ background while the green dashed line shows the real-D_s^s background. The blue curve shows the sum of all these contributions.

4. $D^+ \to \gamma e^+\nu_e$ decays

We present the first search for the radiative leptonic decay $D^+ \to \gamma e^+\nu_e$ in the charm sector based on a double-tag method using a data sample of 2.93 fb^{-1} collected at $\sqrt{s} = 3.773$ GeV. The singly tagged D^- mesons are reconstructed through six specific hadronic decay modes. We search for the signal $D^+ \to \gamma e^+\nu_e$ in the remaining charged tracks and showers recoiling against the single tagged D^- mesons. In the distributions of $U_{\rm miss}$, no significant $D^+ \to \gamma e^+\nu_e$ signal is observed. With a 10 MeV cutoff on the radiative photon energy, the upper limit of the decay BF for $D^+ \to \gamma e^+\nu_e$ is $\mathcal{B}(D^+ \to \gamma e^+\nu_e) < 3.0 \times 10^{-5}$ at the 90% C.L.[3]. The result approaches the theoretical predictions.

5. Semileptonic $D_{(s)}$ decays

5.1. $D^{0(+)} \to Pe^+\nu_e(P = K^-, \pi^-, K^0 \ or \ \pi^0)$

In the SM, neglecting the lepton mass, the differential decay rate for $D^+ \to Pe^+\nu_e(P = K^-, \pi^-, K^0 or \pi^0)$ is given by

$$\frac{d}{dq^2} = X\frac{G_F^2}{24\pi^3}|V_{\rm cd(s)}|^2 p^3 |f_+^P(q^2)|^2, \tag{2}$$

where p is the momentum of the pseudoscalar meson P in the rest frame of the D meson, q^2 is the squared four momentum transfer, i.e., the invariant mass of the electron and neutrino system, and $f_+^P(q^2)$ is the form factor which describes the strong interaction between the final state quarks and is usually parameterized in data analysis. In Eq. (2), X is a multiplicative factor due to isospin, which equals to 1 for modes $D^0 \to K^-e^+\nu_e$, $D^0 \to \pi^-e^+\nu_e$, $D^+ \to K^0e^+\nu_e$ and 1/2 for mode $D^+ \to \pi^0e^+\nu_e$. These semileptonic decays provide an excellent chance to validate the lattice QCD calculations by measuring the form factors at $q^2 = 0$, and to measure the magnitudes of CKM matrix elements $|V_{cs}|$ and $|V_{cd}|$.

Based on 2.93 fb^{-1} data taken at 3.773 GeV, BESIII well suited to study the dynamics of the $D^0 \to K^-e^+\nu_e$ and $D^0 \to \pi^-e^+\nu_e$ decays using double-tag method. The BFs are measured to be $\mathcal{B}(D^0 \to K^-e+\nu_e) = (3.505 \pm 0.014_{\rm stat} \pm 0.033_{\rm syst})\%$ and $\mathcal{B}(D^0 \to \pi^-e+\nu_e) = (0.295 \pm 0.004_{\rm stat} \pm 0.003_{\rm syst})\%$, which are consistent with the previous measurements and are the most precise at present [4]. Similarly, the BFs for $D^+ \to \bar{K}^0e^+\nu_e$ and $D^+ \to \pi^0e^+\nu_e$ are measured to be $\mathcal{B}(D^+ \to \bar{K}^0e^+\nu_e) = (8.60 \pm 0.06_{\rm stat} \pm 0.15_{\rm syst})\%$ and $\mathcal{B}(D^+ \to \pi^0e^+\nu_e) =$

$(0.363 \pm 0.008_{\rm stat} \pm 0.005_{\rm syst})\%$, which are consistent within errors with previous measurements but with better precisions [5]. We also studied the differential decay rates of these two processes. We first divided the semileptonic events into several q^2 bins and measure the partial decay rates in each q^2 bin, then we fit these partial decay rates using different form factor models. Single pole model, modified pole model and series expansion are tried. Figures 3 and 4 show the fit resulsts.

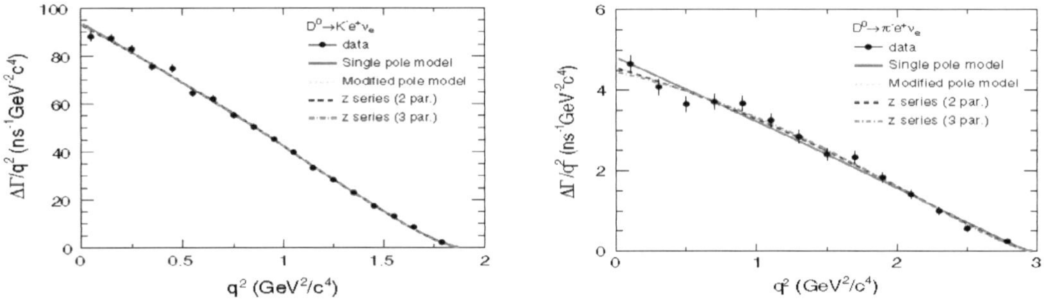

Figure 3. Fits to partial decay widths of $D^+ \to \bar{K}^0 e^+ \nu_e$ (left) and $D^+ \to \pi^0 e^+ \nu_e$ (right).

From the fits, we extract the product $f_+(0)|V_{cs(d)}|$ and other form factor parameters. Using the values for $f_+^{K(\pi)}(0)|V_{cs(d)}|$ from the two-parameter z-series expansion fits and with $f_+^K(0) = 0.747 \pm 0.011 \pm 0.015$ [6] and $f_+^\pi(0) = 0.666 \pm 0.020 \pm 0.021$ [7] calculated in LQCD, $|V_{cs}|$ is obtained to be $0.9601 \pm 0.0033 \pm 0.0047 \pm 0.0239$ ($|V_{cd}| = 0.2155 \pm 0.0027 \pm 0.0014 \pm 0.0094$), where the first uncertainties are statistical, the second ones systematic, and the third ones are due to the theoretical uncertainties in the form factor calculations. The BESIII results are in good agreement with the previous measurements, and with the best precision to date.

5.2. $D_s^+ \to \eta(\eta')e^+\nu_e$

By analyzing the data taken at 4.009 GeV, BESIII studied the semileptonic $D_s^+ \to \eta(\eta')e^+\nu_e$ decays [8]. The 13157 ± 240 singly tagged D_s^- mesons are reconstructed in ten hadronic decay modes. The information of the signal candidate is inferred from the variable $U_{\rm miss}$. Figure 5 shows the distributions of $U_{\rm miss}$. After subtracting the backgrounds, the BFs are determined to be $\mathcal{B}(D_s^+ \to \eta e^+\nu_e) = (2.30 \pm 0.31_{\rm stat} \pm 0.08_{\rm syst})\%$ and $\mathcal{B}(D_s^+ \to \eta' e^+\nu_e) = (0.93 \pm 0.30_{\rm stat} \pm 0.05_{\rm syst})\%$, which are consistent with previous measurements within uncertainties. Combining the BF measurements, we obtain the ratio $B(D_s^+ \to \eta' e^+\nu_e)/B(D_s^+ \to \eta e^+\nu_e) =$

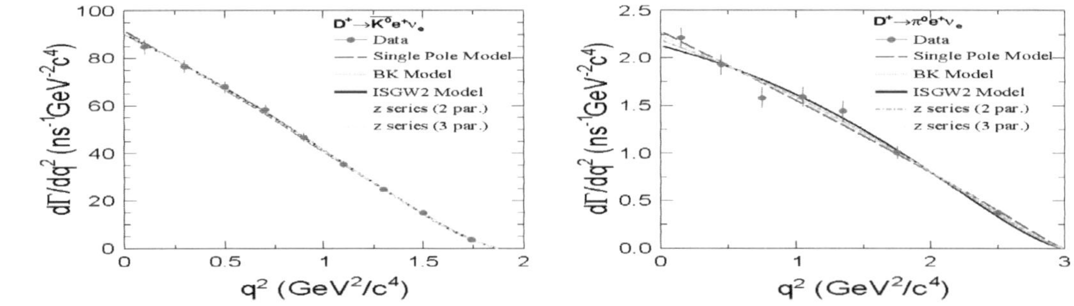

Figure 4. Fits to partial decay widths of $D^+ \to \bar{K}^0 e^+ \nu_e$ (left) and $D^+ \to \pi^0 e^+ \nu_e$ (right).

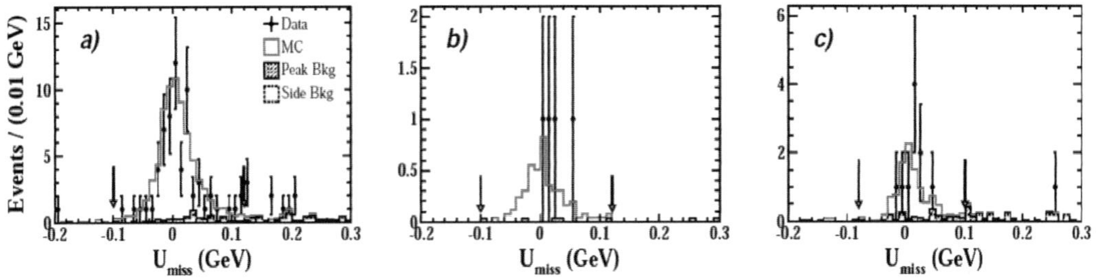

Figure 5. The $U_{\rm miss}$ distributions of the (a) $D^+ \to \mu^+\nu_\mu$, (b) $D_s^+ \to \eta'(\eta\pi^+\pi^-)e^+\nu_e$ and (c) $D_s^+ \to \eta'(\gamma\rho^0)e^+\nu_e$ candidates, where arrows denote the signal regions.

$0.40{\pm}0.14_{\rm stat}{\pm}0.02_{\rm syst}$, which provides complementary information to understand η-η' mixing [8].

6. Handronic D decays

An amplitude analysis of the decay $D^0 \to K^-\pi^+\pi^+\pi-$ has been performed with the data at $\sqrt{s} = 3.773$ GeV. With a nearly background free sample of about 16000 events, we investigate the substructure of the decay and determine the relative fractions and the phases among the different intermediate processes. Our amplitude model includes the two-body decays $D^0 \to \bar{K}^{*0}\rho^0$, $D^0 \to K^-a_1^+(1260)$ and $D^0 \to K_1^-(1270)\pi^+$, the three-body decays $D^0 \to \bar{K}^{*0}\pi^+\pi^-$, $D^0 \to K^-\pi^+\rho^0$, as well as the four-body nonresonant decay $D^0 \to K^-\pi^+\pi^+\pi-$. The dominant intermediate process is $D^0 \to K^-a_1^+(1260)$, accounting for a fit fraction of 54.6%. By using the inclusive BF $\mathcal{B}(D^0 \to K^-\pi^+\pi^+\pi-) = (8.07 \pm 0.23)\%$, taken from the PDG and the fit fraction for the different components $FF(n)$ obtained in this analysis, we calculate the exclusive absolute BFs for the individual components with $\mathcal{B})(n) = \mathcal{B}(D^0 \to K^-\pi^+\pi^+\pi-) \times FF(n)$ [9].

7. Summary

By analysing 2.93 and 0.482 fb^{-1} data taken at $\sqrt{s} = 3.773$ and 4.009 GeV with the BESIII detector, we present a selection of recent BESIII charm results. From the leptonic $D_{(s)}^+$ and semileptonic $D^{+(0)}$ decays we determined the most precise values for the decay constant f_{D^+} the hadronic form factors $f_+^{K(\pi)}(0)$, and the form factor shape $f_+^{K(\pi)}(q^2)$. which provide important test to LQCD calculations, CKM matrix unitary. The study of charmed meson hadronic decays is given with significant improved precision. BESIII have taken more than 3 fb^{-1} $D_sD_s^*$ data at $\sqrt{s} = 4.180$ GeV, many more results are expected.

References

[1] M. Ablikim *et al.* (BESIII Collaboration), Phys. Rev. D **89**, 051104(R) (2014).
[2] M. Ablikim *et al.* (BESIII Collaboration), Phys. Rev. D **94**, 072004 (2016).
[3] M. Ablikim *et al.* (BESIII Collaboration), Phys. Rev. D **95**, 071102(R) (2017).
[4] M. Ablikim *et al.* (BESIII Collaboration), Phys. Rev. D **92**, 072012 (2015).
[5] M. Ablikim *et al.* (BESIII Collaboration), Phys. Rev. D **96**, 012002 (2017).
[6] H. Na *et al.* (HPQCD Collaboration), Phys. Rev. D **82**, 114506 (2010).
[7] H. Na *et al.* (HPQCD Collaboration), Phys. Rev. D **84**, 114505 (2011).
[8] M. Ablikim *et al.* (BESIII Collaboration), Phys. Rev. D **94**, 112003 (2016).
[9] M. Ablikim *et al.* (BESIII Collaboration), Phys. Rev. D **95**, 072010 (2017).

IOP Conf. Series: Journal of Physics: Conf. Series **1024** (2018) 012038 doi:10.1088/1742-6596/1024/1/012038

EnsarRoot: The framework for simulation and data analysis for ENSAR

Pablo Cabanelas[1,2], Héctor Alvarez-Pol[1,2], Elisabet Galiana[1,2] and Yago González-Rozas[1,2] for the SATNuRSE workpackage of ENSAR2

[1] Dpt. de Física de Partículas, Universidade de Santiago de Compostela, E-15782 Santiago de Compostela, Spain

[2] Instituto Galego de Física de Altas Enerxías (IGFAE), Dpt. Física de Partículas, E-15782 Universidade de Santiago de Compostela, Spain

E-mail: pablo.cabanelas@usc.es

Abstract. EnsarRoot is the simulation and analysis framework for the ENSAR project. It provides the software infrastructure and examples to develop an analysis and simulation code for Nuclear Physics experiments. It is written in C++, based on the FairRoot framework and implements data structure and geometry definitions based on ROOT. Dealing with the transport engines is done through the Virtual Monte Carlo concept. The code includes examples of different kind of detectors, specific event generators and event viewers, among other features.

1. Introduction

ENSAR and ENSAR2 (second phase) represent the integrating activity for European scientists who are performing research in Nuclear Physics [1]. It is funded by the European Commission within the HORIZON 2020 Programme, and the core aim is to provide access to large-scale nuclear research facilities in Europe, as GSI/FAIR, CERN/ISOLDE, GANIL and many others. EnsarRoot has been developed under the ENSAR Joint Research Activity SiNuRSE (Simulations for Nuclear Reactions and Structure in Europe), which continues as SATNuRSE (Simulations and Analysis Tools for Nuclear Reactions and Structure in Europe) in ENSAR2.

EnsarRoot is the dedicated framework for the ENSAR project. A software framework is a piece of code, a set of classes, scripts and templates, providing a given functionality that can be selectively completed or modified by additional user-written code, in this case to perform simulations and data analysis for the description of Nuclear Physics experiments. The purpose of a so structured framework is to improve the efficiency of creating new software, increase the reliability of a new application and reduce the programming effort. It is written in C++, it is based on the FairRoot framework libraries [2], which are common to many experiments at GSI-FAIR, and it loads the ROOT [3] libraries providing a ROOT-like output structure. EnsarRoot implements different examples of detectors and experimental setups, and uses the Virtual Monte Carlo concept for running simulations [4].

In the next paragraphs, a general description of the software is presented, and also some words are devoted to implemented cases.

2. The EnsarRoot Framework

EnsarRoot is a suitable framework for simulation and analysis of small and medium scale Nuclear and Particle Physics experiments, publicly available [5] under a GNU LGPL licence [6]. As already mentioned, it is written in C++, and CMake [7] is used as a build system. It runs in many Linux distribution and Mac OSX, so that users could install it in any linux box or select some configurations that have been previously tested. It is based on the FairRoot library to implement a set of base classes. It serves as the core where other developments can be implemented, that is, an user code for specific detectors could be added, based on existing examples, both for the simulation and for the development of additional analysis tasks.

For the simulation, the interface with the transport engines Geant3 and Geant4 [8, 9] is done through the concept of Virtual Monte Carlo, and the implementation we include derives from FairRoot. This concept allows the user to build a simulation independently from the Monte Carlo implementation itself.

The reconstruction is done event by event and is organized in tasks. For each event, several tasks or algorithms are accomplished, such as digitization and hit registration in the active volumes, previously defined in the simulation section. The users can create their own algorithm inheriting from existing example tasks or from the `FairTask` class. Being ROOT based, the initialization of tasks is done with ROOT steering macros and no executable files are needed. The input-output workflow is arranged also in ROOT structures like TFile, TTree or TClonesArray. It makes use of `TGeo` ROOT file format for geometry and navigation.

2.1. Software arrangement

The software package is arranged in modules. There are some common base modules which contain the features to be called by the different implementations:

- *ensarbase*: contains the `EnsarDetector` class. This class inherits from `FairDetector` class, and implements the local reference system and the global positioning of any implemented detector in the framework.
- *ensardata*: contains a list of the active detector systems, as well as classes `EnsarMCStack` and `EnsarMCTrack`, which are responsible for the storage of the output data of the transport engine in ROOT branches.
- *ensargen*: contains the different developed event generators to be called in the simulation initialization (see 2.2).
- *ensarevtdisplay*: contains all the features of a dedicated EVE event display (see 2.3), which are initialized through the class `EnsarEventManager`.

There are another two common modules: *gconfig* and *macros*. The first contains all the needed scripts for the configuration of the selected transport engine (Geant3 or Geant4), and the second, general example scripts for running the simulation and analyzing data.

Making use of these common base modules, the user implements its own system or application in a dedicated module. Each system or application module contains three major submodules:

- The main system class is located in the *detector* submodule and should inherits from the `EnsarDetector` class. This class transform the primary output of the transport engine and register the detector information in the corresponding ROOT TTree branch.
- The *data* submodule contains the dedicated classes for managing the data.
- The corresponding geometry files are stored in the *geometry* submodule.

Besides, the *macros* submodule contains the root scrips for creating the geometry, running simulations and perform the data analysis.

EnsarRoot provides templates for the construction of some of the most commonly used Nuclear Physics detectors, that is, complete system modules for experimental arrangements with NaI and CsI scintillator crystals, plastic scintillators, High Purity Germanium Detector, Silicon and Double-sided Silicon Strip Detectors or Resistive Plate Chambers.

Each system module can be activated and deactivated in the `CMakeLists` file before the compilation.

2.2. Dedicated Event Generators

Some specific event generators has been developed and implemented in EnsarRoot. Some interesting cases are:

- Standard Ion Generator: any given ion can be set as the initial transported particle by calling the `EnsarIonGenerator` class, which has an interface to the `FairIon` class, passing A, Z and charge state as arguments to define an ion.
- Generator for gamma-correlations in proton induced reactions: a dedicated event generator for correlated gamma particle emission has been embedded in the framework through classes `EnsarCascadeGen` and `EnsarSiGenerator`.
- Giant and Pygmy Dipole Resonance Generator: a complete generator has been implemented in the submodule *CascadeGen*, with easy running and parameter handling scripts; the output is produced in a proper ascii EnsarRoot format.
- CRY (Cosmic Ray Air Showers) Generator interface: the output of the CRY Cosmic Rays event generator [10] can be directly plug into the frame by calling the interface `EnsarCryAsciiGenerator`.

2.3. Event Display

It has been implemented an useful event display based on the EVE package of ROOT [11]. It draws not only the general view of detectors geometry and event tracks, but also their projections in different axis using the `TEveProjections` features. Also, a dedicated calorimeter representation has been implemented using the `TEveCalo`, which allows a topological study of the event in an user-friendly environment. Fig. 1 shows a simulated event in a CsI calorimeter using the event display.

3. Implementation Cases

Besides the templates for the most common cases of single particle detectors, some specific complex cases have been implemented in EnsarRoot and are delivered in the package. The two most important cases are presented in the following subsections.

3.1. The CALIFA Demonstrator Experiment

CALIFA [12] is the calorimeter for the detection of gammas and light charged particles of the R3B experiment [13] at FAIR. A dedicated experiment for testing the CALIFA Barrel Demonstrator has been carried out in November of 2016 at the tandem accelerator of the CTN/IST facilities, in Lisbon, Portugal [14]. Two sets of 64 CsI(Tl) crystals each and a HPGe detector were exposed to gamma rays emitted in proton reactions over Al targets. The full setup was simulated and implemented in the module *ctn*. The analysis of the simulated data and its comparison with the real data is being done with EnsarRoot. Figures 2 and 3 show the geometry implementation of the setup detectors.

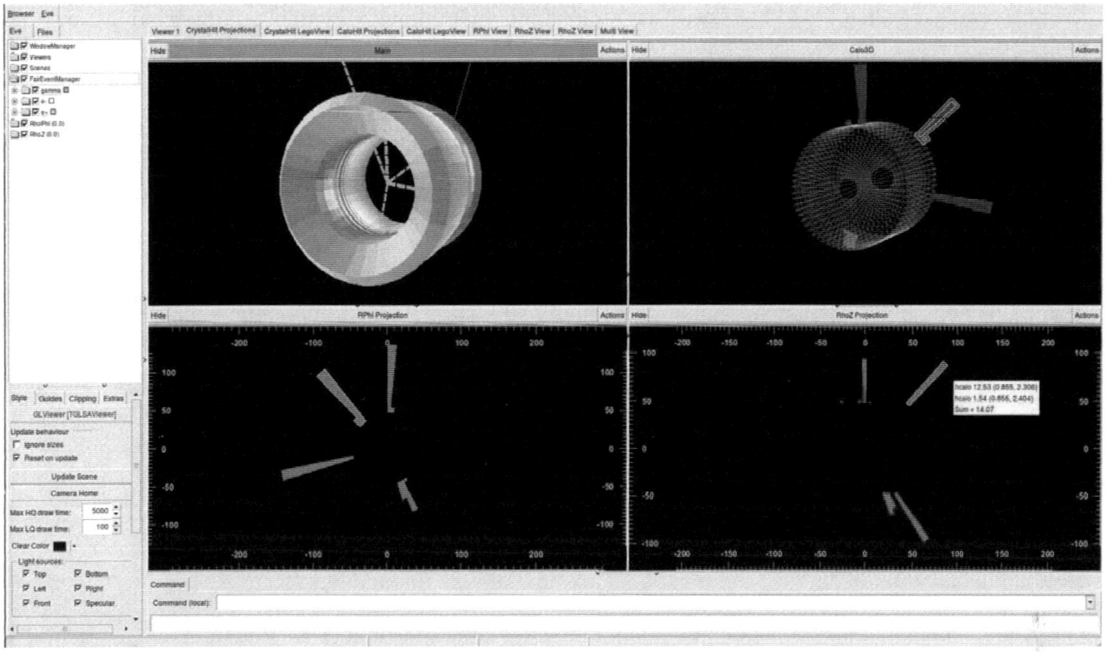

Figure 1. Example of a simulated event in a CsI calorimeter using the implemented event display of EnsarRoot

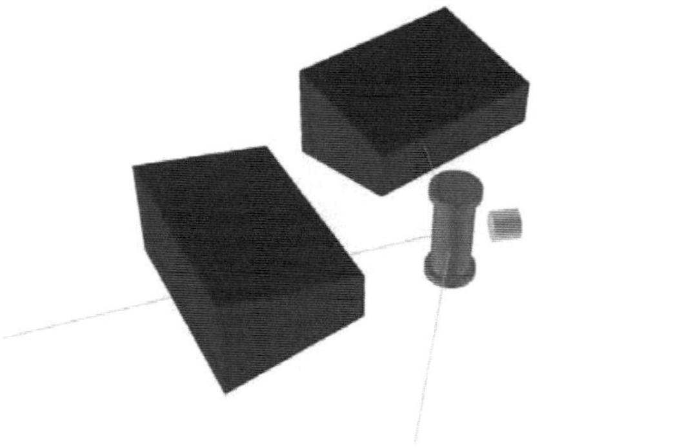

Figure 2. Geometry implementation of the setup for the The CALIFA Demonstrator Experiment, with the two segments of the demonstrator, the HPGe detector and a reaction chamber

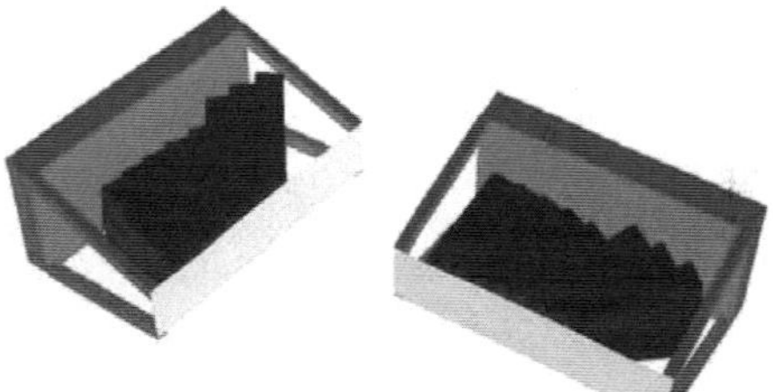

Figure 3. Detail of the inner part of the CALIFA Demonstrator segments. The crystals and the inner structure are visible

3.2. The TRAGALDABAS Detector

The TRAGALDABAS Detector [15] is a cosmic ray telescope based on Resistive Plate Chambers located at the Facultade de Física of the University of Santiago de Compostela, Spain. It is devoted to the understanding of the atmospheric showers arriving to the Earth surface. The system was implemented in the framework in the module *tragaldabas*, and all the simulations, performance analysis, efficiency and acceptance studies are being done with EnsarRoot. A picture of the system is shown in Fig. 4, and Fig. 5 shows a simulated event where the cosmic rays are generated with CRY and are propagated with the `EnsarCryAsciiGenerator` interface.

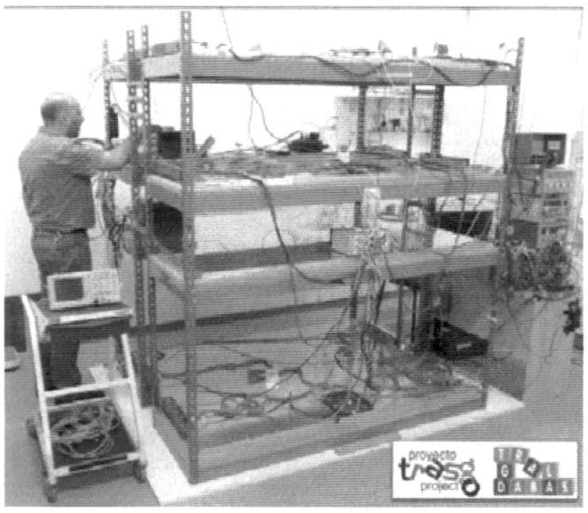

Figure 4. Picture of the TRAGALDABAS Detector System based on four planes of Resistive Plate Chamber detectors

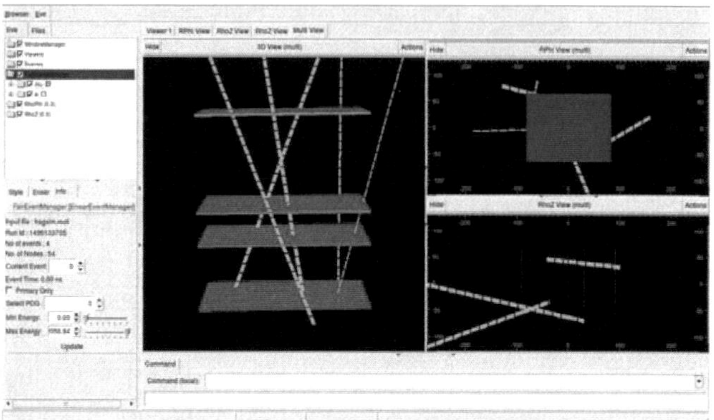

Figure 5. Example of a simulated air shower event over the TRAGALDABAS Detector in EnsarRoot

4. Summary

EnsarRoot, a general purpose simulation and event reconstruction framework for nuclear experiments has been developed. It is based on the ROOT package and the FairRoot library. It provides a set of classes and templates which allows the user to implement its own simulation in an easy way. It uses the Virtual Monte Carlo concept for the transport engines and implements several dedicated event generators. Input and output data for analysis is arranged in ROOT data structures. Different systems and experiments were already implemented in EnsarRoot and are delivered with the package as examples.

Acknowledgments

This project has received funding from the European Union Horizon 2020 research and innovation programme under grant agreement No 654002 (ENSAR2), and from Plan Galego de Investigación, Innovación e Crecemento (I2C) of Xunta de Galicia under project No. POS-B/2016/015. The authors also thanks Enrique Nacher for the development of the Giant and Pygmy Dipole Resonance generator.

References

[1] European nuclear science and applications research - 2 URL `http://www.ensarfp7.eu/`
[2] Bertini D *et al.* 2008 *J. Phys.: Conf. Ser.* **119** 032011
[3] Brun R and Rademakers F 1997 *Nucl. Inst. and Meth. in Physics Research* A **389** 81
[4] Hrivnacova I *et al.* 2003 (*Preprint* `cs.SE/0306005`)
[5] URL `https://github.com/EnsarRootGroup/EnsarRoot`
[6] URL `https://opensource.org/licenses/lgpl-license`
[7] URL `https://cmake.org/`
[8] Brun R *et al.* 1987 GEANT3 CERN-DD-EE-84-1
[9] Agostinelli S *et al.* 2003 *Nucl. Inst. and Meth. in Physics Research* A **506** 250
[10] C Hagmann D L and Wright D 2007 *IEEE Nucl. Sci. Conf. Record* **24** 36
[11] URL `https://root.cern.ch/eve`
[12] Cortina D *et al.* 2014 *Nuclear Data Sheets* **120** 99
[13] URL `http://www.gsi.de/r3b`
[14] Cabanelas P *et al.* 2017 *GSI Scientific Report 2016* **RESEARCH-NUSTAR-KR-9** 219
[15] Garzón J A *et al.* 2017 (*Preprint* `astro-ph.HE/1701.07277`)

IOP Conf. Series: Journal of Physics: Conf. Series **1024** (2018) 012039 doi:10.1088/1742-6596/1024/1/012039

Properties of open and hidden charm mesons in light quark matter

Martin Cleven, Volodymyr K. Magas, and Angels Ramos

E-mail: cleven@fqa.ub.edu,vladimir@fqa.ub.edu,ramos@fqa.ub.edu
Departament de Fisica Quantica i Astrofisica and Institut de Ciencies del Cosmos
Universitat de Barcelona, 08028-Barcelona, Spain

Abstract. In this work we study the implications of light-quark pionic matter at finite temperatures on the properties of open and hidden charm mesons. The meson-meson interactions are described by means of a chiral unitary approach accounting for coupled channels and the finite temperature effects are included using the Imaginary Time Formalism. The key observables presented here are the spectral function and the pion-induced width at finite temperatures for the mesons D, D^* and most importantly J/ψ.

1. Introduction

With a new generation of heavy-ion-collision experiments operational or under construction – LHC, FAIR or NICA– there is a demand for theoretical predictions of hadronic properties at temperatures and densities far from standard nuclear physics scenarios. Among the various subjects, the study of the J/ψ stands out as a candidate to signal deconfinement, according to the Matsui and Satz prediction [1]. A quark-gluon plasma (QGP) produced in the collision would screen the $c\bar{c}$ interaction or ionize the charmonium state. Either way, this would lead to a suppression of events.

However, although in previous experiments such a drop was actually seen [2], it still remains unclear whether this is indeed related to the formation of a QGP. Another explanation, accounted for in the 'co-mover' models, is that inelastic interactions of the J/ψ with the surrounding hadronic medium are responsible for the reduction in J/ψ detection and hence no QGP is necessary to explain it.

The inelastic interactions of the J/ψ with the surrounding hadronic medium, accounted for in the 'co-mover' models, offer alternative mechanisms to explain the drop in the J/ψ production that either do not involve the transition to QGP state or, at least, reduce the number of J/ψ during the evolution in the hadronic phase of reaction, and thus have to be taken into account.

Obtaining the properties of the D and D^* mesons in hot pionic matter, which we consider to be the first level approximation of the matter generated in ultrarelativistic heavy ion collision, will be one of the main topics of the present study. While interesting in itself, the study of these mesons is also closely related to that of the J/ψ through coupled-channel effects. Therefore, the second focus will be on studying the properties of the hidden charm J/ψ meson under the same conditions. Such a combined effort will provide a comprehensive understanding of how the properties of open and hidden charm mesons are modified when interacting with pionic matter at non-vanishing temperatures.

IOP Conf. Series: Journal of Physics: Conf. Series **1024** (2018) 012039 doi:10.1088/1742-6596/1024/1/012039

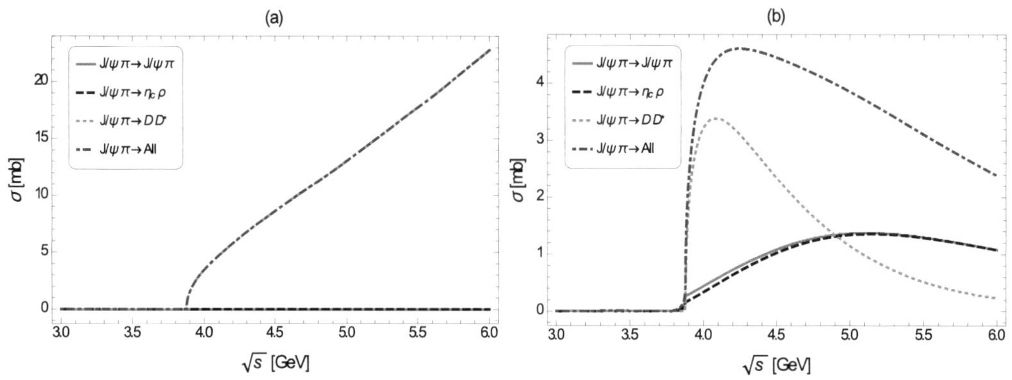

Figure 1. Cross Sections for $J/\psi\pi \to X$ employing the non-unitarized amplitudes (left panel) and the unitarized ones (right panel)

2. Unitarized amplitudes in the vacuum

In this section we will briefly recall the main features of meson-meson scattering in a chiral $SU(4)$ as used in Ref. [3]. For further reading see *e.g.* Refs. [4, 5].

Throughout this work potentials for pseudoscalar-pseudoscalar and pseudoscalar-vector scattering will be derived from the Lagrangian

$$\mathcal{L}_{\mathrm{PPPP}} = \frac{1}{12f^2}\left\langle J^\mu J_\mu + \Phi^4 M \right\rangle \qquad \mathcal{L}_{\mathrm{VPVP}} = -\frac{1}{4f^2}\left\langle J^\mu \mathcal{J}_\mu \right\rangle \ , \tag{1}$$

where the vector currents for pseudoscalar and vector mesons are given by

$$J_\mu = (\partial_\mu \Phi)\Phi - \Phi(\partial_\mu \Phi), \qquad \mathcal{J}_\mu = (\partial_\mu \mathcal{V}_\nu)\mathcal{V}^\nu - \mathcal{V}_\nu(\partial_\mu \mathcal{V}^\nu), \tag{2}$$

respectively with the $SU(4)$ pseudoscalar and vector multiplets Φ and $\mathcal{V}_\mu$, respectively. This allows us to calculate the Weinberg-Tomozawa-like potentials for the coupled channel interactions $D\pi$, $D\eta$ and $D^*\pi$, $D^*\eta$ in the open-charm and $J/\psi\pi, \eta_c\rho, D\bar{D}^* + c.c.$ in the hidden-charm sector. For a detailed discussion of the amplitudes we refer to Ref. [3]. At this point we only state the most striking feature in the hidden-charm sector. All interactions that do not contain the charmed meson pair vanish at the order we are studying including $J/\psi\pi \to J/\psi\pi$. This has important implications as will be discussed further down the line. The $S-$wave projections of these amplitudes V can be used as the kernel for the Lippmann-Schwinger Equation. Using the on-shell formalism this simplifies to a simple algebraic equation that can easily be solved as

$$T = (1 - VG)^{-1} V \vec{\epsilon} \cdot \vec{\epsilon}' \ , \tag{3}$$

for the scattering of vector mesons off pseudoscalar ones, while for the purely pseudoscalar case one simply needs to remove the polarization vectors from the previous equation. The diagonal matrix G contains the two-meson loops calculated using dimensional regularization.

In the following we calculate the cross sections for $J/\psi\pi$ scattering into different final states, elastic and inelastic, using the theoretical framework introduced before. This is instructive for several reasons. First of all, it gives us an idea about the order of magnitude for the interactions of the J/ψ in a pionic medium. The inelastic cross section in particular allows us to gauge the hadronic dissociation of the J/ψ in such a scenario.

Fig. 1 shows the cross sections for the processes $J/\psi\pi \to J/\psi\pi$, $J/\psi\pi \to \eta_c\rho$ and $J/\psi\pi \to DD^*$. The left panel displays the cross sections using non-unitarized amplitudes, the right panel shows the cross section using the unitarized amplitudes according to Eq. (3). When comparing these two we see several noteworthy differences.

First of all, as a result of vanishing tree-level amplitudes for $J/\psi\pi \to J/\psi\pi$ and $\eta_c\rho$ the corresponding cross sections vanish as well. However, once the amplitude is unitarized, these processes can occur through charmed meson loops which give rise to a cross section in the range of 1 mb. This will also become an important feature once we consider in-medium effects as we will see in the following chapter. Notice also that unitarization changes the high-energy behaviour of the cross section from the linear rise to one that drops after peaking around 4 GeV.

3. Interactions at finite temperatures

In the following we will provide an overview of how the meson-meson interactions are modified by a surrounding pion medium at finite temperatures. For brevity we will focuss on the main results and refer to Ref. [3] and references therein for more details. The modifications will be two-fold. First, we will use the Imaginary Time Formalism (ITF) to incorporate the impact of finite temperatures. This will be done by modifying the propagators and through the loops also the unitarized amplitudes. Secondly, we will dress the propagators for the mesons M with the self energy obtained from closing the pion line in the $M\pi \to M\pi$ scattering amplitude. This way the dressed propagators account for the pion bath. As these dressed propagators affect the amplitudes, the process needs to be iterated several times to be self-consistent.

In ITF the zeroth components of the four-momenta are modified using $q_0 \to \mathrm{i}\omega_n = \mathrm{i}2\pi nT$ with the discrete Matsubara frequencies $\omega_n = \mathrm{i}2\pi nT$ and replacing the zeroth component of the loop integrals with a sum over these frequencies. As mentioned above, in IFT the interactions remain unchanged and only the propagators are modified. The meson propagator becomes

$$D_M(\omega_n, \vec{q}; T) = [(\mathrm{i}\omega_n)^2 - \vec{q}^{\,2} - m_M^2 - \Pi_M(\omega_n, \vec{q}; T)]^{-1} \, . \tag{4}$$

Once the sums over the Matsubara frequencies are performed one may analytically continue the result to external frequencies $\omega + \mathrm{i}\varepsilon$. In the case of the two-meson loop at finite temperature we find

$$G_{MM'}(p^0, \vec{p}; T) = \int \frac{\mathrm{d}^3 q}{(2\pi)^3} \int \mathrm{d}\omega \int \mathrm{d}\Omega \frac{S_M(\omega, \vec{q}; T) S_{M'}(\Omega, \vec{p} - \vec{q}; T)}{p^0 - \omega - \Omega + \mathrm{i}\varepsilon} [1 + f(\omega, T) + f(\Omega, T)] \, . \tag{5}$$

Here $f(\omega, T) = [\exp(\omega/T) - 1]^{-1}$ is the meson Bose distribution function at temperature T and $S_M(\omega_n, \vec{q}; T) = -(1/\pi)\mathrm{Im}(D_M(\omega_n, \vec{q}; T))$ is the spectral function of the meson M. For details of the regularization and a more detailed discussion of how the finite temperatures affect the individual loops and consequently through unitarization the amplitudes see Ref. [3].

The second important temperature-dependent quantity is the self energy of a meson M obtained from closing the pion line in the $M\pi \to M\pi$ T-matrix element $T_{M\pi}$. After applying the ITF Feynman rules and carrying out the Matsubara sum we find for real energies p^0

$$\Pi_M(p^0, \vec{p}) = \int \frac{\mathrm{d}^3 q}{(2\pi)^3} \int \mathrm{d}\Omega \frac{f(\Omega, T) - f(\omega_\pi, T)}{(p^0)^2 - (\omega_\pi - \Omega)^2 + \mathrm{i}\varepsilon} \left(-\frac{1}{\pi}\right) \mathrm{Im} T_{M\pi}(\Omega, \vec{p} + \vec{q}; T). \tag{6}$$

In Fig. 2 we show the imaginary parts of the self energies (upper panel) of D, D^* and J/ψ resulting from their interaction with the hot pion bath and the resulting spectral functions (lower panel) as a function of the energy at increasing temperatures $T = 50, 100, 150$ MeV. Naturally, the behaviour of the spectral functions mirrors the behaviour of the self energy. In the case of the charmed mesons we see how the strength spreads considerably when increasing the temperature from 50 to 150 MeV. In other words, the spectral function of the charmed mesons broadens significantly in a finite-temperature pion bath.

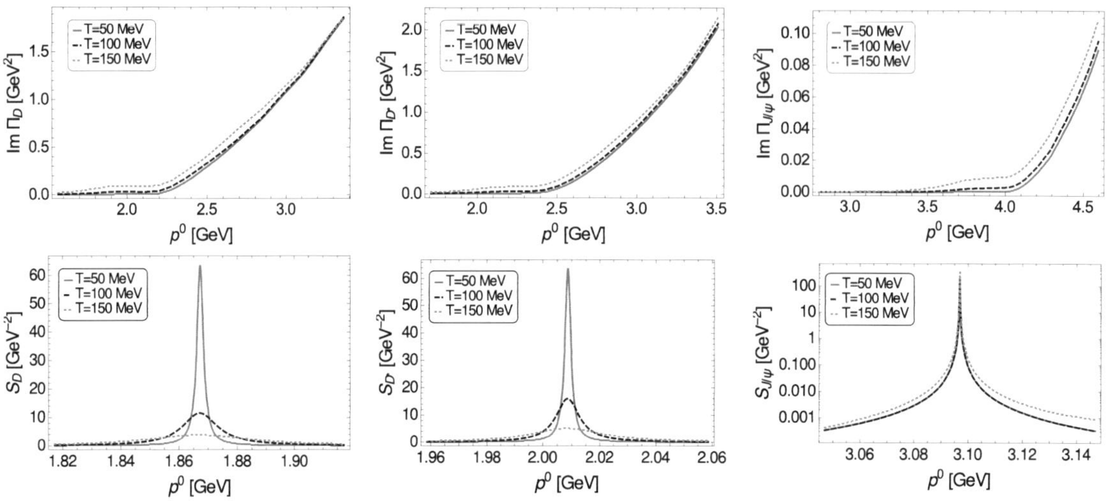

Figure 2. Upper panel: Imaginary part of the pion-induced self energies for D, D^* and J/ψ at temperatures $T = 50, 100, 150$ MeV.
Lower panel: Spectral functions of D, D^* and J/ψ at temperatures $T = 50, 100, 150$ MeV. Notice that for D and D^* we use a linear, for J/ψ a logarithmic plot.

The situation is obviously different for the J/ψ spectral function. Notice that in this case we have also included the J/ψ vacuum width of $\Gamma^{\rm Vac}_{J/\psi} = (93 \pm 3)$ keV [6]. On one hand this makes the numerical determination of the spectral function easier, on the other hand it gives an intuitive measure of the impact of the pion-induced width. The large self-energy for the charmed mesons made this incorporation irrelevant there.

The J/ψ spectral functions at the temperatures $T = 50$ MeV and $T = 100$ MeV are indistinguishable from the vacuum one. One needs to go to temperatures as large as $T = 150$ MeV to start seing noticeable broadening of the J/ψ spectral function compared to its free case shape. The reason for this is basically two-fold. First, all interactions that contribute to the J/ψ are weak which leads to a lower self energy. Secondly, due to the absent leading interaction for $J/\psi\pi$ scattering this process has to go through a charmed meson pair. This, however, corresponds to a significantly larger threshold and has therefore less impact at energies close to the mass of the J/ψ.

While a broadening spectral function for $T = 150$ MeV can be seen from our results, it still remains similar to the vacuum case. From this we can conclude that the J/ψ retains most of its shape in hot pionic matter. A suppession in the detection of J/ψ mesons in a hot pionic medium should therefore not be attributed to the interaction with the surrounding hadrons. Thus, if observed, such a drop might indeed point towards the detection of a QGP.

To demonstrate the impact of increasing temperatures we use a more illustrative quantity, the width of the mesons M obtained from the surrounding pions

$$\Gamma_M = {\rm Im}\Pi_M(p^0 = m_M, \vec{p} = 0)/m_M. \tag{7}$$

Fig. 3 shows the temperature behaviour of the three mesons in question, D, D^* and J/ψ. Let us discuss the final results which correspond to the last iteration at each sub-figure of Fig. 3, namely $n = 3$ for the charmed meson case and $n = 4$ for J/ψ (see Ref. [3] for more details on iterations). In the case of the charmed mesons we see a slow rise in the single MeV region for

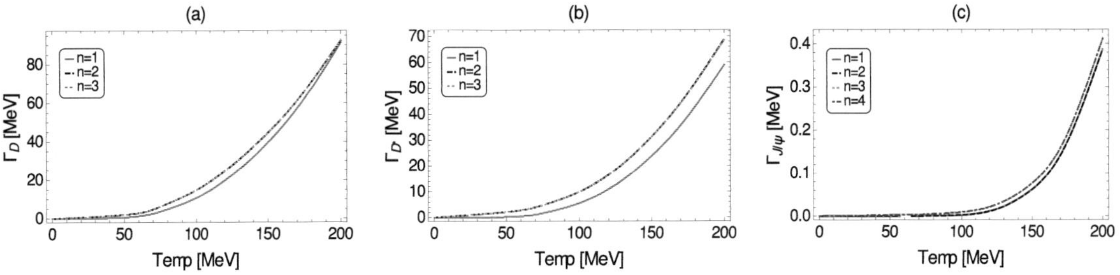

Figure 3. We show the pion-induced width as a function of the temperature. To illustrate the process of achieving self consistent results we show this for different iterations n.

temperatures below 100 MeV and a sharper rise beyond that up to around $70 - 90$ MeV for temperatures approaching 200 MeV. Compared to the hadronic width of the D^* in the vacuum of the order of 100 keV this means an increase of 2-3 orders of magnitude depending on the temperature.

The general behaviour is similar for the J/ψ albeit suppressed by 2-3 orders of magnitude. At temperatures around 150 MeV the pion induced width exceeds the vacuum width $\Gamma^{\text{Vac}}_{J/\psi} = (93 \pm 3)$ keV [6].

4. Summary

We have calculated the behaviour of the charmed mesons and the J/ψ in a pion bath at finite temperatures using unitarized $SU(4)$ chiral amplitudes. Using the ITF formalism we obtained observables that are relevant for understanding the behaviour of mesons containing charm quarks in hot mesonic matter, most notably the induced width and spectral function.

We find that the charmed D and D^* mesons acquire a substantial width, reaching values in the $30 - 40$ MeV range at $T = 150$.

The modification of the J/ψ meson properties by a hot pion bath is substantially more moderate, even at temperatures of around $T = 150$ MeV the J/ψ retains most of its shape in hot pionic matter. Consequently, an observed drop in J/ψ detection should indeed point towards the formation of a QGP, the interaction with the surrounding hadrons being essentially negligible.

Our results can be used as input in theoretical simulations of open and hidden charm meson propagation in a hot pion gas, aiming at understanding the transition from a hadronic to a QGP phase.

Acknowledgements: This work is partly supported by the Spanish Ministerio de Economia y Competitividad (MINECO) under the project MDM-2014-0369 of ICCUB (Unidad de Excelencia 'María de Maeztu'), and, with additional European FEDER funds, under the contract FIS2014-54762-P and the Spanish Excellence Network on Hadronic Physics FIS2014-57026-REDT and from the Generalitat de Catalunya contract 2014SGR-401.

References

[1] Matsui T and Satz H, *Phys. Lett.* B **178** (1986) 416.
[2] Abreu M C *et al.* [NA50 Collaboration], *Phys. Lett.* B **450**, 456 (1999); *Phys. Lett.* B **477**, 28 (2000).
[3] Cleven M, Magas V K and Ramos A, *Phys. Rev.* C **96** (2017) 4, 045201 .
[4] Gamermann D, Oset E, Strottman D and Vicente Vacas M J, *Phys. Rev.* D **76** (2007) 074016.
[5] Gamermann D and Oset E, *Eur. Phys. J.* A **33** (2007) 119.
[6] Patrignani C *et al.* [Particle Data Group], *Chin. Phys.* C **40**, no. 10, 100001 (2016).

Construction of Multi Wire Proportional Chambers for the CBM Transition Radiation Detector

Florian Roether
Goethe-University Frankfurt, Max-von-Laue-Str. 1, 60438 Frankfurt am Main, Germany
E-mail: roether@ikf.uni-frankfurt.de

Abstract. The Compressed Baryonic Matter (CBM) experiment at the future Facility for Antiproton and Ion Research (FAIR) will explore the QCD phase diagram in the region of high net-baryon densities. The Transition Radiation Detector (TRD) with its multi-layer design will provide electron identification for higher momenta and contribute to the fragment identification.

1. Introduction

The main aim of the Compressed Baryonic Matter experiment is the exploration of the QCD phase-diagram in the region of high net-baryon densities by observing high energy nucleus-nucleus collisions. The high event rate of up to 10^7 Hz grants access to rare probes like multi-strange hyperons, hypernuclei and vector mesons. To identify those rare observables an efficient

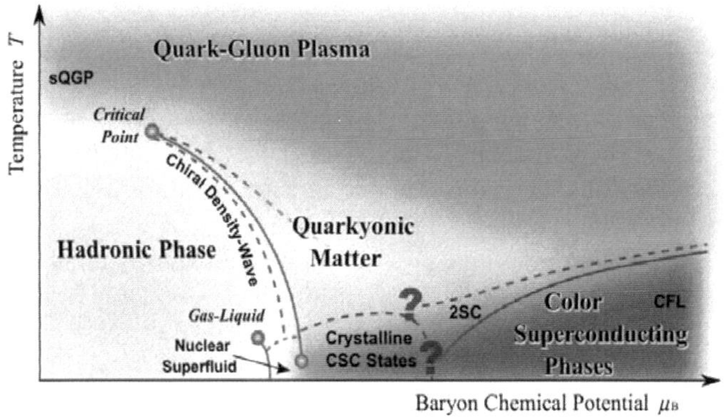

Figure 1. Sketch of the phase diagram containing different predicted phase boundaries and states of matter. [2]

background suppression is needed. Along with the high interaction rate this imposes high demands on the participating detectors in terms of precision, rate capability and radiation hardness. At CBM the task of a precise electron identification will be fulfilled by the Ring

IOP Conf. Series: Journal of Physics: Conf. Series **1024** (2018) 012040 doi:10.1088/1742-6596/1024/1/012040

Imaging Cherenkov Detector (RICH) and the Transition Radiation Detector (TRD). While the RICH is effective for momenta below 6 GeV/c the TRD is the only detector capable of providing electron identification above 7 GeV/c. In addition the TRD can provide fragment identification by measuring the specific energy loss [2]. The overall design of the TRD at SIS100 consist of

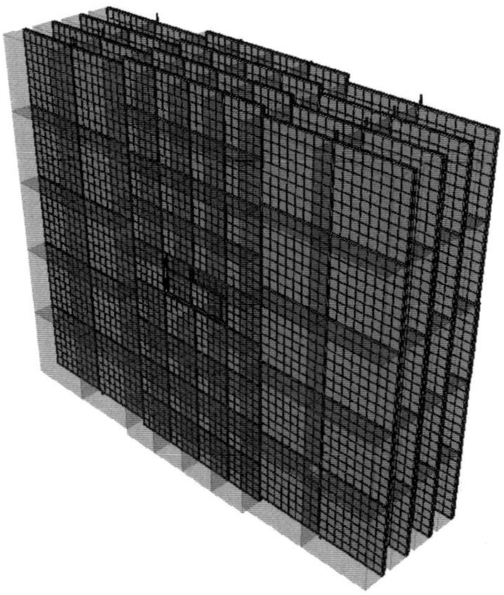

Figure 2. Front side of the TRD with the radiators shown in blue.f [2]

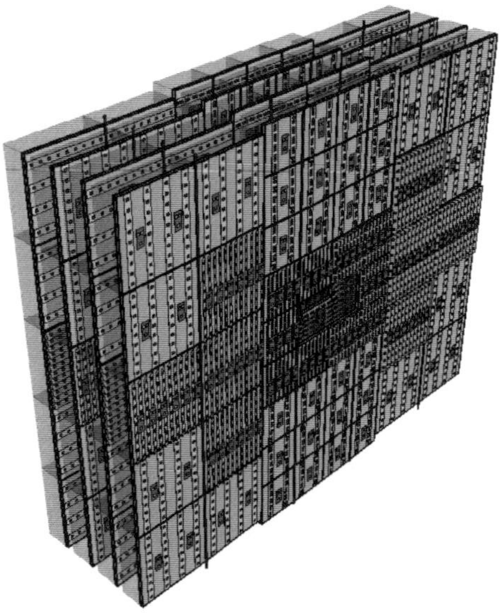

Figure 3. Back side of the TRD. The FEBs are shown in yellow. [2]

one station with four layers. Each layer is made of 20 large chambers measuring $99 \times 99\,\mathrm{cm}^2$ and 34 smaller chambers with outer dimensions of $57 \times 57\,\mathrm{cm}^2$.

2. Transition Radiation

Transition radiation occurs when a charged particle traverses the interface of two media with different dielectric constants [1]. The radiated spectrum depends on the Lorentz factor γ and can be described for relativistic particles by the following expression:

$$\frac{\mathrm{d}^2 W}{\mathrm{d}\omega \mathrm{d}\Omega} = \frac{\alpha}{\pi^2} \left(\frac{\theta}{\gamma^{-2} + \theta^2 + \xi_1^2} - \frac{\theta}{\gamma^{-2} + \theta^2 + \xi_2^2} \right)^2 \tag{1}$$

with $\xi_i^2 = \omega_{Pi}/\omega$, where ω_{Pi} is the plasma frequency of the foil and the medium in between, α the fine structure constant and θ the emission angle. As the yield of a single interface is far too low, radiators are constructed such that a passing particle crosses many interfaces. This can be implemented for example as a stack of foils. As shown in Fig.4 the radiated spectrum depends on the configuration of the radiator and the Lorentz factor of the passing particle. This enables the TRD to distinguish between electrons with a high Lorentz factor and particles with a low Lorentz factor such as pions. While the produced spectrum of regular foil radiators can be well described theoretically, making them a good reference, their construction is time-consuming and unpractical. Furthermore, the radiator needs an outer frame to hold the foils in place. Irregular radiators made from fiber or foam material have been proven to be a good self-sustaining alternative and have been tested at test beam campaigns. As a candidate for the final experiment a foam radiator was chosen which showed a performance close to a foil reference radiator.

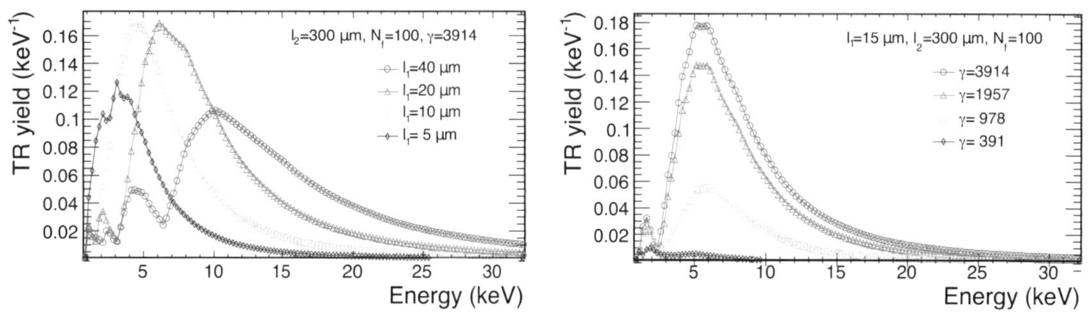

Figure 4. Left: Radiated spectrum produced by a stack of 100 foils with a gap of 300 μm as a function of the foil thickness. Right: Radiated spectrum as a function of the Lorentz factor. [1]

3. Multi Wire Proportional Chamber

To detect the transition radiation a gas detector is used. Gas detectors are a cost-efficient solution to cover a large area while keeping the material budget low. As stated by their name, the active area of these kind of detectors consists of a gas filled volume. Inside an electrical field is created by several electrodes. Passing charged particles ionize the gas and the produced ion-electron pairs get separated by the electrical field. In a multi wire proportional chamber (MWPC) the electrical field is generated by a plane of anode wires and two cathode planes. The primary electrons drift towards the anode wires. The increasing field around the anode wires accelerates the primary electrons to the point where they can also ionize the gas. This effect is called the gas amplification and depends on the configuration of the electrical field. The produced electrical charge can be read out either directly through the anode wires or by the charge which is induced on the cathode planes. By separating the cathode plane into individually read out pads it is possible to get an additional position information. The distribution of the induced charge on the pads is described by the pad response function and depends on the pad width and the distance to the anode wires. Different designs of MWPCs have been tested. To

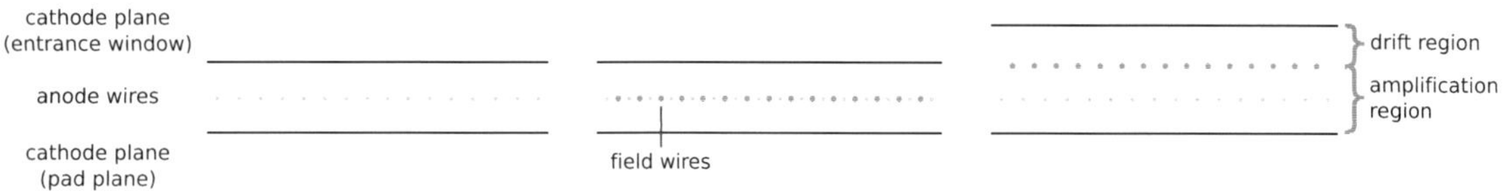

Figure 5. Different types of MWPCs with the classical type on the left, a configuration with alternating wires in the middle, and a design with an additional drift region on the right.

minimize the absorption of the TR-photons before entering the gas volume, one of the cathode planes is implemented as a thin aluminized Mylar or Capton foil, which also serves as the entrance window. The signal collection time arises from the ion-movment. While a classical MWPC (Fig.5 left) is by far the fastest design, its gas amplification correlates strongly with the uniformity of the cathode planes. As the entrance window is made from a thin foil, even moderate pressure differences between the inside of the detector and the environment can cause a deformation which has a huge effect of the electrical field and therefore the gain. One possible

IOP Conf. Series: Journal of Physics: Conf. Series **1024** (2018) 012040 doi:10.1088/1742-6596/1024/1/012040

countermeasure is to place additional ground wires between the anode wires. These so-called field wires alter the electrical field and stabilize the gas gain. For the final design the entrance window is stabilized with a lattice grid and an additional drift region is added. This drift region is separated by the amplification through a cathode wire plane and decouples the electrical field from the entrance window. The thicker gas volume increases also the absorption probability of TR-photons. Even through this design is slower than the ones mentioned before, the signal collection time is within the design goal of 0.3 µs. To achieve the required position resolution in the order of 300 µm, the width of the pads and the distance to the anode wires is tuned in a way that the most central pads get 80% of the charge. This ensures a good position and energy resolution.

4. Front End Electronic

As the CBM experiment does not have a global trigger, a self-triggered front end electronic is used to read out the detector. The front-end boards (FEBs) are mounted directly on the back of each chamber (see Fig.3) and are equipped with special designed readout ASICs, the socalled SPADICs. Each SPADIC has 32 analog input channels [2]. The incoming signals are going first

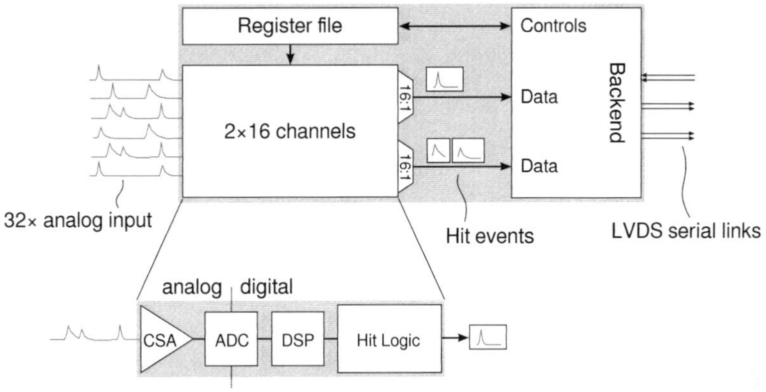

Figure 6. Block diagram of the Self-triggered Pulse Amplification and Digitization ASIC (SPADIC)

through a charge sensitive amplifier (CSA) before they are digitized by a continuously sampling 9-bit ADC. In the digital part a programmable digital filter can be used for tail cancelation and signal shaping. The hit detection logic can trigger on a configurable threshold or the slope of the signal. Furthermore the SPADIC is capable of triggering neighboring channels even if the signals on these pads do not meet the criteria to trigger the readout them self. This ensures that also the induced charge on neighboring pads is registered. Each triggered channel produces a hit message which includes among other information the digitized signal and a timestamp. This information is transferred via low-voltage differential signaling links (LVDS) to the Read Out Board (ROB) where the data transmitted via optical fibers to the entry nodes.

5. Electron identification

In order to identify particles, individual hit messages are getting combined by the cluster finder to individual hits and the information of the position and the total deposited energy gets extracted. The average energy loss for electrons is increased due to the additional energy deposition by the absorbed TR-photon, which causes a second maximum in the electron spectrum (See Fig.7). The separation of pions and electrons for tracks in the TRD will be done by a likelihood algorithm.

IOP Conf. Series: Journal of Physics: Conf. Series **1024** (2018) 012040 doi:10.1088/1742-6596/1024/1/012040

For the likelihood method a clean spectrum of electrons and pions is needed as an input. The likelihood for a given track to be an electron (or a pion) is defined as:

$$L = \frac{P_e}{P_e + P_\pi}, \qquad P_e = \prod_{i=1}^{N} P(E_i|e), \qquad P_\pi = \prod_{i=1}^{N} P(E_i|\pi) \tag{2}$$

with $P(E_i|e)$ and $P(E_i|\pi)$ as the probability that the energy deposit in layer i is caused by an electron or a pion, respectively. This likelihood for electrons and pions for a track with hits in at least three layers is shown in Figure 8. For a given electron efficiency, the fraction of misidentified pions can be directly derived by this method. The performance of a TRD can be determined by this pion efficiency. The design goal of this detector is to achieve a pion misidentification of less than 5% at an electron efficiency of 90%. Simulations based on data acquired at test beam campaigns have shown that this can be achieved with four layers.

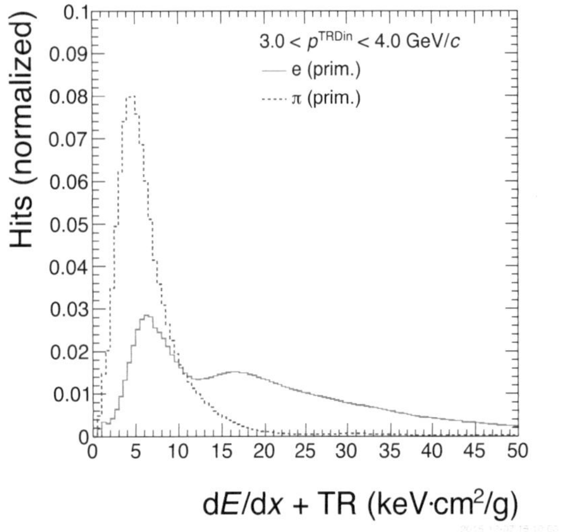

Figure 7. The simulated average energy loss in a single chamber for electrons and pions.

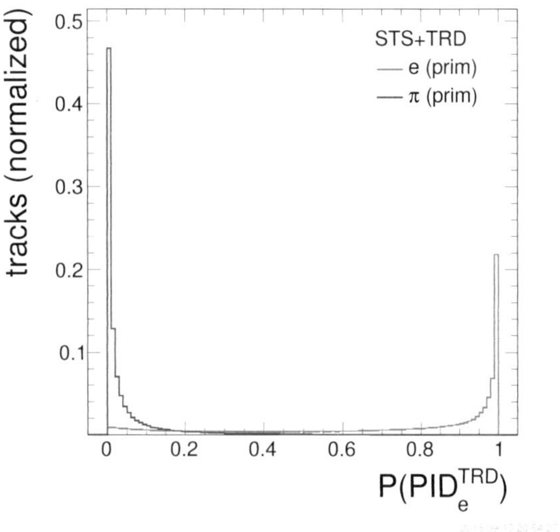

Figure 8. Likelihood for a given track to be an electron (red) or pion (blue).

6. Conclusion

Even though many similar TRDs have been build, each new experiment provides new requirements and challenges. The CBM experiment with its unprecedented interaction rate places enormous demands on the rate capability of all detectors participating at the experiment. After smaller prototype chambers have been successfully tested, new full size detectors have been build and are already being tested.

References

[1] A. Andronic and J. P. Wessels. Transition Radiation Detectors. Nucl. Instrum. Meth., A666:130-147, 2012.
[2] Technical Design Report of the TRD (to be published)

IOP Conf. Series: Journal of Physics: Conf. Series **1024** (2018) 012041 doi:10.1088/1742-6596/1024/1/012041

Polarization reconstruction algorithm for a Compton polarimeter

M Vockert[1,2], G Weber[2], U Spillmann[3], T Krings[4] and Th Stöhlker[1,2,3]

[1] Institute for Optics and Quantum Electronics, Friedrich Schiller University Jena, Max-Wien-Platz 1, 07743 Jena, Germany

[2] Helmholtz Institute Jena, Fröbelstieg 3, 07743 Jena, Germany

[3] GSI Helmholtzzentrum für Schwerionenforschung GmbH, Planckstraße 1, 64291 Darmstadt, Germany

[4] Institut für Kernphysik, Forschungszentrum Jülich, Wilhelm-Johnen-Straße, 52425 Jülich, Germany

E-mail: Marco.Vockert@uni-jena.de

Abstract. We present the technique of Compton polarimetry using X-ray detectors based on double-sided segmented semiconductor crystals that were developed within the SPARC collaboration. In addition, we discuss the polarization reconstruction algorithm with particular emphasis on systematic deviations between the observed detector response and our model function for the Compton scattering distribution inside the detector.

1. Introduction

The study of particle and photon polarization phenomena occurring in the interaction of fast ion and electron beams with matter is of particular relevance for the understanding of cosmic and laboratory plasmas, where high temperatures, high atomic charge states and high field strengths prevail. In addition, polarization-sensitive studies of radiation emitted by processes in highly-charged, heavy ions may provide detailed insights into both relativistic particle dynamics and atomic structure properties at extreme electromagnetic field strengths [1, 2]. In recent years, polarimetry of hard X-rays was also proven to be a unique tool to study subtle spin-dependent effects in the realm of atomic physics at high field strengths [3, 4]. With high-precision X-ray polarimetry even quantum electrodynamic and parity nonconservation effects may become accessible [5, 6]. Moreover, X-ray polarimetry is proposed as a tool for the diagnosis of spin-polarized ion beams for future experiments at the Facility for Antiproton and Ion Research (FAIR) [7].

A significant part of the recent hard X-ray polarization measurements were performed within the Stored Particle Atomic Research Collaboration (SPARC) [8] using polarimeters based on double-sided segmented semiconductor crystals. These devices employ a detector technology that has been continuously improved over the last two decades [9, 10, 11]. In this report we will briefly present the technique of polarization measurements with such detectors and then focus on the present status of the analysis of possible systematic uncertainties introduced by our reconstruction method for the incident X-ray (linear) polarization.

IOP Conf. Series: Journal of Physics: Conf. Series **1024** (2018) 012041 doi:10.1088/1742-6596/1024/1/012041

2. Compton polarimetry

The angular differential cross section of the Compton scattering process is described by the Klein-Nishina equation (Eq. 1):

$$\frac{d\sigma}{d\Omega} = \frac{1}{2} r_0^2 \left(\frac{E'}{E}\right)^2 \left(\frac{E'}{E} + \frac{E}{E'} - 2\sin^2\theta\cos^2\varphi\right) \tag{1}$$

where E denotes the incident photon energy and $E'(\theta)$ the photon energy after the scattering with the corresponding polar scattering angle θ, while φ is the azimuthal scattering angle and r_0 being the classical electron radius. Note that the scattered photon is preferably emitted perpendicular to the incident photon electric field vector, whereas emission in the parallel direction is less probable. Thus, the degree of linear polarization as well as the orientation of the polarization axis of an incident photon beam can be obtained from the azimuthal emission pattern of Compton scattered photons. A detailed discussion of this technique can be found in Ref. [12].

Within the SPARC collaboration several double-sided segmented X-ray detectors were constructed as dedicated Compton polarimeters [10, 11, 13]. They provide good detection efficiency, energy and time resolution together with millimeter to submillimeter two-dimensional position resolution and a large detection area. When an incident photon undergoes Compton scattering inside such a detector, the scattered photon may be detected at a different position within the same detector crystal. Thus each segment of the detector crystal serves as both a scatterer and an absorber for the scattered X-rays, resulting in a high polarimeter efficiency due to a large active area and a good coverage of the azimuthal scattering angle. Combining energy, position and time information of each interaction then allows a reconstruction of the Compton scattering events [14]. A series of recent measurements demonstrated the detector ability for precise and efficient linear polarization studies [4, 15, 16, 17]. Depending on the crystal material and thickness, which can reach up to about 2 cm, polarimeters of the aforementioned type can be tailored to incident photon energies ranging from a few dozen keV up to the MeV regime.

In the following we will focus on data associated with a recently commissioned Si(Li) Compton polarimeter (Fig. 1 a). This instrument consists of a 9 mm thick lithium-drifted silicon crystal,

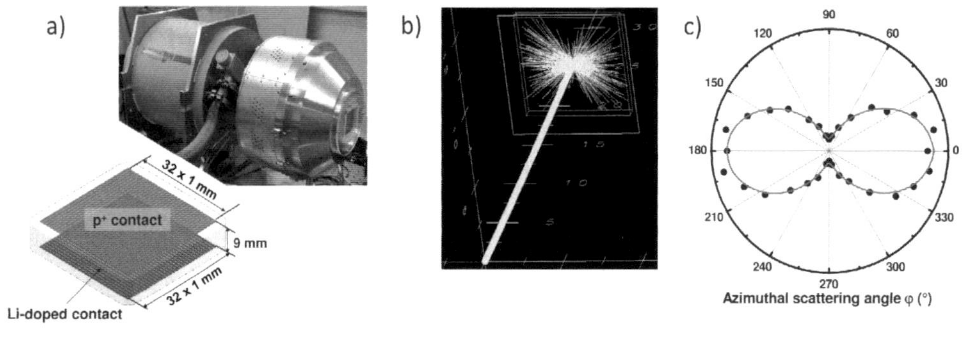

Figure 1. a) Photograph of a recently comissioned Compton polarimeter developed within the SPARC collaboration, together with a schematic drawing of the double-sided segmented detector crystal. b) MC simulation of Compton scattering of a 100 % polarized photon beam impinging on the detector crystal. c) Azimuthal scattering distribution for highly polarized K-REC photons generated at the ESR storage ring [13] (black dots) plotted together with the model curve (red line) shown in Eq. 2, being adjusted to the experimental data.

segmented on each side into 32 strips of 1 mm width each, and features an improved energy

IOP Conf. Series: Journal of Physics: Conf. Series **1024** (2018) 012041 doi:10.1088/1742-6596/1024/1/012041

resolution compared to polarimeters developed previously, see [13] for details. The Monte Carlo-(MC-)simulated scattering distribution of a 100 % polarized beam impinging on the detector crystal is illustrated in Fig. 1 b), exhibiting a pronounced anisotropy as expected from Eq. 1. Finally, Fig. 1 c) displays the azimuthal scattering distribution of nearly 100 % polarized X-ray radiation recorded in a commissioning experiment of the aforementioned Si(Li) polarimeter. These photons result from radiative electron capture (REC) into the projectile K shell (K-REC) during the collision of bare xenon ions at an energy of 30.93 MeV/u with hydrogen molecules at the internal gas target of the experimental storage ring (ESR) of the GSI facility at Darmstadt, Germany [18], and are known to exhibit a high degree of linear polarization [19, 20]. To extract the degree of linear polarization, a model function based on the Klein-Nishina equation is adjusted to the experimental data. This function is of the form

$$f(\varphi) = C\left(A - 2B\left(\frac{1}{2}(1 - \widetilde{P_L}) + \widetilde{P_L}\cos^2(\varphi + \varphi_0)\right)\right). \tag{2}$$

Therein the parameters A, B and the normalizing factor C are defined as:

$$A = \frac{1}{I}\sum_{\theta=\theta_0}^{\theta_1} I_\theta\left(\frac{E'(\theta)}{E} + \frac{E}{E'(\theta)}\right) \quad ; \quad B = \frac{1}{I}\sum_{\theta=\theta_0}^{\theta_1} I_\theta \sin^2\theta \quad ; \quad C = \frac{I}{N_B}\frac{1}{A - B}$$

wherein I_θ is the number of acquired Compton events at the scattering angle θ (being already corrected for geometric effects due to the angular binning and finite position resolution of the detector), I is the sum of all reconstructed Compton events over the interval $[\theta_0, \theta_1]$ (usually selected as $(90 \pm 15)^\circ$ [11]) and N_B is the number of angular bins. The energies E and $E'(\theta)$ are defined as in Eq. 1. The reconstructed degree of linear polarization $\widetilde{P_L}$ and the orientation of the polarization vector with respect to the detector axis φ_0 are free parameters that are adjusted to the experimental data by means of a least-squares optimization. Note that the degree of linear polarization of the incident radiation is obtained from $\widetilde{P_L}$ after correcting for the polarimeter quality and other effects as explained in Ref. [14]. However, this is beyond the scope of the present work.

3. Systematic deviation of detector response function and model function

When drawing conclusions from the output of the reconstruction algorithm, it is crucial to quantify to what extent the model reproduces the real detector response to incident polarized X-rays. This is of particular importance for precision experiments aiming for an accuracy of better than 1 % in the determination of the degree of polarization. In order to determine the level of statistical uncertainty below which systematic deviations between the detector response and our model function emerge as the dominant contribution to the overall uncertainty of the reconstructed polarization characteristics, we performed a series of MC simulations based on the EGS5 (Electron-Gamma Shower) package [21]. It was previously found that such simulations are able to reproduce all relevant detector features encountered in real-world measurements [14, 22]. As a test case we used the experimental setup and incident radiation characteristics of the recent commissioning measurement for the new Si(Li) polarimeter. From the measured data we obtained a preliminary value of the degree of linear polarization of (0.860 ± 0.008), based on about 17 000 reconstructed Compton events (without applying any corrections to account for polarimeter quality, misidentified Compton events, etc.). With our simulation routine we replicated the experimental conditions and used incident REC radiation data obtained from the REC calculator (RECAL) code [23, 24] to produce data sets with the number of reconstructed Compton events ranging from 10^3 to 10^6. Two of these data sets, illustrating the difference between low and high numbers of reconstructed Compton events, are presented in Fig. 2. Here the azimuthal scattering distribution is plotted for the simulated data together with the model function, shown in Eq. 2, which using the degree of linear polarization as a free parameter models these data sets. A significant deviation between the data set and our model is indicated

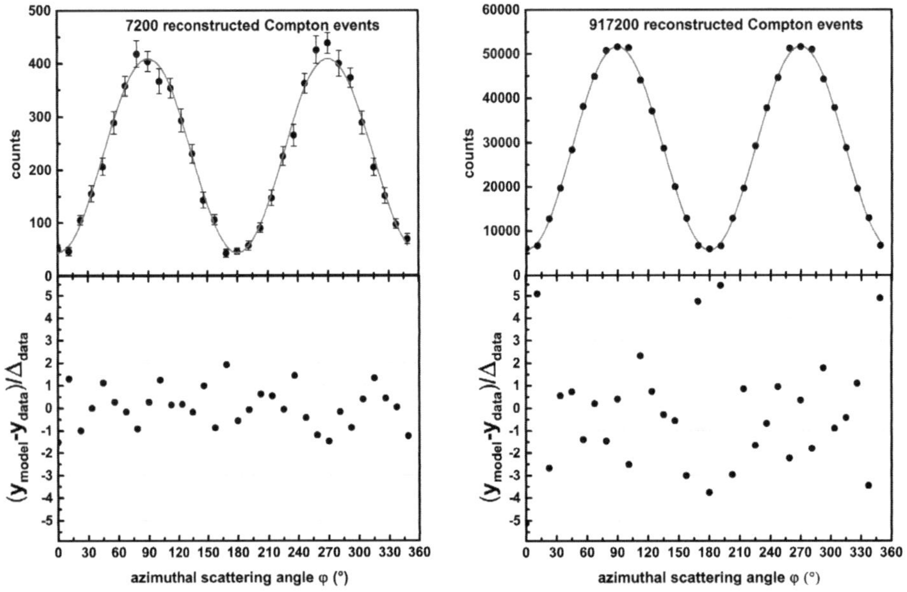

Figure 2. Top pair: Azimuthal scattering distribution (black dots) including error bars indicating the 1σ statistical uncertainty, together with fitted curve using Eq. 2 (red line) for low (left) and very high statistics (right). Bottom pair: Difference between obtained data points and the fitted curve, relative to the individual errors, for data sets with low (left) and very high statistics (right).

by a high number of data points deviating more than the standard deviation (being determined by the statistical uncertainty of the data) from the model. A systematic deviation is clearly found for the 'high statistics' data set shown on the right side of Fig. 2. To quantify this deviation for a whole data set we used the reduced chi-squared $\chi^2_{\text{red}} = \frac{\chi^2}{N}$ (χ^2 being the weighted sum of the squared difference between the data points and the associated fitted curve and N being the number of data points minus the number of free parameters), which is plotted as a function of the number of reconstructed Compton events in Fig. 3. As a rule of thumb, a χ^2_{red} value $\gg 1$ indicates a model that does not capture all significant features of a given data set. This is the case for the data sets consisting more than about 30 000 reconstructed Compton events. In this region the use of Eq. 2 to reconstruct the polarization of the incident radiation is questionable as the χ^2_{red} is rapidly increasing with decreasing statistical uncertainty of the data. Nonetheless, the amount of data obtained in our commissioning experiment is in the range of about 17 000 reconstructed Compton events and therefore still well within the reliability range of our model. The reconstructed degree of linear polarization is also presented in Fig. 3, with error bars corresponding to the statistical uncertainty. It is found that, while all polarization values are in agreement with each other, there is a slight trend towards lower polarization values with decreasing statistical uncertainty. This hints at the presence of a systematic effect that becomes significant once the statistical uncertainty drops below a certain level.

IOP Conf. Series: Journal of Physics: Conf. Series **1024** (2018) 012041 doi:10.1088/1742-6596/1024/1/012041

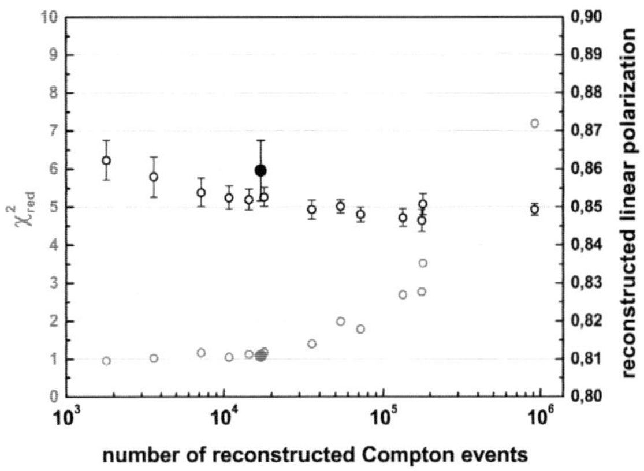

Figure 3. Values of the reduced chi-squared χ^2_{red} (red circles) and reconstructed linear polarization (black circles) for multiple simulated data sets with different statistical uncertainties. In addition, points in the plot are displayed for comparison with experimental data points (filled dots), taken from the commissioning experiment at ESR [13].

4. Summary

The current status of our analysis indicates that our simple model based on the Klein-Nishina equation apparently works for small statistics, but the equation is not able to describe fully the azimuthal scattering distribution of reconstructed Compton events at high statistics. To further improve our analysis and understanding of the data, studies on higher statistics are ongoing and alternative reconstruction algorithms are taken into account.

References

[1] Stöhlker Th et al. 2009 *Eur. Phys. J.-Spec. Top.* **169** 5–14
[2] Weber G et al. 2015 *J. Phys. Conf. Ser.* **583** 012041
[3] Tashenov S et al. 2011 *Phys. Rev. Lett.* **107** 173201
[4] Märtin R et al. 2012 *Phys. Rev. Lett.* **108** 264801
[5] Marx B et al. 2013 *Phys. Rev. Lett.* **110** 254801
[6] Fratini F, Trotsenko S, Tashenov S, Stöhlker Th and Surzhykov A 2011 *Phys. Rev.* A **83** 052505
[7] Surzhykov A, Fritzsche S, Stöhlker Th and Tashenov S 2005 *Phys. Rev. Lett.* **94** 203202
[8] Stöhlker Th et al. 2011 *AIP Conf. Proc.* **1336** 132–7
[9] Kroeger R A, Johnson W N, Kurfess J D and Phlips B F 1999 *Nucl. Instrum. Meth.* A **436** 165–9
[10] Spillmann U et al. 2008 *Rev. Sci. Instrum.* **79** 083101
[11] Weber G et al. 2010 *J. Instrum.* **5** C07010
[12] Lei F, Dean A J and Hills G L 1997 *Space Sci. Rev.* **82** 309–88
[13] Vockert M et al. 2017 *Nucl. Instrum. Meth.* B **408** 313–6
[14] Weber G et al. 2015 *J. Phys.* B **48** 144031
[15] Weber G et al. 2010 *Phys. Rev. Lett.* **105** 243002
[16] Blumenhagen K H et al. 2016 *New J. Phys.* **18** 103034
[17] Hess S et al. 2009 *J. Phys. Conf. Ser.* **163** 012072
[18] Vockert M et al. to be published 2017
[19] Hess S et al. 2009 *J. Phys. Conf. Ser.* **194** 012025
[20] Tashenov S et al. 2006 *Phys. Rev. Lett.* **97** 223202
[21] Hirayama H, Namito Y, Bielajew A F, Wildermann S J and Nelson W R 2005 *SLAC report* SLAC-R-730
[22] Weber G, Bräuning H, Märtin R, Spillmann U and Stöhlker Th 2011 *Phys. Scripta* **T144** 014034
[23] Weber G, Ding H, Herdrich M O and Surzhykov A 2015 *J. Phys. Conf. Ser.* **599** 012040
[24] Herdrich M O, Weber G, Gumberidze A, Wu Z W and Stöhlker Th 2017 *Nucl. Instrum. Meth.* B **408** 294–300

Inverse magnetic catalysis in nonlocal chiral quark models

V P Pagura[1], D Gómez Dumm[2,3], S Noguera[1] and N N Scoccola[3,4,5]

[1] Departamento de Física Teórica and IFIC, Centro Mixto Universidad de Valencia-CSIC, E-46100 Burjassot (Valencia), Spain
[2] IFLP, CONICET – Departamento de Física, Fac. de Cs. Exactas, Universidad Nacional de La Plata, C.C. 67, (1900) La Plata, Argentina
[3] CONICET, Rivadavia 1917, (1033) Buenos Aires, Argentina
[4] Physics Department, Comisión Nacional de Energía Atómica, Av. Libertador 8250, (1429) Buenos Aires, Argentina
[5] Universidad Favaloro, Solís 453, (1078) Buenos Aires, Argentina

E-mail: Valeria.Pagura@uv.es

Abstract. We study the behavior of strongly interacting matter under an external magnetic field in the context of nonlocal Polyakov–Nambu–Jona-Lasinio (PNJL) -like models. We find that at zero temperature the condensates display the well-known Magnetic Catalysis effect, showing a good quantitative agreement with lattice QCD results. Moreover, when extended to finite temperature we find that the Inverse Magnetic Catalysis effect is naturally incorporated.

1. Introduction

Over the last years the understanding of the behavior of strongly interacting matter under extremely intense magnetic fields has attracted increasing attention, due to its relevance for subjects such as the physics of magnetars [1], the analysis of heavy ion collisions at very high energies [2] or the study of the first phases of the Universe [3]. Consequently, considerable work has been devoted to study the structure of the QCD phase diagram in the presence of an external magnetic field [4]. From most low-energy effective models of QCD it was generally expected that, at zero chemical potential, the magnetic field would lead to an enhancement of the chiral condensate ("magnetic catalysis"), independently of the temperature of the system. However, LQCD calculations [5, 6] show that, whereas at low temperatures one finds indeed such an enhancement, close to the critical chiral restoration temperature light quark condensates exhibit a nonmonotonic behavior as functions of the external magnetic field, which results in a decrease of the transition temperature when the magnetic field is increased. This effect is known as inverse magnetic catalysis (IMC). Although many scenarios have been considered in the last few years to account for the IMC [7–22], the mechanism behind this effect is not yet fully understood. With this motivation, we study the behavior of strongly interacting matter under an external magnetic field in the framework of nonlocal chiral quark models, theories that are proposed as nonlocal extensions of the well-known (P)NJL model, intending to go a step further towards a more realistic effective approach to QCD [23–30].

2. Theoretical formalism

Let us start by stating the Euclidean action for our nonlocal NJL-like two-flavor quark model,

$$S_E = \int d^4x \left\{ \bar{\psi}(x) \left(-i\partial\!\!\!/ + m_c\right) \psi(x) - \frac{G}{2} j_a(x) j_a(x) \right\} . \tag{1}$$

Here m_c is the current quark mass, equal for u and d quarks. The nonlocal currents are given by $j_a(x) = \int d^4z\, \mathcal{G}(z)\, \bar{\psi}(x+\frac{z}{2})\, \Gamma_a\, \psi(x-\frac{z}{2})$, where $\Gamma_a = (\mathbb{1}, i\gamma_5\vec{\tau})$, and $\mathcal{G}(z)$ is a nonlocal form factor. In order to study the influence of an external magnetic field we introduce in Eq. (1) a coupling to an electromagnetic gauge field $\mathcal{A}_\mu$. This can be done by performing appropriate changes in the covariant derivative and in the nonlocal currents (see Refs. [32, 33] for details). We restrict to the case of a constant and homogeneous magnetic field along the 3-axis. Next we perform a standard bosonization, introducing scalar and pseudoscalar fields $\sigma(x)$ and $\vec{\pi}(x)$. Within the mean field approximation (MFA), pseudoscalar field vacuum expectation values (VEVs) vanish, and we assume the VEV of the scalar field, $\bar{\sigma}$, to be homogeneous in coordinate space. In this way, following the Ritus eigenfunction method [31] we find the corresponding action, $S_{\rm bos}^{\rm MFA}$ [32].

We extend the analysis of the model to a system at finite T by using the standard Matsubara formalism. In order to account for confinement effects, we include the coupling of fermions to the Polyakov loop (PL) assuming that quarks move on a constant color background field $\phi = ig\,\delta_{\mu 0}\, G^\mu_a \lambda^a/2$, where G^μ_a are the SU(3) color gauge fields. We will work in the so-called Polyakov gauge, in which the matrix ϕ is given a diagonal representation $\phi = \phi_3\lambda_3 + \phi_8\lambda_8$, taking the traced Polyakov loop $\Phi = \frac{1}{3}{\rm Tr}\, \exp(i\phi/T)$ as an order parameter of the deconfinement transition. We include effective gauge field self-interactions through a PL potential. In this work we consider two alternative forms proposed in the literature: a potential given by a polynomial function based on a Ginzburg-Landau ansatz (Poly) [35], and the so-called "improved" PL potential proposed in Ref. [36], in which the full QCD potential is related to a Yang-Mills potential (Poly Imp). Then, the MFA thermodynamical potential $\Omega^{\rm MFA}_{B,T}$ and the associated gap equations can be obtained. It is seen that $\Omega^{\rm MFA}_{B,T}$ turns out to be divergent, we regularize it following Ref. [26]. By deriving $\Omega^{\rm MFA}_{B,T}$ with respect to m_c we get the magnetic field dependent quark condensate for each flavor

$$\begin{aligned}
\langle\bar{\psi}_f\psi_f\rangle^{\rm reg}_{B,T} &= -\frac{|q_fB|\,T}{\pi}\sum_c \int\frac{dp_3}{2\pi}\sum_{k=0}^{\infty}\sum_{n=-\infty}^{\infty} \\
&\quad \left\{ \frac{M^{-,f}_{k,p_{\|nc}}\left[p^2_{\|nc} + 2k|q_fB| + (M^{+,f}_{k,p_{\|nc}})^2\right] + (+\leftrightarrow -)}{\left(2k|q_fB| + p^2_{\|nc} + M^{-,f}_{k,p_{\|nc}} M^{+,f}_{k,p_{\|nc}}\right)^2 + p^2_{\|nc}\left(M^{+,f}_{k,p_{\|nc}} - M^{-,f}_{k,p_{\|nc}}\right)^2} \right. \\
&\quad \left. - \frac{2m_c}{p^2_{\|nc} + 2k|q_fB| + m_c^2} \right\} \\
&\quad -\frac{N_c m_c^3}{4\pi^2}\left[\frac{\ln\Gamma(x_f)}{x_f} - \frac{\ln 2\pi}{2x_f} + 1 - \left(1 - \frac{1}{2x_f}\right)\ln x_f\right] \\
&\quad +\frac{|q_fB|}{\pi}\sum_c\sum_{k=0}^{\infty}\alpha_k \int\frac{dp}{2\pi}\frac{m_c}{\epsilon^f_{kp}\left[1+\exp(\epsilon^f_{kp}/T + i\phi_c)\right]} ,
\end{aligned} \tag{2}$$

where the color index c runs over r, g, b and the color background fields are $\phi_r = -\phi_g = \phi_3$, $\phi_b = 0$. We have defined $p_{\|nc} = (p_3, [(2n+1)\pi T + \phi_c])$ where n is associated to the Matsubara frequencies, $x_f = m_c^2/(2|q_fB|)$, $\alpha_k = 2 - \delta_{k0}$ and $\epsilon^f_{kp} = \sqrt{2k|q_fB| + p^2 + m_c^2}$. We have also defined

$$M^{\lambda,f}_{k,p_{\|nc}} = (-1)^{k-\frac{1-\lambda s_f}{2}} \int_0^\infty dr\, r\, \exp(-r^2/2)\left[m_c + \bar{\sigma}\, g\left(\frac{|q_fB|}{2}r^2 + p^2_{\|nc}\right)\right] L_{k-\frac{1-\lambda s_f}{2}}(r^2). \tag{3}$$

Here $s_f = \text{sign}(q_f B)$, while $g(p^2)$ is the Fourier transform of $\mathcal{G}(z)$, $L_k(x)$ are the Laguerre polynomials, and the index k labels the Landau levels.

Finally, to make contact with the LQCD results quoted in Ref. [6] we define the normalized quark condensate

$$\Sigma^f_{B,T} = \frac{2\,m_c}{S^4}\left[\langle\bar\psi_f\psi_f\rangle^{\text{reg}}_{B,T} - \langle\bar\psi_f\psi_f\rangle^{\text{reg}}_{0,0}\right] + 1\ , \tag{4}$$

where the scale S is given by $S = (135 \times 86)^{1/2}$ MeV. We also introduce the definitions $\Delta\Sigma^f_{B,T} = \Sigma^f_{B,T} - \Sigma^f_{0,T}$ and $\Delta\bar\Sigma_{B,T} = (\Delta\Sigma^u_{B,T} + \Delta\Sigma^d_{B,T})/2$, which correspond to the subtracted normalized flavor condensate and the normalized flavor average condensate, respectively.

For definiteness we consider the case of a Gaussian form factor $g(p^2) = \exp(-p^2/\Lambda^2)$. Thus the model parameters are m_c, G, and Λ, that we fix so as to reproduce empirical values of f_π, m_π and a given value of the quark condensate at zero T and B, $\Phi_0 \equiv (-\langle\bar\psi_f\psi_f\rangle^{\text{reg}}_{0,0})^{1/3}$ (details can be found in Ref. [27]), in addition to the parameters related to the PL potentials considered.

3. Numerical Results

Numerical results at $T = 0$ are shown in Fig. 1. In the left panel we quote the predictions for $\Delta\bar\Sigma_{B,0}$ as function of eB for various model parametrizations, found to be very similar for all cases considered and in very good agreement with LQCD results from Ref. [6]. The results for $\Sigma^u_{B,0} - \Sigma^d_{B,0}$, shown in the right panel, present an overall good agreement with LQCD calculations, however, the dependence on the model parametrization is somewhat larger.

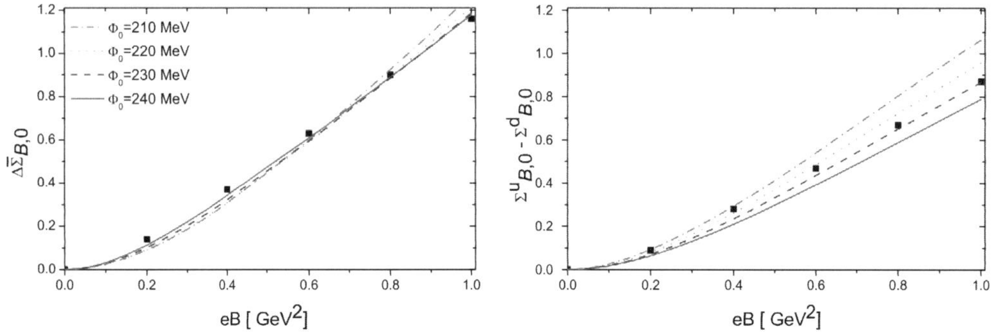

Figure 1. Normalized condensates as functions of eB at $T = 0$. The curves correspond to different parametrizations identified by Φ_0. Full square symbols correspond to LQCD results of Ref. [6]. Left panel: subtracted flavor average; right panel: flavor difference.

We turn now to our results for the case of finite temperature. In Fig. 2 we quote the values obtained for $\Delta\bar\Sigma_{B,T}$ as a function of eB, for some representative values of T. All values correspond to the parametrization leading to $\Phi_0 = 230$ MeV considering the polynomial PL potential, yet qualitatively similar results are found for the other parametrizations. In contrast to what happens at zero temperature, the quantity $\Delta\bar\Sigma_{B,T}$ does not display a monotonous increase with eB when one approaches the chiral transition temperature [for this parameter set one has $T_c(eB = 0) = 179$ MeV]. In fact, the curves reach a maximum after which $\Delta\bar\Sigma_{B,T}$ starts to decrease with increasing eB, implying that the present nonlocal model naturally exhibits the IMC effect found in LQCD. This feature can also be seen from the results displayed in Fig. 3. In the upper pannel we show $(\Sigma^u_{B,T} + \Sigma^d_{B,T})/2$ and Φ as a functions of T for some selected values of eB, while in the lower pannel we present the associated susceptibilities defined as $\chi_{cond} = \partial[(\Sigma^u_{B,T} + \Sigma^d_{B,T})/2]/\partial T$ and $\chi_\Phi = \partial\Phi/\partial T$. As expected, for all values of eB it is found a crossover transition from the chiral symmetry broken phase to the (partially) restored one as the temperature increases. However, contrary to what happens e.g. in the standard local NJL model [4], it is seen that within the present model the transition temperature decreases as the magnetic field increases. Moreover, both chiral restoration and deconfinement transitions are observed simultaneously, as predicted by PNJL-like models at $eB = 0$ [28].

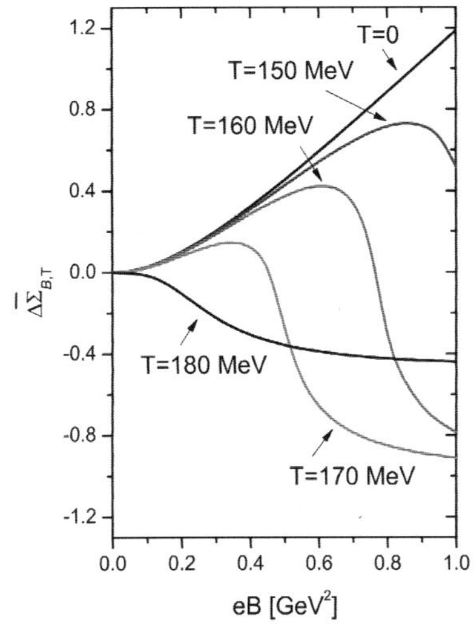

Figure 2. Subtracted normalized flavor average condensate as a function of eB for different representative temperatures. All results correspond to $\Phi_0 = 230$ MeV and polynomial PL potential, also in Fig. 3.

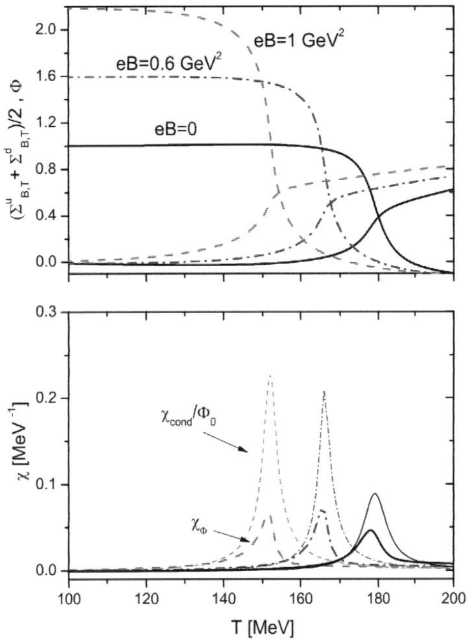

Figure 3. Normalized flavor average condensate and PL (up) and corresponding susceptibilities (down) as functions of T for representative values of eB.

To be more specific, let us define the critical temperature as the value of T at which χ_{cond} reaches a maximum. In Fig. 4, the relative quantity $T_c(B)/T_c(0)$ is displayed as a function of eB together with LQCD results from Ref. [6]. It is seen that all parameter sets considered here lead to a decrease of the critical temperature when eB gets increased, i.e. in all cases the IMC effect is observed. On the other hand, the strength of the IMC effect is rather sensitive to the parametrization. In the case whitout coupling to the PL the decrease of the T_c is small compared with LQCD estimates, however, this is cured once the PL is included.

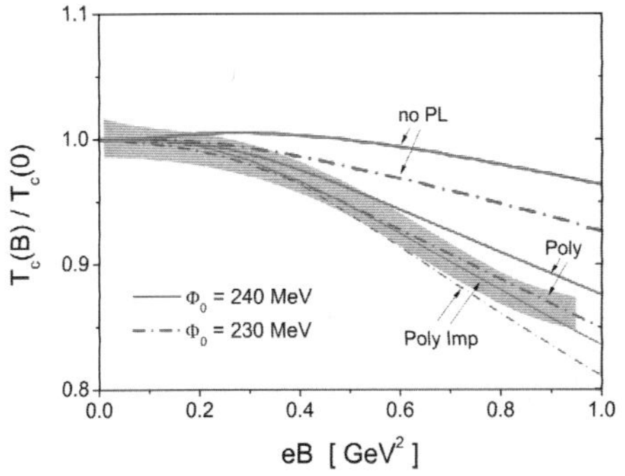

Figure 4. Normalized chiral restoration temperatures as functions of eB for various model parametrizations for the two PL potentials considered. For comparison results obtained excluding the coupling to the PL are also displayed. LQCD results of Ref. [6] are indicated by the grey band.

Acknowledgments
This work has been supported in part by CONICET and ANPCyT (Argentina), grants PIP14-492, PIP12-449, and PICT14-03-0492, by the UNLP (Argentina), Project No. X718,

by the Mineco (Spain), contract FPA2013-47443-C2-1-P, by Centro de Excelencia Severo Ochoa Programme grant SEV-2014-0398, and by Generalitat Valenciana (Spain), grant PrometeoII/2014/066.

References

[1] Duncan R C and Thompson C 1992 *Astrophys. J.* **392** L9
Kouveliotou C *et al* 1998 *Nature* **393** 235
[2] Kharzeev D E, McLerran L D and Warringa H J 2008 *Nucl. Phys.* A **803** 227
Skokov V, Illarionov A Y, and Toneev V 2009 *Int. J. Mod. Phys.* A **24** 5925
Voronyuk V, Toneev V, Cassing W, Bratkovskaya E, Konchakovski V, and Voloshin S 2011 *Phys. Rev.* C **83** 054911
[3] Vachaspati T 1991 *Phys. Lett.* B **265** 258
Enqvist K and Olesen P 1993 *Phys. Lett.* B **319** 178
[4] Kharzeev D E, Landsteiner K, Schmitt A and Yee H U 2013 *Lect. Notes Phys.* **871** 1
Andersen J O, Naylor W R and Tranberg A 2016 *Rev. Mod. Phys.* **88** 025001
Miransky V A and Shovkovy I A 2015 *Phys. Rept.* **576** 1
[5] Bali G S, Bruckmann F, Endrodi G, Fodor Z, Katz S D, Krieg S, Schafer A and Szabo K K 2012 *JHEP* **1202** 044
[6] Bali G S, Bruckmann F, Endrodi G, Fodor Z, Katz S D and Schafer A 2012 *Phys. Rev.* D **86** 071502
[7] Fraga E S, Noronha J and Palhares L F 2013 *Phys. Rev.* D **87** 114014
[8] Bruckmann F, Endrodi G and Kovacs T G 2013 *JHEP* **1304** 112
Bali G S, Bruckmann F, Endrodi G, Gruber F and Schaefer A 2013 *JHEP* **1304** 130
[9] Fukushima K and Hidaka Y 2013 *Phys. Rev. Lett.* **110** 031601
[10] Chao J, Chu P and Huang M 2013 *Phys. Rev.* D **88** 054009
[11] Fraga E S, Mintz B W and Schaffner-Bielich J 2014 *Phys. Lett.* B **731** 154
[12] Ferreira M, Costa P , Menezes D P, Providência C and Scoccola N 2014 *Phys. Rev.* D **89** 016002
[13] Farias R L S, Gomes K P, Krein G I and Pinto M B 2014 *Phys. Rev.* C **90** 025203
[14] Ayala A, Loewe M and Zamora R 2015 *Phys. Rev.* D **91** 016002
Ayala A, Cobos-Martnez J J, Loewe M, Tejeda-Yeomans M E and Zamora R 2015 *Phys. Rev.* D **91** 016007
[15] Fayazbakhsh S and Sadooghi N 2014 *Phys. Rev.* D **90** 105030
[16] Andersen J O, Naylor W R and Tranberg A 2015 *JHEP* **1502** 042
[17] Mueller N and Pawlowski J M 2105 *Phys. Rev.* D **91** 116010
[18] Ferrer E J, de la Incera V and Wen X J 2015 *Phys. Rev.* D **91** 054006
[19] Braun J, Mian W A and Rechenberger S 2016 *Phys. Lett.* B **755** 265
[20] Rougemont R, Critelli R and Noronha J 2016 *Phys. Rev.* D **93** 045013
[21] Ayala A, Dominguez C A, Hernandez L A, Loewe M and Zamora R 2016 *Phys. Lett.* B **759** 99
[22] Mao S 2016 *Phys. Lett.* B **758** 195
[23] Noguera S and Scoccola N N 2008 *Phys. Rev.* D **78** 114002
[24] General I, Gomez Dumm D and Scoccola N N 2001 *Phys. Lett.* B **506** 267
Gomez Dumm D and Scoccola N N 2002 *Phys. Rev.* D **65** 074021
[25] Scarpettini A, Gomez Dumm D and Scoccola N N 2004 *Phys. Rev.* D **69** 114018
[26] Gomez Dumm D and Scoccola N N 2005 *Phys. Rev.* C **72** 014909
[27] Gomez Dumm D, Grunfeld A G and Scoccola N N 2006 *Phys. Rev.* D **74** 054026
[28] Contrera G A, Gomez Dumm D and Scoccola N N 2008 *Phys. Lett.* B **661** 113
Contrera G A, Gomez Dumm D and Scoccola N N 2010 *Phys. Rev.* D **81** 054005
Contrera G A, Orsaria M and Scoccola N N 2010 *Phys. Rev.* D **82** 054026
Carlomagno J P, Gomez Dumm D and Scoccola N N 2013 *Phys. Rev.* D **88** 074034
[29] Gomez Dumm D, Noguera S and Scoccola N N 2011 *Phys. Lett* B **698** 236
Gomez Dumm D, Noguera S and Scoccola N N 2012 *Phys. Rev.* D **86** 074020
[30] Blaschke D, Buballa M, Radzhabov A E and Volkov M K 2008 Yad. Fiz. **71**, 2012 (2008) [Phys. Atom. Nucl. **71**, 1981 (2008)].
[31] Ritus V I 1978 *Sov. Phys. JETP* **48** 788
[32] Pagura V P, Gomez Dumm D, Noguera S and Scoccola N N 2017 *Phys. Rev.* D **95** no. 3 034013
[33] Gomez Dumm D, Izzo Villafañe M F, Noguera S, Pagura V P and Scoccola N N (in press) *Phys. Rev.* D (*Preprint* hep-ph/04742)
[34] Dumitru A, Pisarski R D and Zschiesche D 2005 *Phys. Rev.* D **72** 065008
[35] Ratti C, Thaler M A and Weise W 2006 *Phys. Rev.* D **73** 014019
[36] Haas L M, Stiele R, Braun J, Pawlowski J M and Schaffner-Bielich J 2013 *Phys. Rev.* D **87** 076004

IOP Conf. Series: Journal of Physics: Conf. Series **1024** (2018) 012043 doi:10.1088/1742-6596/1024/1/012043

Magneto-hydrodynamic simulations of Heavy Ion Collisions with ECHO-QGP

G Inghirami[1,2,3,4], L Del Zanna[5,6,7], A Beraudo[8], M Haddadi Moghaddam[9], F Becattini[5,7] and M Bleicher[1,2,3,4]

[1] Frankfurt Institute for Advanced Studies, Ruth-Moufang-Straße 1, 60438 Frankfurt am Main, Germany.
[2] Institute for Theoretical Physics, Goethe-Universität, Max-von-Laue-Straße 1, 60438 Frankfurt am Main, Germany.
[3] GSI Helmholtzzentrum für Schwerionenforschung GmbH, Planckstraße 1, 64291 Darmstadt, Germany.
[4] John von Neumann Institute for Computing, Forschungszentrum Jülich, 52425 Jülich, Germany.
[5] Dipartimento di Fisica e Astronomia, Università di Firenze, Via G. Sansone 1, I-50019 Sesto F.no (Firenze), Italy.
[6] INAF - Osservatorio Astrofisico di Arcetri, L.go E. Fermi 5, I-50125 Firenze, Italy.
[7] INFN - Sezione di Firenze, Via G. Sansone 1, I-50019 Sesto F.no (Firenze), Italy.
[8] INFN - Sezione di Torino, Via P. Giuria 1, I-10125 Torino, Italy.
[9] Department of Physics, Hakim Sabzevari University, P. O. Box 397, Sabzevar, Iran.

E-mail: inghirami@fias.uni-frankfurt.de

Abstract. It is believed that very strong magnetic fields may induce many interesting physical effects in the Quark Gluon Plasma, like the Chiral Magnetic Effect, the Chiral Separation Effect, a modification of the critical temperature or changes in the collective flow of the emitted particles. However, in the hydrodynamic numerical simulations of Heavy Ion Collisions the magnetic fields have been either neglected or considered as external fields which evolve independently from the dynamics of the fluid. To address this issue, we recently modified the ECHO-QGP code, including for the first time the effects of electromagnetic fields in a consistent way, although in the limit of an infinite electrical conductivity of the plasma (ideal magnetohydrodynamics). In this proceedings paper we illustrate the underlying 3+1 formalisms of the current version of the code and we present the results of its basic preliminary application in a simple case. We conclude with a brief discussion of the possible further developments and future uses of the code, from RHIC to FAIR collision energies.

1. Introduction

Heavy ion collisions represent an excellent method to study strongly interacting matter at high energy densities [1, 2]. In turn, relativistic hydrodynamics has proved to be a reasonable model of the expanding Quark Gluon Plasma fireball that it is expected to form at high collision energies [3, 4]. In order to provide a better explanation of certain experimental observables, in particular the collective anisotropic flows, viscous effects have been taken into account [5, 6, 7, 8] and recent developments involve also anisotropic hydrodynamics [9, 10, 11]. However, only in the last years it has been realized that the magnetic fields generated by the fast moving charges contained in the colliding nuclei might be strong enough [12, 13, 14] to produce measurable effects, like the chiral magnetic effect [15], the chiral magnetic wave [16, 17], the chiral vortical effect [18], changes in the QGP EoS [19] and modifications of the collective flows [20, 21, 22]. In the first studies of their influence on the dynamical evolution of the QGP fireball, the magnetic fields have been modeled as external entities not coupled with the medium [23, 24, 25]. To overcome this limitation, we modified the ECHO-QGP code [26] to consistently take into account the interplay between the plasma and the electromagnetic field, although in the approximation of an infinite electrical conductivity of the medium (ideal relativistic magnetohydrodynamics). In this proceedings paper we will briefly review the formalism and present some preliminary results.

2. Essential theory and numerical implementation

2.1. The basic formalism

We consider a system composed by a dominant species determining a main *fluid* current, associated to the net-baryon current, and by a secondary species at the origin of the conduction current [27, 28]. We impose the standard conservation laws for this fluid current N^μ and for the *total* (matter and fields) energy-momentum tensor of the plasma $T^{\mu\nu}$:

$$d_\mu N^\mu = 0, \tag{1}$$
$$d_\mu T^{\mu\nu} = 0, \tag{2}$$

with d_μ being the covariant derivative. Then, we consider that the electromagnetic field obeys Maxwell's equations:

$$d_\mu F^{\mu\nu} = -J^\nu \quad (d_\mu J^\mu = 0), \tag{3}$$
$$d_\mu F^{\star\mu\nu} = 0, \tag{4}$$

where $F^{\mu\nu}$ is the Faraday tensor and $F^{\star\mu\nu} = \frac{1}{2}\epsilon^{\mu\nu\lambda\kappa}F_{\lambda\kappa}$ is its dual. We neglect any possible polarization and magnetization effect of the plasma and we assume to have local thermal equilibrium. With all these assumptions, we can define a single *fluid four-velocity* u^μ ($u_\mu u^\mu = -1$) and, in its comoving frame, the electric and magnetic fields as: $e^\mu = F^{\mu\nu}u_\nu$, ($e^\mu u_\mu = 0$) and $b^\mu = F^{\star\mu\nu}u_\nu$, ($b^\mu u_\mu = 0$). In general, the Ohm law which relates the current with the fields has a tensorial structure, nevertheless in our ideal MHD approximation we assume that the plasma has an infinite electrical conductivity and, in this special case, it is possible to show that $e^\mu = 0$ is implied and that the electromagnetic energy-momentum tensor can be expressed as

$$T^{\mu\nu} = (e + p + b^2)u^\mu u^\nu + (p + \tfrac{1}{2}b^2)g^{\mu\nu} - b^\mu b^\nu \tag{5}$$

where e is the energy density, p the thermal pressure, $b^2 = b_\mu b^\mu$ (all measured in the comoving frame of the fluid) and $g^{\mu\nu}$ is the metric tensor (we use either Minkwoski or Milne coordinates). Eventually, the system of ideal RMHD equations which we solve is

$$d_\mu(nu^\mu) = 0, \tag{6}$$
$$d_\mu[(e + p + b^2)u^\mu u^\nu + (p + \tfrac{1}{2}b^2)g^{\mu\nu} - b^\mu b^\nu] = 0, \tag{7}$$
$$d_\mu(u^\mu b^\nu - u^\nu b^\mu) = 0, \tag{8}$$

plus an equation of state (EoS) to close it, in the unknowns n, e, p, u^μ, and b^μ.

2.2. The RMHD module in ECHO-QGP

To numerically solve the system of ideal RMHD equations, we need to separate the time and space components (3+1 split) [29] and rewrite the equations in *conservative form* [30]. We refer

IOP Conf. Series: Journal of Physics: Conf. Series **1024** (2018) 012043 doi:10.1088/1742-6596/1024/1/012043

to other articles [31, 32, 7, 26] to get more details about the procedure, here we report just the final form of the equations:

$$\partial_0 \mathbf{U} + \partial_i \mathbf{F}^i = \mathbf{S}, \tag{9}$$

where

$$\mathbf{U} = |g|^{\frac{1}{2}} \begin{pmatrix} \gamma n \\ S_j \equiv T^0_j \\ \mathcal{E} \equiv -T^0_0 \\ B^j \end{pmatrix}, \ \mathbf{F}^i = |g|^{\frac{1}{2}} \begin{pmatrix} \gamma n v^i \\ T^i_j \\ S^i \equiv -T^i_0 \\ v^i B^j - B^i v^j \end{pmatrix}, \ \mathbf{S} = |g|^{\frac{1}{2}} \begin{pmatrix} 0 \\ \frac{1}{2} T^{ik} \partial_j g_{ik} \\ -\frac{1}{2} T^{ik} \partial_0 g_{ik} \\ 0 \end{pmatrix} \tag{10}$$

are respectively the set of *conservative variables*,the *fluxes* and the *source terms*.
The components of $T^{\mu\nu}$ shown in the previous equations are:

$$S_i = (e+p)\gamma^2 v_i + \varepsilon_{ijk} E^j B^k, \tag{11}$$

$$T_{ij} = (e+p)\gamma^2 v_i v_j + (p + u_{\text{em}}) g_{ij} - E_i E_j - B_i B_j, \tag{12}$$

$$\mathcal{E} = (e+p)\gamma^2 - p + u_{\text{em}}, \tag{13}$$

where we have defined the electromagnetic energy density $u_{\text{em}} = \frac{1}{2}(E^2 + B^2)$.
The v^i are the contravariant 3-vector components of the fluid velocity which are related to the previous 4-vector by $u^\mu = (\gamma, \gamma v^i)$, where $\gamma = (1 - v^2)^{-1/2}$ is the Lorentz factor of the bulk flow and $v^2 = v_k v^k$. The electric and magnetic field are now 3-vectors measured in the laboratory frame which are related to the previous 4-vectors by $e^\mu = (\gamma v_k E^k, \gamma E^i + \gamma \varepsilon^{ijk} v_j B_k)$ and $b^\mu = (\gamma v_k B^k, \gamma B^i - \gamma \varepsilon^{ijk} v_j E_k)$, where ε_{ijk} is the Levi-Civita pseudo-tensor of the spatial three-metric, namely $\varepsilon_{ijk} = |g|^{\frac{1}{2}}[ijk]$, with $g = \det\{g_{\mu\nu}\} = -\det\{g_{ij}\} < 0$ and $[ijk]$ the usual alternating symbol of three-dimensional space with values ± 1 or 0. In this $3D+1$ scheme, the ideal Ohm law $e^\mu = 0$ translates to $E_i = -\varepsilon_{ijk} v^j B^k$. This means that the components of the electric field are not independent anymore, but they are derived from the v^i and B^i dynamical variables. From a numerical point of view, it is also necessary to enforce the solenoidal (i.e. "null-B divergence") condition $\partial_i(|g|^{\frac{1}{2}} B^i) = 0$. This problem can be managed in different ways [30]; for its simplicity, we adopt the "general Lagrange multipliers" (known also as Dedner's) method [33, 34].
ECHO-QGP is based on finite difference schemes, we invite to read Refs. [32, 7, 26] for details. Here we just mention that we use the HLL (Harten-Lax-Van Leer) [35] approximate solver to solve the Riemann problem and a second or third order Runge Kutta algorithm to perform time integration.

3. A preliminary application

3.1. Setup

After testing the code [26], we applied it in a quite typical and simple case of heavy ion collisions. We performed a $2D+1$ simulation of two gold nuclei colliding at $\sqrt{s_{\text{NN}}} = 200\text{GeV}$ with an impact parameter $b = 10\text{fm}$, adopting geometrical Glauber initial conditions[36] to determine the spatial energy density distribution at $\tau = 0.4\text{fm/c}$. We computed the spatial distribution of the magnetic field as in Ref. [14], assuming that the nuclei moved in a medium with constant electrical conductivity $\sigma = 5.8\text{MeV}$ until the system reached the condition of local thermal equilibrium at $\tau = 0.4\text{fm/c}$, when we started the simulation. During the magnetohydrodynamical evolution we assumed that the medium had infinite electrical conductivity, consistently with our model, and we used the massless particles EoS $p = e/3$. We chose 150MeV/fm^3 as freeze-out energy density and we computed the elliptic flow of positive pions using the Cooper-Frye prescription [37].

3.2. Results

In Fig. 1, we compare the time evolution of the magnetic field in the center of mass of the system in the ideal RMHD approach (i.e. fluid with an infinite electrical conductivity, coupled with the magnetic field) versus the evolution of the magnetic field in a medium with constant electrical conductivity, decoupled from the dynamics of the fluid. We can notice that the expansion on the transverse plane makes the magnetic field to decay faster compared to the Bjorken flow case (line a vs line e), nevertheless the decay is still much slower than in the case of a magnetic field generated by electric charges moving in the void (line c).
In Fig. 2, we compare the final $v_2(p_T)$ of pions with and without an initial magnetic field. In the context of our setup, we don't observe any significant difference between these two cases.

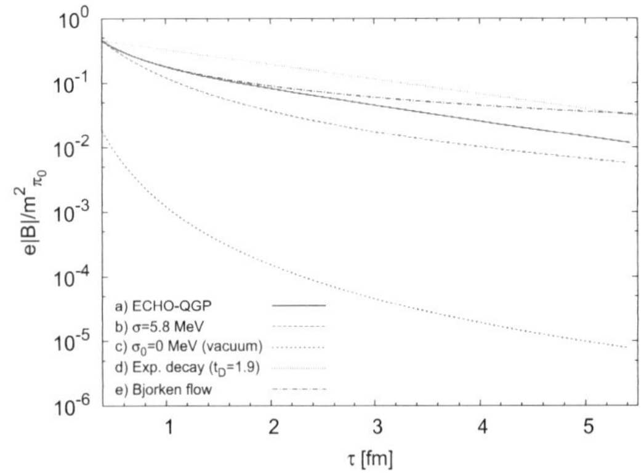

Figure 1: Comparison of the time evolution of the magnitude of the magnetic field (in neutral pion mass units squared) at the center of the grid in five different cases: a) ECHO-QGP 2D+1 RMHD evolution b) time evolution of the magnetic (as in Ref. [14]) in a medium with uniform and constant electrical conductivity $s = 5.8$ MeV c) same as in case b), but assuming zero electrical conductivity $s = 0$ MeV (vacuum) d) exponential decay of magnetic field as modeled in Ref. [25], with $t_D = 1.9$ e) Bjorken flow [38]. The parameters are for the reaction Au+Au, b=10 fm at $\sqrt{s_{NN}}$=200 GeV.

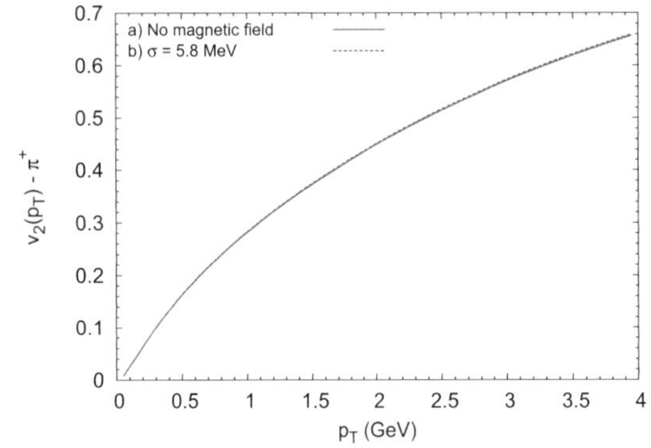

Figure 2: v_2 of π^+ in two cases: a) Without magnetic field, b) With an initial magnetic field computed assuming $\sigma = 5.8$ MeV. The parameters are for the reaction Au+Au, b=10 fm at $\sqrt{s_{NN}}$=200 GeV.

4. Conclusions and future perspectives

The present work represents a necessary, but limited step toward a realistic dynamical description of the Quark Gluon Plasma coupled with electromagnetic fields. Short term plans involve 3D+1 simulations with a lattice QCD + Hadron gas EoS, studying collisions at different energies and impact parameters, while long term plans will probably focus on resistive RMHD [39, 40, 41] and magnetic fields of chiral origin [42]. Hadronic interactions and particle decays at the end of the hydro phase will have to be taken into account, as well. Some effects that might be partially due to magnetic fields [43], like the splitting between the polarization of Λ and $\overline{\Lambda}$ [44], are more evident at low collision energies, close to the regime of the upcoming FAIR CBM experiment. Unfortunately, at such low collision energies hydrodynamic simulations can provide a realistic description of the QGP only for a short and delayed time, especially when focusing on the most peripheral events, in which magnetic fields are expected to be strongest. Nevertheless, even in the case of CBM, a consistent treatment of the magnetic fields in the hydro phase will certainly contribute to provide a more accurate and complete physical model of the QGP evolution in heavy ion collisions.

Acknowledgments
G. Inghirami was supported by a GSI grant in cooperation with the J. von Neumann Institute for Computing; he also acknowledges support from the H-QM and HGS-HIRe schools. M. Haddadi Moghaddam thanks the ministry of science and technology of Iran and the Physics Dept. and the INFN Turin. The computational resources were provided by the INFN Florence, by the FIAS and by the CSC Frankfurt. This work was supported by the Florence Univ. grant "Fisica dei plasmi relativistici: teoria e applicazioni moderne" and by COST Action CA15213 "THOR".

References
[1] Csernai L P 1994 *Introduction to relativistic heavy ion collisions*
[2] Vitev I 2005 *Int. J. Mod. Phys.* **A20** 3777–3782
[3] Maruhn J A 1991 *Int. Rev. Nucl. Phys.* **5** 81–178
[4] Kolb P F, Sollfrank J and Heinz U W 2000 *Phys. Rev.* **C62** 054909
[5] Romatschke P and Romatschke U 2007 *Phys. Rev. Lett.* **99** 172301
[6] Bozek P 2012 *Phys. Rev.* **C85** 034901
[7] Del Zanna L, Chandra V, Inghirami G, Rolando V, Beraudo A, De Pace A, Pagliara G, Drago A and Becattini F 2013 *Eur. Phys. J.* **C73** 2524
[8] Karpenko I, Huovinen P and Bleicher M 2014 *Comput. Phys. Commun.* **185** 3016–3027
[9] Martinez M, Ryblewski R and Strickland M 2012 *Phys. Rev.* **C85** 064913
[10] Tinti L 2015 *Phys. Rev.* **C92** 014908
[11] Molnar E, Niemi H and Rischke D H 2016 *Phys. Rev.* **D93** 114025
[12] Kharzeev D E, McLerran L D and Warringa H J 2008 *Nuclear Physics A* **803** 227 – 253 ISSN 0375-9474
[13] Tuchin K 2013 *Adv. High Energy Phys.* **2013** 490495
[14] Tuchin K 2013 *Phys. Rev.* **C88** 024911
[15] Fukushima K, Kharzeev D E and Warringa H J 2008 *Phys. Rev. D* **78**(7) 074033
[16] Kharzeev D E and Yee H U 2011 *Phys. Rev. D* **83**(8) 085007
[17] ALICE Collaboration (ALICE Collaboration) 2016 *Phys. Rev. C* **93**(4) 044903
[18] Kharzeev D, Liao J, Voloshin S and Wang G 2016 *Progress in Particle and Nuclear Physics* **88** 1 – 28 ISSN 0146-6410
[19] Bali G S, Bruckmann F, Endrődi G, Katz S D and Schäfer A 2014 *JHEP* **08** 177
[20] Gürsoy U, Kharzeev D, Marcus E and Rajagopal K 2016 *Nucl. Phys.* **A956** 389–392
[21] Moghaddam M H, Azadegan B, Kord A F and Alberico W M 2017 (*Preprint* `1705.08192`)
[22] Das A, Dave S S, Saumia P S and Srivastava A M 2017 (*Preprint* `1703.08162`)
[23] Deng W T and Huang X G 2012 *Phys. Rev.* **C85** 044907
[24] Roy V and Pu S 2015 *Phys. Rev.* **C92** 064902
[25] Pang L G, Endrődi G and Petersen H 2016 *Phys. Rev.* **C93** 044919
[26] Inghirami G, Del Zanna L, Beraudo A, Moghaddam M H, Becattini F and Bleicher M 2016 *Eur. Phys. J.* **C76** 659
[27] Bekenstein J D and Oron E 1978 *Phys. Rev. D* **18**(6) 1809–1819
[28] Anile A M 1989 *Relativistic fluids and magneto-fluids : with applications in astrophysics and plasma physics* (Cambridge ; New York : Cambridge University Press) ISBN 0521304067
[29] Gourgoulhon E 2007 (*Preprint* `gr-qc/0703035`)
[30] Font J A 2008 *Living Rev. Rel.* **11** 7
[31] Del Zanna L, Bucciantini N and Londrillo P 2003 *Astronomy & Astrophysics* **400** 397–414
[32] Del Zanna, L, Zanotti, O, Bucciantini, N and Londrillo, P 2007 *Astronomy & Astrophysics* **473** 11–30
[33] Dedner A, Kemm F, Kröner D, Munz C D, Schnitzer T and Wesenberg M 2002 *Journal of Computational Physics* **175** 645 – 673 ISSN 0021-9991
[34] Mignone A and Tzeferacos P 2010 *Journal of Computational Physics* **229** 2117 – 2138 ISSN 0021-9991
[35] Harten A, Lax P D and van Leer B 1983 *SIAM Rev.* **25** 35–61 ISSN 0036-1445; 1095-7200/e
[36] Glauber R J 1959 *Lectures in Theoretical Physics* (New York : WE Brittin and LG Dunham)
[37] Cooper F and Frye G 1974 *Phys. Rev. D* **10**(1) 186–189
[38] Roy V, Pu S, Rezzolla L and Rischke D 2015 *Physics Letters B* **750** 45 – 52 ISSN 0370-2693
[39] Dionysopoulou K, Alic D, Palenzuela C, Rezzolla L and Giacomazzo B 2013 *Phys. Rev.* **D88** 044020
[40] Bucciantini N and Del Zanna L 2013 *Mon. Not. Roy. Astron. Soc.* **428** 71
[41] Del Zanna L, Papini E, Landi S, Bugli M and Bucciantini N 2016 *Mon. Not. Roy. Astron. Soc.* **460** 3753–3765
[42] Hirono Y, Hirano T and Kharzeev D E 2014 (*Preprint* `1412.0311`)
[43] Becattini F, Karpenko I, Lisa M, Upsal I and Voloshin S 2017 *Phys. Rev.* **C95** 054902
[44] Adamczyk L *et al.* (STAR) 2017 (*Preprint* `1701.06657`)